COUPLED
BIOLUMINESCENT
ASSAYS

COUPLED BIOLUMINESCENT ASSAYS

Methods, Evaluations, and Applications

MICHAEL J. COREY

A John Wiley & Sons, Inc., Publication

Published by John Wiley & Sons, Inc., Hoboken, New Jersey

Published simultaneously in Canada

For general information on our other products and services or for technical support, please contact our Customer Care Department within the United States at (800) 762-2974, outside the United States at (317) 572-3993 or fax (317) 572-4002.

Wiley also publishes its books in a variety of electronic formats. Some content that appears in print may not be available in electronic formats. For more information about Wiley products, visit our web site at www.wiley.com.

Library of Congress Cataloging-in-Publication Data:

ISBN 978-0-470-10883-3

Printed in the United States of America

10 9 8 7 6 5 4 3 2 1

"There are two ways of spreading light:
to be the candle or the mirror that reflects it."

–Edith Wharton

CONTENTS

PREFACE

Every scientist with a career of sufficient length has a favorite story of serendipity. In the 1990s, while studying acceleration of the complement system by antibodies directed against its regulators, I became dissatisfied with the available means of measuring cell death. Either I had to load up the cells with a radioactive label, hope it would not perturb the results, and then clean it up; or I had to wait a long time for a weak signal from lactate dehydrogenase released by the dying cells. It was then that my mentor Bob Kinders said, "Why don't you come up with a luminescent assay?" Five minutes after Bob posed his question, I was racing upstairs to try the method that eventually became aCella-TOX^{TM}, a *coupled bioluminescent* (CB) general cytotoxicity assay. After much help from many people, I obtained my own patent (an experience every life scientist should have at least once), published, found partners, got it on the market, and finally wrote this book.

Today, it is not difficult to come up with a new CB assay, but in a sense, all of the developments derive to some degree either from Arne Lundin's ideas as expressed in the late 1970s and 1980s regarding the many possible ways of expanding the use of luciferase in coupled assays or from the important work of Shimomura in isolating and characterizing the first photoproteins. Other academic workers and, notably, vendors such as Promega have driven the field far beyond the initial applications, to the point where there is probably a CB alternative for more than half of all biochemical measurements proposed. The activities of enzymes that are critical in today's medicine, including kinases, phosphatases, proteases, acetylcholinesterase, nitric oxide synthase (NOS), and many others, may be measured rapidly and with extraordinary sensitivity by using this approach. In other cases, such as reporter assays, luciferase activity simply serves as a robust quantitative marker. The power of CB methods to provide information with extraordinary speed and sensitivity is growing and expanding, yet apart from reporter assays, they are virtually unknown to many life scientists.

Because of the general lack of familiarity with CB technology, I begin the book with what I hope is a solid introduction to bioluminescence in general, including the tools available, as well as the biochemical and enzymatic principles that make these assays possible and useful. To those who have skipped or forgotten their thermodynamics, it may appear paradoxical that CB reactions (and other coupled enzymatic reaction series) can be run *even if the reaction series is endothermic*, that is, even if it absorbs energy at the level of the individual molecule. This is just one of a series of illuminating surprises that await the scientist who engages this field for the first time.

The reader will note the presence in this book of a great deal of material related to competing nonluminescent methods, especially in chapters dealing with major drug discovery targets, such as kinases and G-protein-coupled receptors (GPCRs). This is both essential and inevitable. The advent of CB assays for many applications is still upon us, or in the future, and the amount of available material relating to them is therefore limited, whereas we know a great deal about fluorometric assays, to give a prominent example. Therefore, it may be highly useful to the assay developer, as well as the scientist seeking to understand the alternatives, to have information about both the CB options and the other choices together in the same chapter. In some cases, the majority of space in the chapters is occupied by the other methods, to allow the developer to assess the full range of strategies he/she can employ. Thus, the intent is both to introduce and describe the CB possibilities and to enable readers to evaluate other methods if they are more suitable.

The book mostly focuses on assay target (kinases, proteases), but there is necessarily some overlap. Calcium, an essential cofactor in aequorin luminescence, is also a major player in the biochemistry of the GPCRs, which are currently the most important class of targets in drug development. Kinases and phosphatases also play roles in GPCR signal transduction, and so on. Finally, I have endeavored to suggest applications that are not directly related to the drug discovery juggernaut, including food safety testing and bioburden measurements, as well as environmental applications. These chapters are arranged differently.

Although every effort has been made to keep the information up-to-date, inevitably a number of methods will have been inadvertently omitted, while others are just appearing on the market as the book goes to press. There is no better evidence of the vitality of the field of CB assays than the range of new assay types now being marketed. Among those that have appeared too recently to treat in depth herein are several fascinating offerings from Promega, including P450-Glo^{TM} for the cytochrome P450s so crucial in drug metabolism; MAO-Glo^{TM} for the monoamine oxidases, which are increasingly important targets for psychoactive pharmaceutical development; GSH-Glo^{TM} for glutathione quantification; Proteasome-Glo^{TM}; and Pgp-Glo^{TM} for assessing the activity of the P-glycoprotein ATPase. These fields of study and their nexus with CB assay development are worthy of their own chapters in a future volume.

I owe thanks to many for help and support. Among them are my outstanding mentors Jack Kirsch, Bob Kinders, Bob Vessella, and Phil Maples; Helen Landicho, who advised me on regulatory issues; technical associates Kathy Schaffer, Connie Ave-Teel, and Caroline Babcock; and all-around Document Wizard Anna Schneider. Tomas Corey and Eva Corey have offered excellent assistance in preparing figures. Lukas Corey assisted with indexing. Sumant Dhawan and Sanjeet Thadani of Cell Technology, Inc. have been invaluable allies in developing and marketing CB products, such as the first CB detection system for nerve gas.

Of the many joys of my scientific career, one of the high points was knowing the great Daniel E. Koshland, Jr., of the University of California, Berkeley. I wish to dedicate this book to his memory.

Michael J. Corey
April 8, 2008

PART I

BACKGROUND TO COUPLED BIOLUMINESCENT ASSAYS

1

INTRODUCTION

1.1 INTRODUCTION TO COUPLED BIOLUMINESCENT ASSAYS

The phenomenon of bioluminescence, the emission of internally generated light by living organisms, holds special fascination for the scientist. We are used to living things that *use* light: they may have light organs for detection and perception of their environment or nearby objects, light transducers that regulate circadian or seasonal rhythms, or membranes capable of harvesting light for conversion to chemical energy. But to the scientist who enters the field of bioluminescence with the usual requisite skepticism, the process whereby animals, fungi, and bacteria *make* their own light seems almost like a pointless biological extravagance that occurs merely for the benefit of our research. Despite all the sensible arguments about interorganism communication and photolocation (1, 2), the whole business has the flavor of an undeserved gift.

And a remarkable gift it is. Though the range of chemistries employed in light-producing reactions is limited, the applications span nearly the entire breadth of the life sciences. Hundreds of researchers who have probably never heard of "coupled bioluminescence" regularly perform reporter assays with recombinant luciferase genes. Thousands of workers in the food industry use a bioluminescent reaction to detect unwanted biological contamination. Scientists in fields as diverse as genetics, environmental monitoring, enzymology, biowarfare prevention and response, and especially drug discovery now have coupled bioluminescent alternatives to numerous slow, insensitive, and expensive procedures. In many cases, however, these scientists may not be aware of the possibilities; hence this book.

Our focus in this book is on the applications of coupled bioluminescence (CB). The discussion presented here of the physical phenomena of luminescence and bioluminescence is intended for the scientist with a general background, including knowledge of chemistry at some level, but no prior experience with luminescence or fluorescence. The material is necessarily limited in extent; for an in-depth introduction to the chemistry and biology of bioluminescence, the reader is encouraged to consult two classic and highly seminal works: *Bioluminescence: Chemical Principles and Methods*, by Shimomura, discoverer of aequorin (3); and *Bioluminescence*, by Harvey, who began working and publishing on the phenomenon of bioluminescence nearly four decades before he finally wrote this "bible" of the field (4). What follows is a survey of the kinds of bioluminescent reactions that can occur and how they may be harnessed to the service of coupled bioluminescent assays. The second half of the chapter is largely a review of the available molecular tools: the many interesting bioluminescent organisms and the relevant characteristics of the luciferases and photoproteins they offer. That section may be of special interest to the scientist considering cutting-edge development of novel coupled bioluminescent techniques. A thorough discussion of how to design, perform, and test these reactions appears in Chapter 2.

1.2   LUMINESCENT TECHNOLOGIES OF THE LIFE SCIENCES

We begin with a definition. *Luminescence* is a critical term for our purposes, and it is desirable to establish its precise meaning from the outset. The most accurate definition in general use is expressed with a sort of negative twist: luminescence is emission of light at temperatures below those that give rise to incandescence. In physical terms, luminescence consists of the emission of photons by electrons that have previously been excited to appropriate energy levels by any means *except* heating. Thus, fluorescence, phosphorescence, chemiluminescence, and bioluminescence all fall within this category.

Exept phosphorescence, all of these kinds of luminescence are in widespread use throughout the life sciences; in fact, the only important category of quantitative liquid-phase assays that does not employ one of these kinds is the venerable and nearly obsolete method known as spectrophotometry (known in the trade as "UV/spec," jargon referring to the fact that spectrophotometric measurements are frequently made in the ultraviolet range). However, these days even spectrophotometers may employ light sources other than incandescent lamps, rendering even this last bastion of the 1950s just another form of luminescence.

At this point, we introduce the two phenomena that are the simplest to explain, though not necessarily to work with, *chemiluminescence* and *bioluminescence*. Though distinct, these two processes have a critical point in common: in each case, light is generated within the sample by chemical reactions. Figure 1.1 illustrates the simple sequence of events underlying assays based on these principles: chemical energy transduced to light within the reaction vessel is radiated *anisotropically* (i.e., in all directions equally) and part of it is captured and measured by a detecting device (the nature of the device is discussed below).

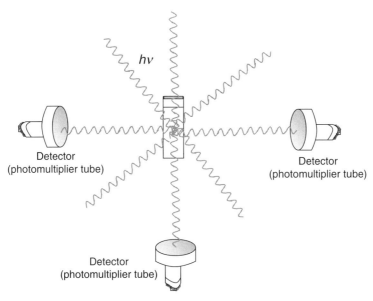

FIGURE 1.1   Schematic representation of autoluminescence. Light ($h\nu$) is emitted equally in all directions by the luminous sample, impinging on one or more detectors, which may be photomultiplier tubes, charge-coupled devices, or other types. No lamp is required. (See the color version of this figure in the Color Plates section.)

This situation may be fruitfully contrasted with the situation depicted in Fig. 1.2, that of spectrophotometry, in which light from an *instrument* source, such as a lamp or light-emitting diode, is *absorbed* by molecules (or supermolecular aggregates, in the case of turbidity measurements) within the sample; the attenuated light intensity is then measured by a device that may be identical to that used for measuring chemiluminescence and bioluminescence. A critical feature of this system is that the intensity

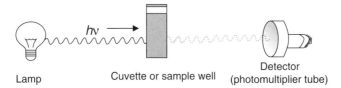

FIGURE 1.2   Schematic representation of spectrophotometry. The lamp illuminates the sample, which absorbs a portion of the light, depending on the concentration of the absorbing species. The remaining light is transmitted through the sample vessel and impinges on the detector. The full intensity of the lamp is known separately from either careful calibration or beam splitting. The fraction of light remaining indicates the concentration to be determined, assuming the absorbance coefficient is known. Multiple concentrations may be measured simultaneously if all absorbance coefficients are known, data from multiple wavelengths are collected, and independent spectra of the absorbing substances are available. (See the color version of this figure in the Color Plates section.)

of the detected light can never exceed that of the incident light from the lamp. Perfect transmittance, which is equivalent to zero absorbance, therefore results in an intensity ratio of 1 in comparison with the blank reading.

The final case we consider here is that of fluorescence (or phosphorescence, which differs from fluorescence for practical purposes only in its timescale and is not treated further here). Fluorometry, or the controlled measurement of fluorescence, is one of the most useful techniques ever to appear in science. It partakes of both the physical complexity of chemiluminescence and bioluminescence and the conceptual complexity (at least) of spectrophotometry. Fluorescent phenomena require a light source, as does spectrophotometry, but, in sharp contrast to the latter method, fluorometry paradoxically requires that as few photons as possible from the source reach the detector. The reason for this can be inferred from Fig. 1.3. Only photons *emitted by sample molecules* are useful in fluorometry. These photons are not emitted as a direct result of chemical reactions or physical interactions of the molecules. Instead, they are the products of the decay of the electronic excitation energy imparted by the light source. In other words, the electrons within the sample are excited by the source, and the process whereby this excitation energy is released by the electrons in returning to the ground state produces the photons to be measured. The reader should note especially the 90° angle between the light path leading from the lamp to the sample and that leading from the sample to the detector. This arrangement is intended to minimize

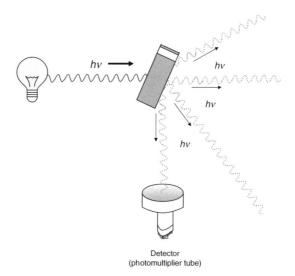

Detector
(photomultiplier tube)

FIGURE 1.3   Schematic representation of fluorometry. Illumination by the lamp in a specific range of wavelengths is absorbed as electronic excitation by sample molecules and reemitted at longer wavelengths. Emission is in all directions, but to avoid receiving light from the lamp, the detector is mounted at an angle to the incident beam. The detector is also tuned by filters or a monochromator to accept only emitted wavelengths. (See the color version of this figure in the Color Plates section.)

the degree of contamination of the sample fluorescence by light shed directly by the source. However, this is sometimes accomplished in other ways, such as by situating the detector proximate to the light source, thus requiring the emission to occur at a 180° from the incident light. The latter instrument geometry is common, for example, in fluorometric microplate assays, where the lamp must deliver light to a single tiny well and the detector must receive reemitted light exclusively from the same well to avoid cross talk.

The fact that the emitted light comes from electronic decay events in the sample molecules has one very important consequence for the nature of the light: its wavelength is virtually always shifted from that of the incident light, and the shift is always in the direction of longer wavelengths (lower energy). The reason for this shift is that the process of capturing and reemitting light is not 100% efficient. Energy is lost to vibration and rotation, to collisions with other molecules, and to the shifting charge dipoles of the excited electron's environment. Fortunately, the amount of change of the wavelength, known as the Stokes shift, is highly consistent and provides a probe of the electronic state and surroundings that is nearly as useful as the fluorescence intensity. The Stokes shift itself may be perturbed by pH, ligand association, and many other phenomena.

At this point, fluorometric assays are the main competitor of CB methods. The astonishing utility and versatility of fluorometry derives from the fact that both excitation and emission characteristics of fluorophores depend strictly on both the chemistry and the environment of the specific molecule. Fluorophores may behave differently depending on diverse phenomena such as pH, ionic strength, hydrophobicity, temperature, spatial separation from other fluorophores, delays between excitation and detection, and many aspects of molecular chemistry. Substrates may be designed specifically to yield fluorophores in response to the activity of enzymes. Fluorophores may be introduced into biological membranes to study their motions. Although fluorometric techniques have a number of inherent drawbacks, including the need for complex collimation and a costly source of illumination, the necessity of irradiating potentially labile samples, and the requirement for precise methods of selecting both excitation and emission wavelengths, it must be emphasized that the technique, or rather a set of techniques, is extraordinarily powerful, and it would be imprudent to assign limits to what fluorometry can do and will do.

Fluorescence may be defined in very different ways elsewhere in the literature; for example, some consider photon liberation in an electronic decay from a singlet state to be fluorescence if spin is preserved (5).

1.3 VARIETIES OF FLUOROMETRIC ASSAYS

Modern fluorometry has advanced far beyond the simple process depicted in Fig. 1.3 of shining light on a sample and measuring fluorescence at an angle. The temporal and spatial dependence of the intensity of emitted light can be exploited in a surprising variety of ways. Here, we describe three fluorometric technologies, all of which have their origins in prior decades but are still yielding new and useful methods:

time-resolved fluorescence (TRF), fluorescence resonance energy transfer (FRET), and fluorescence polarization (FP).

1.3.1 Time-Resolved Fluorescence

All fluorescence is time dependent, in the sense that unless continuous illumination is provided, the electronic excitation energy that gives rise to fluorescence will soon be exhausted, generally within nanoseconds. Moreover, even if the lamp never stops shining, photobleaching (oxidation of the sample induced by the radiant energy) will eventually attenuate the signal and alter the sample itself. However, the time dependence of certain fluorescent phenomena can be made to yield valuable information about the state of the fluorophore. In particular, fluorescence due to the label can be readily distinguished from autofluorescence or other adventitiously fluorescent molecules within the sample, since these sources of fluorescence decay much more rapidly than the emission of a fluorophore tuned for TRF through clever chemistry. Thus, TRF is made possible by the very long fluorescence lifetimes of the species employed for the purpose.

The lanthanide metals have so far proven to be by far the most useful fluorophores for TRF in the life sciences, with fluorescence lifetimes in the hundreds of microseconds, compared to roughly 1–100 ns for most fluorophores. By waiting to measure the fluorescence signal until light emission by the ordinary fluorophores has decayed, therefore, one can obtain a readout from the lable that is very nearly quantitatively pure. Among lanthanides, the Eu^{3+} (europium) ion is most commonly used, but Tb^{3+} (terbium), Sm^{3+} (samarium), and Dy^{3+} (dysprosium) also show the phenomenon and, fortunately, exhibit well-separated emission wavelengths, allowing multiplexing. In other words, separate molecules within the sample can be labeled with each of these lanthanides, which can then be quantified via fluorescence at different wavelengths.

Among a number of web-based resources further explicating the phenomenon of TRF, the excellent PerkinElmer material at the following URL is strongly recommended to the reader: http://las.perkinelmer.com/content/relatedmaterials/brochures/bro_delfiaresearchreagents.pdf.

1.3.2 Fluorescence Resonance Energy Transfer

Fluorescence resonance energy transfer, also known as Förster resonance energy transfer, is a process wherein one donor fluorophore in a sample emits a photon that, rather than proceeding out of the sample solution, excites an electron in a nearby acceptor molecule. The acceptor may or may not reemit the photon at a still higher wavelength; when it does not, the acceptor is known as a quencher. Because FRET very strongly depends on the separation distance between the fluorophores (in fact, it depends directly on the orbital overlap integral, which falls off with the sixth power of the separation), it can be used as a means of measuring intramolecular or intracomplex distances, although it is only sensitive within a small range that depends on the particular pair of fluorophores being used. Figure 1.4 presents a schematic diagram of the FRET process.

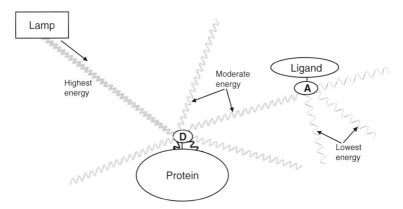

FIGURE 1.4 Schematic representation of FRET. Low-wavelength (high-energy) light from the lamp excites the donor fluorophore (D), which reemits light of moderate wavelength and energy. The acceptor (A) fluorophore is not excited by light at the wavelength of the lamp emission, but can accept fluorescent energy from the donor. After doing so, it in turn reemits light of a still longer wavelength (lowest energy). This process occurs only when the donor and acceptor are proximate, leading to emission at the lowest expected wavelength; when they are distant, only light of the moderate wavelength is observed. (See the color version of this figure in the Color Plates section.)

Over the years, FRET has enjoyed a productive career as a means of measuring separations of amino acid residues within proteins (6), assessing protease activity (e.g., with the EnzChek® Peptidase/Protease Assay Kit from Invitrogen), characterizing conformational changes in biomolecules (7), elucidating membrane dynamics (8), and studying protein–protein interactions (9). This is only a partial list of the assay methods that are possible with this versatile technique.

1.3.3 Fluorescence Polarization

As FRET is a means of obtaining information about molecular separations, so fluorescence polarization (formerly known as fluorescence anisotropy) is a way of measuring molecular size. This is accomplished by attaching a fluorophore to (or allowing it to associate with) the molecule or multimolecular complex under study. The sample solution is then irradiated with polarized light, the molecule or complex rotates, and the excitation energy is reemitted in a different direction (see Fig. 1.5). Because the average rate of molecular rotation depends inversely on the square of the size, it is precisely the time dependence of the emission that is informative as to the size of the fluorescent species. A large molecule rotates slowly (if a few dozen nanoseconds can be considered slow), yielding an emission signal later than a smaller fluorophore. It will occur to the reader that if the fluorophore is very small and can be made to associate with a large molecule or complex, the difference between the two FP signals will be correspondingly great. That is in fact the case, although FP can generally be

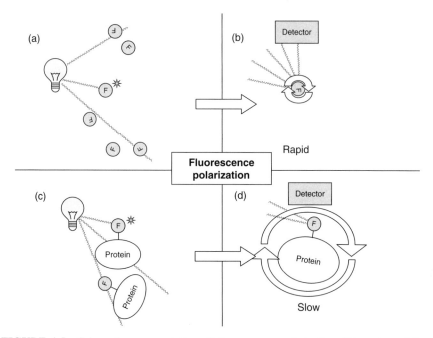

FIGURE 1.5 Schematic representation of fluorescence polarization. The unbound fluorophore of panel (a) is excited by polarized light from the lamp only if its excitation dipole is properly aligned. Other fluorophores are not excited. (b) The excited fluorophore rotates rapidly because of its small size, emitting light that reaches the detector, which is mounted at a 90° angle from the lamp beam. (c) The protein-bound fluorophore is also excited by the polarized lamp beam only if its dipole is properly aligned. The protein itself and other fluorophores are unaffected. (d) Because the rotation of the complex is slow, the fluorescence decays before the fluorophores dipole is aligned with the detector. Little or no fluorescence is seen. Thus the fluorescence reading depends on the size of the fluorescing complex. (See the color version of this figure in the Color Plates section.)

used to obtain binding information about two molecules of roughly the same size as well.

FP has been utilized in numerous recent studies of intermolecular associations and membrane phenomena, of which only a small number can be cited here (10–12). The method is highly versatile and often appears in surprising variations, such as a widely used kit available from Abbott Laboratories for assessment of homocysteine in plasma (13).

1.4 CHEMILUMINESCENCE AND BIOLUMINESCENCE

We return to our main focus to address these two forms of *autoluminescence* and explain the distinctions between them. The most prominent point is that while both fluorometric and spectrophotometric measurements rely on a lamp as the source of energy, luminescent reactions by definition supply their own energy in chemical form. The nature of the energy-containing molecule(s) and the means by which the energy is transduced to light are of high interest to workers in the field and have profound implications for the utility of the various light-generating reactions in life science research.

It is possible to be fairly specific about the varieties of chemistries generally seen in these reactions. Both chemiluminescence and bioluminescence involve the emission of a photon accompanying the decay of a hyperoxidized intermediate. The difference is in how the intermediate is generated: in chemiluminescence it is the product of an oxidative chemical reaction, without obligatory enzymatic involvement, while in bioluminescence an enzyme (or photoprotein, such as aequorin) plays a critical role in forming the intermediate. Each of these processes is described in more detail below.

1.4.1 Chemiluminescence

Chemiluminescence has been well described, sometimes in concert with bioluminescence, from the points of view of both chemistry and applications in several useful volumes (14–16). Like bioluminescence, chemiluminescence is currently enjoying a rapid expansion of the range of its possible applications, but in the case of chemiluminescence, this is due almost exclusively to the progress in the relevant chemistries, while in the bioluminescence field, molecular biological tricks play an active role. It has become a routine to expect that if a fluorogenic substrate is on the market for assays of a particular enzyme, a chemiluminescent substrate is or will soon become available.

Among the simplest chemiluminescent reactions from a conceptual point of view is the spontaneous interaction of luminol with hydrogen peroxide (H_2O_2) in the presence of a second oxidant (such as iron or copper) to produce 3-aminophthalate (3-APA) and an impressive amount of bluish light, according to the mechanism shown in Fig. 1.6.

This reaction is exploited in a number of important *coupled chemiluminescent* reaction series, including the one involving horseradish peroxidase (HRP). The action of HRP greatly increases the luminescence of the reaction. The detection method involves covalent conjugation of the HRP enzyme to a ligand of interest. Typically, the ligand associates with a protein or other species adsorbed or covalently linked to an insoluble matrix (e.g., a Western blot membrane). Free ligand is washed away, an HRP-conjugated antibody directed to the ligand is added, more washing occurs, and finally the substrates are added to yield the luminescent signal. Although the components of the luminol/hydrogen peroxide reaction are easy to list, the role of the second oxidant and the process whereby HRP enhances the luminescence are not yet clear.

The luminol/hydrogen peroxide reaction is useful with certain categories of defined reagents, but it does not constitute a general system for chemiluminescent detection of enzyme activity. The major trend of this kind is toward molecules that incorporate the simple structure depicted in Fig. 1.7. The first stable dioxetane was synthesized in 1968, and the molecule was found to luminesce upon thermal decomposition (17).

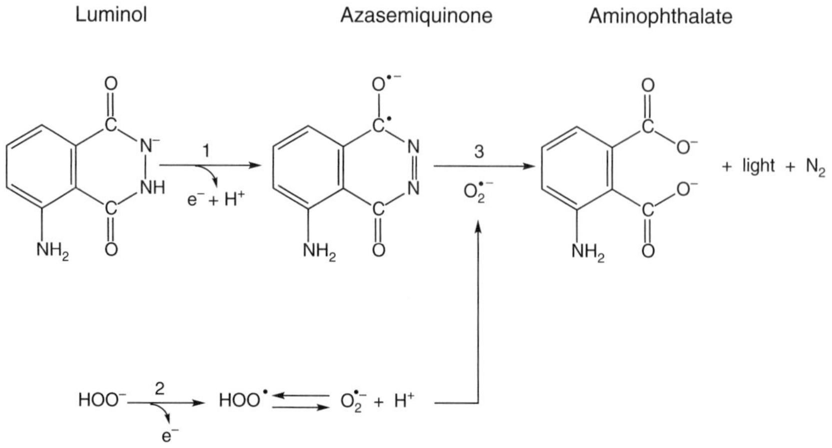

Luminol   Azasemiquinone   Aminophthalate

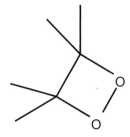

FIGURE 1.6 Luminol/peroxide light-generating reaction series. Luminol undergoes spontaneous reorganization to the azasemiquinone radical. The superoxide anion also generated by spontaneous processes from H_2O_2 attacks the electrophilic carbons to form a dicarboxylic acid, releasing N_2 and generating a photon of visible light. Reproduced from http://www.liv.ac.uk/Electrochemiluminescence/lusemiq.htm by permission of Dr. Robert Wilson.

Subsequent work on less stable 1,2-dioxetanes led to the issuance of a broad patent in 1994 covering "Enzymatically cleavable chemiluminescent 1,2-dioxetane compounds capable of producing light energy when decomposed" (18). One of the major advances of this work was the provision of water-soluble 1,2-dioextane substrates. Other advantages of the new admantylidene dioxetane compounds, including a more rapid approach to steady-state light emission, were claimed. Alkaline phosphatase, β-galactosidase, and (by implication) esterases, acyl transferases, glycosidases, amidases, and a wide variety of other hydrolases were among the enzyme groups named as actual or potential assay targets for these chemiluminescent substrates.

FIGURE 1.7 Common structure of the 1,2-dioxetanes, the predominant luminescent moieties of both synthetic and natural luminescent molecules. The light-generating process involves collapse of the dioxetane ring with liberation of CO_2.

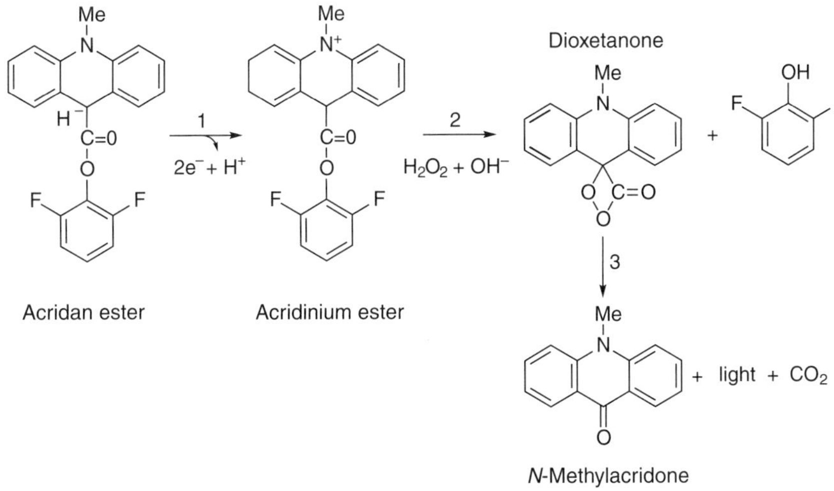

FIGURE 1.8 Common structure of the acridinium and acridan esters. Shown at the left is an acridan ester, which can lose hydride to become an acridinium ester under conditions that are subject to control. The acridinium ester reacts further with H_2O_2 and the hydroxyl ion, yielding a dioxetane structure that decomposes to yield light. The development of the acridan esters allows the chemiluminescent even to be initiated by an electrical signal; hence the term "electrochemiluminescence." Reproduced from http://www.liv.ac.uk/ Electrochemiluminescence/acmech.htm by permission of Dr. Robert Wilson.

Another development was the advent of the acridinium esters (Fig. 1.8), molecules that undergo rearrangements to yield 1,2-dioxetanes. This allowed greater control of the luminogenic process. Another advance was realized with the principle of *electrochemiluminescence* (ECL), which was also called "electrogenerated chemiluminescence" in early publications (19–21). This allows an even higher level of control of the timing and position of the chemiluminescent event. For example, the acridan ester 2′,6′-difluorophenyl 10-methylacridan-9-carboxylate decays to an acridinium ester in response to an electrical impulse when an appropriate oxidant is present, and the decay event can be timed to occur following an interval subsequent to a diffusion event, enzyme activity, or addition of another reaction component (such as coated magnetic beads) (22). This level of temporal tuning enables very sensitive antibody-based assays for small molecules such as *p*-nitrophenol and trinitrotoluene (TNT) (23). The reader is encouraged to visit Dr. Wilson's fascinating web site for more details regarding chemiluminescence and ECL: http://www.liv.ac.uk/Electrochemiluminescence/home1.htm.

The availability of chemiluminescent enzymatic substrates is expanding rapidly.

1.4.2 Bioluminescence

The central players in bioluminescence are of course the luciferase or photoprotein, which promotes the transformation of the prosthetic group that leads to light emission, and the prosthetic group itself. Both types of molecules come in a number of varieties, and the degree to which their reactions are understood also varies considerably, although the majority have been well characterized.

By definition, bioluminescence involves a protein that performs an essential catalytic role in forming the luminogenic intermediate. However, the term "catalytic" must be qualified in the context of photoproteins, since regeneration of the active photoprotein can be very slow; Shimomura (3) reports that several hours under strictly defined conditions are required *in vitro*. This is in keeping with the stoichiometric light generation observed in these systems, as described below.

The fascinating phenomenon of bioluminescence is found in many diverse classes of organisms and in fact spans the kingdoms. As Shimomura expressed it, "There is no obvious rule or reason in the distribution of luminous species among microbes, protists, plants, and animals." He goes on to quote Harvey: "It is as if the various groups had been written on a blackboard and a handful of damp sand cast over the names. Where each grain of sand strikes, a luminous species appears." The analogy is a particularly apt one. Part of a clump of wet sand disintegrates into individual grains, while part remains in larger fragments; the luminous species are found in similar manner, some phylogenetically isolated, others with many luminous evolutionary neighbors.

Energy is of course required to produce light, but it is perhaps surprising that in some cases of bioluminescence, no conventional energy-storing molecule such as ATP or nictonamide adenine dinucleotide (NADH) is directly consumed during the reaction. Energy is stored instead within the photoprotein/prosthetic group system for release as luminescence in response to a triggering event. Despite their single-turnover characteristics, some of these proteins, notably aequorin, are proving to be extraordinarily useful in the laboratory, and for CB reactions in particular. Such critical matters of detail pervade the study of bioluminescence at the level of the organism, and as a result the classification of bioluminescent systems presented below is organized roughly according to taxonomic considerations, beginning with the enzymes that are most useful or promising in CB assays.

1.5 COMMON BIOLUMINESCENCE SYSTEMS

1.5.1 Firefly Luciferase

Insects of three orders carry out bioluminescent reactions using the same prosthetic group (luciferin) and luciferases that are probably homologous to each other. These orders are Collembola, Hemiptera, and Coleoptera; of these, Coleoptera includes the *Lampyridae* (fireflies), luminous worms, and beetles in which the luciferases have been characterized to some extent. Firefly luciferase, which will be our focus for the remainder of this section, was purified and partially characterized in 1956 (24). The activity of the luciferase of the American firefly, *Photinus pyralis*, was the first to be demonstrated *in vitro* (25); the diffusing energy source, ATP, was identified three decades later (26). Whether coincidentally or otherwise, this system, the first to be characterized biochemically, has also proven to be by far the most useful in the life sciences. An important reason for this is the "glow" kinetics of the reaction. The luminescence of firefly luciferase (in the presence of an adequate ATP concentration and other needed reagents, including the luciferin cofactor, which is the actual luminogenic prosthetic group, as well as coenzyme A and the magnesium ion) lasts for hours, and its lifetime can be extended still further by the use of reagents that protect the enzyme from oxidative inactivation.

1.5.1.1 Firefly Luciferase Reaction Mechanism Figure 1.9 is a simplified reaction scheme of firefly luciferase. In the first step, the energy of ATP hydrolysis is transduced to form the dioxetane derivative of luciferin, which subsequently undergoes chemical decay, with concomitant emission of yellow-green light. The mechanism of the light emission step and the exact identity of the luminogenic intermediate, however, have been extraordinarily difficult to study, and it is not clear whether consensus has yet been reached. Until about 1990, the emitting species was considered to be a dioxyanion intermediate resulting from decarboxylation subsequent to the hydrolysis of the adenylate ester bond (27); however, later work appeared to show that two distinct monoanionic oxyluciferin derivatives were responsible for both the natural

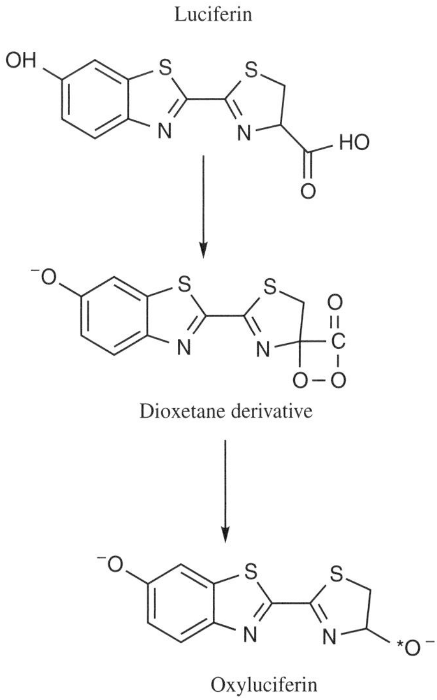

FIGURE 1.9 Simplified series of transformations of the luciferin prosthetic group during the light-generating reaction of luciferase. Luciferin first reacts with molecular oxygen to form a 1,2-dioxetane in a step that utilizes the energy of hydrolysis of ATP. This species then undergoes decarboxylation to generate an excited oxyanion, leading to luminescence.

luminescence and an observed emission mode in the red (28). Those with special interest can consult the first chapter of Shimomura's volume (3) for a valuable discussion of the mechanistic issues. It is worth noting that a proposed intermediate in the light emission is a dioxetanone, which undergoes decarboxylation (29) in a reaction strikingly similar to some of the chemiluminescent schemes. Virtually all luciferases appear to have mechanisms involving decarboxylations of similar structures (30).

1.5.1.2 Means of Enhancing Firefly Luciferase Luminescence   The luminescence of the reaction as shown in Fig. 1.9 reaches its maximum intensity within half a second, followed by rapid decay over the ensuing 2–3 s and a much slower diminution over several minutes or hours. The cause of this kinetic behavior has been proposed to be a by-product of the reaction, dehydroluciferyl adenylate, which is held to strongly inhibit luciferase; according to this model, its buildup early in the reaction process is responsible for the initial rapid decay. The addition of coenzyme A alleviates this potential problem, since the coenzyme displaces the adenylate moiety, forming a luciferin derivative that is only very mildly inhibitory. With reductants supplied in excess, therefore, the long-lasting glow kinetics that have proven so useful in laboratory work are observed (31, 32). The commercial world was initially rather slow in adopting this idea, but Promega Corporation, a leader in the field of bioluminescence, introduced a "glow"-type luciferase assay kit in 1991. However, the Promega workers also challenged the model of the observed inhibition, attributing the effect of coenzyme A to its reducing power, rather than the given displacement reaction, and, indeed, other reductants supplied in high concentrations appear to convert the reaction to the "glow" type (33).

In practice, several parameters enter into both signal intensity and signal lifetime, including the concentrations of the buffer constituents and the source and treatment of the luciferase itself. Some commercial assay systems are designed for experiments requiring very long lifetimes of the glow reaction, while for others, flash kinetics are ideal and coenzyme A is not necessary or desirable. The author has found one or another of the commercial kits to be suitable for development of most CB assay methods, but the trade-off of sensitivity and lifetime suggests the benefits of evaluating more than one kit for the needs of a particular application.

1.5.1.3 Genetically Engineered Firefly Luciferases   Since the widespread dissemination of genetic engineering techniques, cloned firefly luciferase has been modified by genetic means in various ways to enhance expression and reduce posttranslational modification (34), *increase* the K_m of the enzyme to alter the regime of ATP quantification (35), tune the wavelength of maximal light emission (36, 37), improve thermal stability (38), and even introduce cleavage sites to enable a novel CB protease assay (39) (see Chapter 10). Generally, these mutant enzymes are available from specific vendors, especially Promega.

1.5.2 *Renilla* Luciferase

The luciferase enzyme from the sea pansy *Renilla* (40, 41) presents an informative and fascinating contrast with the firefly enzyme. Despite the fact that the prosthetic group is the same *coelenterazine* (2-(*p*-hydroxybenzyl)-6-(*p*-hydroxyphenyl)-8-benzylimidazo[1,2-*a*]pyrazin-3-(7*H*)-one) found in association with photoproteins such as aequorin (see Section 1.5.4), this is a true luciferase, as opposed to the single-turnover luminescent proteins found in other coelenterates (see Section 1.5.5). The *Renilla* enzyme exhibits a highly respectable optimum turnover number of nearly 2 s^{-1}.

1.5.2.1 Renilla Luciferase Reaction Scheme and Emission Characteristics The
reaction scheme of *Renilla* luciferase (Fig. 1.10) bears similarities to those of both the beetle enzymes and the coelenterate photoproteins. Like these photoproteins, the enzyme acts on coelenterazine and the activity is under the control of calcium, but the coelenterazine, instead of being sequestered by the protein for use, is activated by a different system. The inactivated form is a sulfated derivative of coelenterazine; the conversion is catalyzed by the enzyme luciferin sulfokinase in a reaction that requires $3',5'$-diphosphoadenosine as an effector (42). The presence of the nucleotide as an effector that is not consumed in the activating reaction stands in fascinating contrast to the requirement of the firefly enzyme for the related molecule ATP, which is of course consumed when fireflies glow.

The emission maximum of *Renilla* luciferase is 480 nm. The separation of this wavelength from the firefly maximum of 562 nm suggests the possibility of dual-mode detection, and this possibility has been exploited quite successfully, as we shall see. Another complexity, however, is introduced by the presence of the equally interesting protein known as "green fluorescent protein" (GFP) in the same organism. *Renilla*, like many luminous marine organisms, actually fluoresces in the green, rather than in the blue, as one would anticipate from the 480 nm emission maximum of the luciferase. This is due to nonradiative transfer of the excitation energy to GFP in the phenomenon known as Förster-type resonance energy transfer or, in the more modern term, fluorescence resonance energy transfer. In essence, the energy of the excited coelenterazine molecule is transferred with high efficiency to GFP, which radiates the energy as green light.

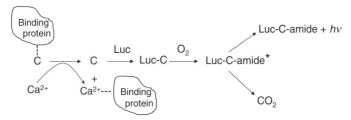

FIGURE 1.10 *Renilla* luciferase reaction scheme. The luminescent prosthetic group coelenterazine is displaced from its binding protein by Ca^{2+}, releasing it to react with the luciferase (Luc). The reaction with molecular oxygen produces the excited state of the complex of the photoprotein with coelenteramide (C-amide), followed by the liberation of the oxidized coelenteramide, CO_2, and light (*hv*).

1.5.2.2 Commercial Development and Practical Applications of Renilla Luciferase Luminescence   *Renilla* luciferase was cloned and sequenced in the early 1990s (43) and was soon in use as a marker of gene expression in diverse cell types (44). Promega's introduction of the Dual-Luciferase Reporter Assay in the 1990s has subsequently allowed this firm, already an outstanding player in CB technologies, to enjoy a very strong market position in two-enzyme luminescent assays, although other firms are making effective counterthrusts, as we shall see in Chapter 12. The *Renilla* enzyme enjoys no major advantages over firefly luciferase on its own, but the fact that its substrate requirements and emission spectrum differ from those of the firefly enzyme gives rise to this very effective two-reporter system, which is useful even in intricate, high-resolution studies of promoter structure–function relationships (45). (However, an alternative dual-reporter system employing firefly luciferase and aequorin was developed and used academically in the same time frame (46).)

1.5.3   Bacterial Luciferases

Before the development of the germ theory, the luminance of bacteria was long attributed to other entities, such as phosphorescence of the ocean, glowing cadavers, and so on. Bacteria were identified as the luminous source in the nineteenth century, and the requirement for long-chain aldehydes of fatty acids was discovered in the 1950s (47). Dihydroflavin mononucleotide $FMNH_2$, a powerful reducing cofactor that participates in the activity of a range of redox-active enzymes, is also required for the reaction to occur. Bacterial luciferase has been used to quantify another important reducing cofactor, nicotinamide adenine dinucleotide, but the role of NADH in the luminescent reaction is merely to supply reducing equivalents for the essential reduction of flavin mononucleotide to its reduced form, a reaction catalyzed by FMN-reductase (48). NADH is much more stable to spontaneous oxidation than $FMNH_2$; hence, the reaction series has evidently evolved to produce $FMNH_2$ *in situ* when needed.

1.5.3.1 Mechanism of Action of Bacterial Luciferases   Elucidation of the mechanism of bacterial luciferases has been a challenging problem, and many scientists have made contributions to our understanding of the very complex reaction series (49–53). Bacterial luciferases may be viewed as mixed-function oxidases that utilized molecular oxygen to oxidize two substrates in a single turnover: $FMNH_2$ and the long-chain aldehyde, which is oxidized to a carboxylic acid. The reaction appears to begin with the formation of a peroxide derivative of $FMNH_2$ by the attack of molecular oxygen, followed by the engagement of the free aldehyde by the peroxide oxygen and subsequent departure of the acid, leaving a hydrated FMN cofactor in an excited state. The decay of this state generates the photon, followed by a dehydration to regenerate FMN for possible recycling by FMN-reductase.

The essential involvement of the $FMNH_2$ cofactor in the mechanism gives rise to the main difficulty in developing CB assays with bacterial luciferase, which is the "flash" nature of the luminescence. $FMNH_2$ is so unstable in the presence of oxygen that any amount introduced into the solution is oxidized to FMN within seconds. This leads to a rapid decay of the luminance, typically with a half-life of approximately

20 s. This may be adequate for "inject-and-read" assays, but for most CB schemes, it introduces an unacceptable constraint, since it is impossible to follow most dynamic processes effectively within this time frame. Provision of NADH, FMN, and FMN-reductase in the reaction solution prolongs the half-life to some degree, but the reaction still does not acquire the useful "glow" characteristics of properly engineered firefly luciferase reactions. Poor solubility of the fatty aldehydes can also become an issue, and the process is further complicated by the organic solvents employed to dissolve the substrates.

1.5.3.2 Coupled Bioluminescent Assays Employing Bacterial Luciferases
Despite the difficulties described, CB assay methods incorporating bacterial luciferases have been successfully developed for specialized research applications (54), enzyme assays (55, 56), and even quantification of small molecules (57–59). The total number of reports of this nature is modest but adequate to show that the technical problems can be surmounted when there is a clear need for a CB assay using these luciferases.

1.5.4 Aequorin

The discovery and study of the jellyfish photoprotein aequorin have contributed greatly to the vitality of research into CB assays in recent years. While some writers lump luciferases along with aequorin-like proteins into the general category of "photoproteins," we use the latter term here to refer to those proteins in which light generation is roughly stoichiometric with the protein quantity in the seconds-to-minutes timescale; in other words, proteins in which rapid catalytic turnover is impossible. (We also exclude fluorescent proteins such as GFP from this category.) By this definition, aequorin is the first-discovered photoprotein and remains by far the most important of this class.

  Aequorea aequorea is a jellyfish found very widely in the Atlantic, Pacific, and Mediterranean. The organism looks like a veined blue umbrella, with the light organs containing the aequorin located at the outer fringe. F. H. Johnson built a machine for cutting jellyfish to separate the light organs; several hundred specimens are required to obtained a milligram quantity of active aequorin (60).

1.5.4.1 Biochemical Properties of Aequorin   It is impossible to improve on Shimomura's description: "Aequorin … is an unusual protein that holds a large amount of energy. The energy can be released in the form of light through an intramolecular reaction triggered by calcium ions" (3). The direct source of the energy is the coelenterazine prosthetic group, which undergoes oxidative decarboxylation during the light generation step, yielding coelenteramide. Coelenterazine appears to be bound to aequorin in the form of a hydroperoxide. Mechanistic studies are still underway to elucidate the unusual process whereby the calcium ion evidently induces conformational changes that enable the decarboxylation step, an elegant example of a "mechanical" triggering of luminescence within a protein. However, much of the recent work has been performed with the photoprotein obelin (61–63), which is closely related to aequorin.

Aequorin responds to the calcium concentration in a manner entirely consistent with its nature as a single-turnover protein. Concentrations of calcium of 10^{-5} M or higher cause a rapid rise in luminance, accompanied by exhaustion of the active aequorin in ~ 10 s or less, while the linear phase with 10^{-6} M calcium persists for 10–20 s, and submicromolar concentrations yield a luminescent signal that is roughly constant for several minutes. Stoichiometric work suggests that two calcium ions are needed to induce aequorin luminescence.

1.5.4.2 Assays with Aequorin   The simplest assay employing aequorin that is generally useful in the life sciences is the straightforward quantification of calcium. Unless one is prepared to undertake sophisticated quantum counting with meticulously calibrated reagents and instruments, a calcium standard whose concentration has been determined by another method is needed. Because of the flash nature of aequorin luminance (due in turn to the single-turnover behavior discussed above), the dynamic range of such an assay is limited. Linearity can therefore be a major concern.

Strictly speaking, this method does not meet our definition of a "coupled bioluminescent" assay, because only a single protein is required for all the events to occur. (In the same sense, a direct measurement of ATP concentration using firefly luciferase is also excluded.) However, CB assays employing these principles are readily imagined. Calcium concentrations are perturbed by a range of physiological and biochemical events, of which one of the most important in current drug discovery is the activity of the many G-protein-coupled receptors (GPCRs) under investigation. The use of aequorin in CB assays in this context is discussed in Chapters 9 and 11. Companies such as PerkinElmer do not market aequorin directly so much as cell lines that express recombinant aequorin, providing a means of quantifying calcium *within* living cells and cellular compartments.

However, active aequorin can be purchased from chemical supply firms such as Sigma–Aldrich, enabling the assay developer to invent novel assays involving calcium detection and quantification. Aequorin is also currently touted as an "anti-aging" dietary supplement and may be available over the counter from corresponding sources, although its activity should not be taken for granted.

1.5.4.3 Recombinant Variants of Aequorin   Like firefly luciferase, aequorin has been subjected to genetic engineering in various ways, originally to facilitate purification (64), but especially in the present day to produce generic chimeric (fusion) proteins with multiple functions, one of which is, of course, calcium sensing (65–68). Many constructs have been tested successfully and it is clear that this protein is highly robust to these manipulations.

1.5.5   Other Luciferases and Photoproteins

The variety of light-generating proteins and their chemical mechanisms is so great that future adaptation of novel members of these classes to the needs of the research and drug development enterprises is virtually certain to occur. We mention here primarily the systems with mechanisms that appear to be biochemically distinct from those of the firefly, *Renilla*, bacteria, and *Aequorea*, to aid those ambitious scientists who wish to bring new reaction schemes into the family of CB assays. As always when bioluminescence is concerned, Shimomura's work provides an excellent reference if more detail is desired (3).

The so-called railroad worms of Central and South America generate light by a mechanism that involves firefly luciferin and depends on ATP and the magnesium ion, but different isozymes yield different wavelengths in separate parts of the worms' bodies (69, 70), providing potential material for structure–function studies. Similarly, isozymes cloned from the Japanese click beetle yielded luciferases that generated several colors, although the wavelengths were all within 50 nm of each other (71).

Certain glowworms of the Appalachians luminesce at wavelengths that are the shortest known for insects (72). Evidence obtained to date indicates that the luminescence does not depend on ATP, but is stimulated by dithiothreitol, a strong biological reductant. It is as yet unclear whether this kind of luciferase can be adapted for use as a bioluminescent redox sensor.

Many of the minute crustaceans known as ostracods are bioluminescent. Among these is *Cypridina hilgendorfii* Muller, which has played an important role in the elucidation of mechanisms of bioluminescence. The Japanese army actually ordered the collection of this species for military purposes during World War II, but the use never occurred. Subsequently, intrepid researchers at Princeton (in the 1920s) and, later, at Nagoya University made progress in purification and characterization of the luciferase and luciferin from this system (73–75), despite the paucity of information on the nature of bioluminescent reactions in general. The power source of *C. hilgendorfii* luciferase is molecular oxygen, which is believed to engage the luciferin of the same animal in forming one of the highly unstable dioxetanone species that are becoming familiar to us. Light emission accompanies decarboxylation. The degree of convergence of these highly diverse luminescent systems toward very similar mechanisms gives a hint as to the remarkable nature of this class of reaction schemes. While *Cypridina* luciferase has not received as much attention as the firefly enzyme has received in recent years, CB assays utilizing the ostracod enzyme have appeared (76–78). There are other oxygen-dependent luciferases among the crustaceans that await full characterization (79), as well as enzymes employing coelenterazine as the prosthetic group along with oxygen (80). Many other marine organisms, including mollusks, also express oxygen-dependent luciferases (3). Shimomura also describes many luciferases of various phyla (including, e.g., luminous fungi) whose mechanisms may involve radical intermediates and/or the participation of excited oxygen species such as superoxide. Some of their reactions do not involve decarboxylation, and some are stimulated by hydrogen peroxide, with or without molecular oxygen. Such mechanistic concerns, while of great interest to enzymologists, are discussed herein only when the molecular requirements of the luminescent reaction suggest a possible role in development of CB assays.

The photoprotein symplectin is obtained from the squid *Symplectoteuthis oualaniensis* of the Indian and Pacific oceans. The luminescent reaction of this protein employs coelenterazine and requires molecular oxygen, but also a monovalent cation (81), suggesting the possibility of CB detection schemes for these ions. Distinct from this is the luminescence of *Odontosyllis*, the Bermuda fireworm, of which the females provide a spectacular mating display in the evenings following a full moon. The males also give light as they approach the females. The reaction of the fireworm luciferase requires the magnesium ion, along with molecular oxygen. Larger dications are generally inhibitory, suggesting the possible value of the protein as a magnesium sensor.

1.6   A COUPLED BIOLUMINESCENT REACTION

With the tools in hand, it remains to investigate an example of a CB reaction, and there can be no better choice to begin with than Arne Lundin's original development, the CB assay of creatine kinase (82) (Fig. 1.11). Each of the two steps of this scheme is eminently reasonable from every point of view, including thermodynamics. In this series, creatine kinase is doing no more than producing ATP from creatine phosphate and adenosine diphosphate (as well as creatine as a by-product), which is precisely its role in the body. ATP is of course the body's most accessible and immediate source of chemical energy, but to maintain homeostasis, mechanisms have evolved to keep the concentration of ATP fairly constant. Creatine phosphate is the proximal backup molecule that can provide ATP units with great rapidity through the action of creatine kinase on creatine phosphate. This rapid response is made possible by the

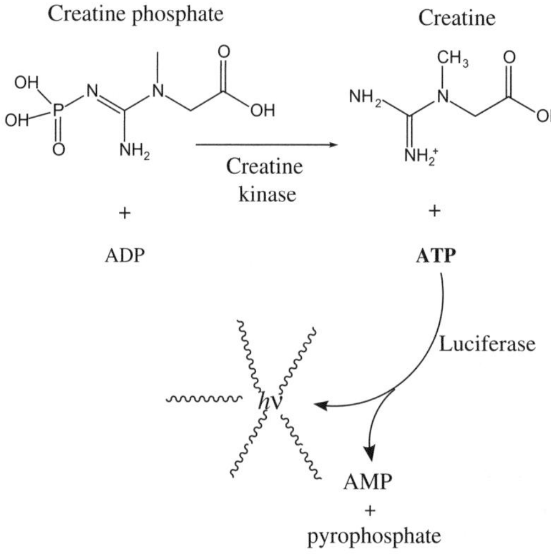

FIGURE 1.11   Lundin's coupled bioluminescent assay for creatine kinase activity. This, the first coupled bioluminescent assay published, is based on the simple concept that ATP evolved by the kinase-catalyzed reaction can be used by luciferase to produce visible light in proportion to the concentration of the kinase. Creatine kinase catalyzes the transfer of the phosphate group from creatine phosphate to ADP, generating ATP and enabling the light reaction.

fact that the transfer of the phosphate group to ATP is thermodynamically favorable (i.e., the terminal phosphodiester bond in ATP is lower in free energy, and therefore more thermodynamically stable, than the corresponding bond in creatine phosphate), and this same fact was likely the reason that the reaction was chosen as the first published CB assay, one in which both the steps are exergonic (thermodynamically favorable). This, however, is the opposite of the direction of action of most kinases from a thermodynamic point of view. ATP is usually the energy *source*, rather than the *sink*, and the process of transferring the phosphate group to a protein or other target is nearly always energetically favored; the inescapable corollary is that the desirable formulation of the CB assay, in which ATP should be formed, is thermodynamically disfavored. This might appear to preclude the use of CB assays in measuring the activity of kinases, but such is not the case, as we shall see in Chapter 4.

Although Lundin successfully patented his creatine kinase assay, there were only a modest number of developments in the field of CB assays over the ensuing 17 years. In 1997, however, this author and his colleagues published the first CB cytotoxicity assay (83), and the field has shown signs of flourishing since that time, although it is still limited in scope, largely because most scientists are not even aware that the methods exist. It is the author's hope to rectify that situation with this book.

While he did not publish separate studies on every possibility, Lundin envisioned long ago the potentially wide range of CB assays that has appeared on the commercial scene only recently (84–86). It is delightful to find that Dr. Lundin has still been contributing to luciferase research quite recently, for example, with an examination of the relationships between the lifetimes of various luciferase reaction formulations and their abilities to quantify extremely low levels of ATP (87).

1.7 SUMMARY

In this chapter, we have been exposed to the physical principles that distinguish CB assays from spectrophotometry and fluorometry, the two most important competing methods. We have also seen many of the enzymes that may be adapted to use in CB assays and looked at the characteristics of their luminogenic reactions. In the next chapter, we will address the question of how to develop, perform, test, and interpret a CB assay and examine key parameters in the decision of whether to use this approach to measure one's phenomenon of choice.

2

COUPLED BIOLUMINESCENT REACTIONS IN PRACTICE

2.1 PRINCIPLES OF COUPLED BIOLUMINESCENT REACTIONS

In a bioluminescent reaction, a small molecule in an excited state is made to emit light in the presence of a protein that facilitates the reaction. Since the initial excitation requires energy input, the reaction depends (or can be made to depend) on the presence of either an energy-containing molecule, whose energy can be transduced to yield the excited state, or a molecule that triggers the release of chemical and/or conformational energy. A *coupled* bioluminescent (CB) reaction differs from this only in that the presence of the initiating molecule is itself the culmination of a separate event or series of events that is of interest to the investigator. Thus, quantification of ATP by luciferase is a bioluminescent assay, but not a CB assay by our definition, since there is no coupled process. Measurement of cell death by such an ATP assay, however, or assessment by such a method of a separate phenomenon that produces or consumes ATP, clearly falls within the category of CB assays.

2.1.1 Requirements for Successful Coupled Bioluminescent Assays

Here, we discuss the conditions that must be fulfilled for the successful performance of CB assays, with the caveat that certain other conditions are desirable but not strictly essential. For example, even in cases where the biochemical conditions for the two coupled steps are necessarily different, it is often possible to develop a CB reaction by adding appropriate reagents, usually in concentrated form, to adjust conditions prior to the luminescent reaction. However, this is required only in a surprisingly small

Coupled Bioluminescent Assays: Methods and Applications, Michael J. Corey
Copyright © 2009 by John Wiley & Sons Inc.

proportion of systems. Of roughly 20 CB assays developed by the author, it has proven possible to run all of them without separate steps by identifying buffers compatible with the various coupled processes, and in only one case (that of calcineurin; see Chapter 5) was a clear advantage in signal observed by providing separate conditions. Thus, the assay developer should carefully consider the potential advantages to the operation of a single-addition assay and evaluate its cost and feasibility.

2.1.1.1 Biochemical Compatibility The principle of biochemical compatibility, while somewhat tautological, is essential to understand and consider when designing and performing these assays. To observe light, one must of course provide conditions that are compatible with its generation. Potential sources of interference with the light reaction range from the obvious excesses of heat, acidity/alkalinity, destructive proteolysis, protein denaturants, and redox status to those pertaining to the specific requirements of the individual luciferases and photoproteins. The beetle luciferases, for example, hydrolyze ATP as an essential step in their mechanisms; these enzymes therefore possess ATP-binding sites, and chemical analogs of ATP may well interfere with these sites and inhibit the luminescence process. It is less obvious that other phosphate-containing small molecules may lead to analogous interference. A notable example is pyrophosphate, which can not only serve as a competitive inhibitor of ATP but also help to initiate the backward reaction of luciferase (*generation* of ATP in the presence of light, pyrophosphate, and AMP), greatly complicating interpretation of the assay results.

The actual species that participates in the luminescent reaction is not ATP, but a complex of ATP with the Mg^{2+} ion. The cation is essential, and reagents that chelate or sequester this ion, such as EDTA and its various cousins or apoenzymes, may shut down the light reaction. Note that EDTA is, however, frequently found in formulations of CB reactions, where its usual purpose is to exert precise control of the final concentration of Mg^{2+} or other dications.

Many luciferases and photoproteins are sensitive to oxidative conditions. In the case of firefly luciferase, this sensitivity is an aspect of its essential nature, and the enzyme is known to undergo spontaneous inactivation through oxidation by a process that is greatly accelerated by enzymatic turnover of the enzyme; in short, this is a form of mechanism-based inactivation. The provision of adequate reductants to protect the enzyme from these effects greatly improves the stability of luciferase and the signal it generates (33).

2.2   INSTRUMENTATION AND EQUIPMENT FOR COUPLED BIOLUMINESCENT ASSAYS

UV/spectrophotometers, gamma counters, fluorometers and multipurpose readers, all of these are really luminometers, since all of them measure the impingement of photons on one or more detectors. Most of these can in fact be used for measurement of bioluminescent and CB reactions, although this should not be assumed without trial. An obvious requirement of luminescent measurements is that the detector should not be flooded with useless (for our purposes) light from the instrument's internal lamp. All of the signals should come from the reaction. It is incredible to this author that many reading devices (microplate readers intended for dedicated, single-purpose fluorometric readouts are among the worst culprits) *do not allow a reading with the lamp off.* Indeed, even blocking the light source with opaque paper or other materials fails as a work-around, since the software senses the fact that no photons from the lamp are reaching the detector in the absence of a sample, concludes that the lamp is broken or burnt out, and refuses to perform. All one can do in such a case is to make a firm resolution never again to purchase an instrument that assumes it is wiser than its user.

Fortunately, with the above-noted exception, virtually all of these instruments can be adapted to use in CB assays. Gamma counters are often marketed as dual-capability instruments, and multimode microplate readers, capable of UV/spec, fluorometry, luminometry, and even exotic fluorescence modes such as fluorescence resonance energy transfer (FRET), fluorescence polarization (FP), time-resolved fluorescence (TRF), and even time-resolved fluorescence resonance energy transfer (TR-FRET), are becoming common. FRET, FP, and TRF are described in Chapter 1.

However, it is important to note that not all instruments capable of reading luminescence can do it equally well. Unfortunately, the marketplace is too dynamic and the evolution of new features too rapid to allow a useful comparison of individual instruments in a volume of this nature, but certain general principles may help the assay developer in making a decision as to whether to invest in a dedicated luminometer or a multimode or other instrument.

First, consider the old-fashioned single-tube luminometer, compared with its single-sample spectrophotometric and fluorometric cousins. With a read time of one to several seconds for each sample, the tube in the luminometer could be lowered or otherwise enshrouded in a detection chamber separate from all other samples. Apart from the sample, the chamber could be completely dark. Every photon represented signal. There is a sharp contrast between this scenario and a UV/spectrophotometry cuvette chamber, where the whole light path had to be precisely collimated so that every photon impinging on the sample from the lamp had an equal opportunity to pass through the solution and enter the detector. Reflections represented noise, as did any light that found a path to the detector without entering the sample. With fluorometry the situation was more complex, since all of the incident beam from the lamp represented noise; only the Stokes-shifted light reemitted by the sample could be used to generate signal. Even with a 90° separation between the excitation and emission light paths, light-scattering materials (such as dust or precipitate) in the sample could divert incident photons to the detector, generating noise. Finally, of course, the fluorescent signal represented only a tiny fraction of the incident light energy, since it was reduced not only by the quantum yield of the sample, but also by the limitations of the geometry, which allowed only photons reemitted along the particular detection pathway vector to be interpreted as signal.

With modern microplate readers, the situation is completely different. The requirements of the various detection modes are entirely distinct. Reading UV/ spectrophotometric signals in these instruments requires transparent plates, since the detector must be collinear with the light source. Fluorometric readings require a different geometry, and the approach that is most commonly used is to illuminate the sample from above and read fluorescence at approximately the same place. The fluorescing molecules detected are therefore those that undergo a rotation of roughly 180° and reemit their photons in the opposite direction. Rotation of any molecule or complex of a size less than that of a cell is extremely rapid relative to the reading time, rendering *anisotropy* of the radiation (i.e., emission with an angular dependence) irrelevant in ordinary fluorometry. However, the phenomenon of anisotropy can be exploited in instruments with special detection timing and geometry via the technique known as fluorescence polarization, or the time-dependent appearance of a fluorescence signal due to molecular rotation. The situation in an instrument that reads at 180° is little different from that of the old 90° collimated fluorometers, except that one would expect to wait approximately twice as long for a given fluorophore to rotate twice as far (although random walk statistics enter this calculation).

Luminescence measurements in microplates have distinct requirements. Since the sample can no longer be carried away from other samples, it must be optically isolated in some other way. Opaque plates are absolutely essential for microplate luminescence measurements. Moreover, unlike the cases of UV/spec and fluorometry, luminescent samples are expected to have luminescent neighbors; thus, the detector geometry must be such as to accept only light from the sample being read and reject the photons from several samples only millimeters away, which may be shining quite brightly. Naturally, this problem is addressed with a combination of optical shielding (generally via close physical coupling of the detector to the well circumference), filters and mirrors, collimation of the light path, and various kinds of signal processing. The advent of the (CCD) charge-coupld device camera and the possibility of obtaining high-resolution images of plates for digital processing have improved the resolution of adjacent signals and accelerated the trend toward simultaneous reading of all the wells even of high-density plates. The reader with technical interest in these matters is encouraged to consult relevant patents, as well as other documents cited therein (88–90). For the purposes of choosing a luminometer for use in practical performance of CB assays, however, the following considerations should be addressed.

2.2.1   Instrument Testing

Many issues enter into the suitability of an instrument for CB assays, some of which are not at all obvious. There are modern instruments based on such technologies as diode arrays and CCDs that can capture entire plate images in a matter of seconds, but if one is to settle for an instrument that reads the wells in a serial fashion, one must consider the cycle time of this process and whether it is rapid enough to capture the luminance information one desires. When flash luminance is the readout, this problem is often eliminated by the simple expedient of the dispense-and-read protocol, which is frequently used with chemiluminescent reactions; however, many important CB reactions are "glow" in nature, and a delay is necessarily interposed between reaction initiation and reading to allow the signal to develop. Glow reactions are actually superior to flash reactions from several points of view, including the opportunity to take multiple readings, the possibility of integrating several readings separated in time (which may allow suppression or robust handling of reaction or instrument artifacts), and the ability to read and interpret changes in the luminance signal and their relationship to the phenomena under study. However, if serial reading of such reactions is to be done, then the user must consider, and preferably test, the process in question with the instrument of interest.

One approach that must be avoided is the use of the "statistical cutoff" read approach employed by some scintillation-counting devices. Here, the length of the read is not measured; instead, the counting process terminates when some predetermined threshold of statistical significance is exceeded. While this may be effective with the very time-constant process of radioactive decay, it is obviously inappropriate for dynamic luminance signals.

Assuming the instrument can read the wells quickly enough to answer the important questions, its sensitivity should be tested. The limit of detection (defined as the level three standard deviations above the background signal) of ATP should be determined, as well as the dependence, if any, on position in the microplate. This is a parameter that is constantly improving with new technology, but as of this writing, there is no need to settle for a limit of detection of greater than 10 amol per well of a 96-well or 384-well plate, and a subattomole limit may well be achieved.

Next, cross talk should be tested. Reagents generating the highest anticipated signal should be measured in wells adjacent to blank wells, and the effect on the blank readings should be determined. (This should be done for all positions in the plate.) The volume of reagent used should be the maximum that is anticipated for regular use; for example, a 250 μL volume in wells of a 96-well plate should be enough to reveal any cross talk due to this effect. It may or may not be possible to achieve zero cross talk under these conditions, but the degree should be known and a decision made as to whether it is acceptable and can be accounted for. Black plates are somewhat less likely to yield cross talk than white plates.

If the instrument is to be used in a drug discovery enterprise, formal validation of its capabilities in the context of the assays to be performed will be required at some point. Some information on this topic is provided in Chapter 13, but validation as a topic is generally outside the scope of this volume. Appropriate expertise is essential for successful assay validation, and the assay developer should acquire suitable training and consult the relevant documents of the FDA and/or other regulatory authorities.

2.2.2 Instrument Features

2.2.2.1 Plate Shaking   Plate shakers can be useful in UV/spectrophotometry, where a submicroliter's worth of contamination has few serious consequences, but in CB assays they are a mixed blessing. It is often possible to engineer the order of addition and physical means of reagent transfer so that mixing of the reaction by shaking is not needed. This is especially true if the final step of reagent addition involves introduction of a considerable volume (at least 20 μL, preferably 50 or more μL) of reagent in such a way that turbulent flow is produced. In solutions of normal viscosity, a good technician or a well-designed injector (see below) can mix a full well of a 96-well plate simply by transferring a volume of this magnitude. The first readout of a glow CB reaction is usually taken after some delay in any case, allowing the turbulence to do its work. Even an injection of 5 μL can cause effective mixing of a volume 10–20 times as great if the injector is so engineered as to deliver angular momentum to the solution.

Nevertheless, there will be instruments and assay conditions that require shaking of plates for adequate mixing. If this is the case, two types of tests should be performed. The first is a gross test for visible cross-contamination, most conveniently carried out by alternately loading wells of clear buffer and wells of dye solution, staggered down the plate. Shaking should be as vigorous as the user anticipates will ever be needed, and of course no cross-contamination should occur at ordinary loading volumes. If this test is passed, the more rigorous version should be performed: wells loaded with high ATP concentrations should be alternated with blank wells and shaken, followed by addition of firefly luciferase. This test is necessary because extremely small volumes of solution, such as those transmitted by invisible aerosols, can have dramatic effects on the results of CB reactions, simply because the methods are so sensitive (see Section 2.3.4). It should go without saying that if the user considering the purchase of an instrument observes cross-contamination of the ATP signals in this second test, the issue of whether the spurious signals are due to aerosol formation through shaking or simply to light leakage from the high-signal wells should be resolved (by performing the test with and without shaking).

2.2.2.2 Injectors

2.2.2.2 Injectors   At least one injector is very nearly a necessity for successful development and performance of CB assays. The reason is that the results of most of these assays are strongly time dependent, and only maintaining a constant interval between the time of reaction initiation and the readout can assure the user of consistent data. A narrow exception may be made if the user is willing to perform time-linear fits of the data in all cases; if so, then it may be possible in some systems to load the plates manually, with a multichannel pipette, for example, and account for the uncertainty and/or inconsistency in reaction initiation times by measuring the rate of increase of the luminescent signal, rather than its absolute level. This method is discussed below in Section 2.4.1, and its use is not restricted to those cases in which automated injection is impossible or undesirable. In any case, the availability of one or more injectors greatly simplifies the performance of these assays.

The issue of what to put in the injector is not a trivial one. In part, the question is related to the strategy of reaction initiation, which is discussed in Section 2.3.3.5. However, one disadvantage of automated injectors is that the environment of the injector reservoir may not be ideal for the reagents stored there. Both heat and (less obviously) light may have undesirable effects on these chemicals, both by causing chemical reactions and, in the case of light, by inducing light-dependent enzymatic reactions if luciferases or photoproteins are stored in the reservoir. Other issues with injectors are deadline losses of potentially precious reagents, cleaning protocols, and volume assurance. If an injector is used, any run of a critical nature should include some kind of control or standard to establish that the delivery volume of the injector is within error tolerance of nominal.

Two injectors are often better than one, especially for dual-mode assays (such as the dual-mode CB cytotoxicity assay described in Chapter 3). Even for single-readout assays, two injectors may provide more flexibility in planning the reaction-initiation step, for example, by allowing the separation of enzymes and substrates until the last possible moment. It is also possible to inject a cell suspension, assuming there is some means of maintaining its homogeneity, such as a magnetic stirrer.

2.2.2.3 Software   Virtually every luminometer workstation comes with software to control the instrument and help analyze the data, but capabilities vary widely, and personal taste enters into one's impressions of software. Those who live in the GLP/drug development world have strict regulatory requirements for validation of all software, as well as ongoing assurance of performance. Often the software is validated by the manufacturer, and then must be validated again on a particular instrument for a specific application by the scientist/user in the context of the whole assay system.

Whether or not a particular worker's instrument falls under regulatory scrutiny, certain features are highly desirable in software for control and analysis of CB assays. The operator should perform an instrument check routine every day when critical work is planned, and it should be possible to initiate this routine with minimal effort and complexity. The software should enable selection of any rectangular subset of the plate wells for reading. Along with this feature, a very important capability of both software and instrument is the ability to predict, or at least test, the exact cycle time of both the individual reads and the entire experimental run. This stands in marked contrast, for example, to the case of ELISA readouts, in which time-critical changes are usually not occurring within seconds. Most CB assays are like true kinetics experiments, even if only a single-point readout is desired. To take a simple example, suppose an assay protocol is designed in which the instrument automatically injects an initiating reagent into each well and then goes back and reads each well. Unless the software and the instrument can be configured to perform the injections and the readings with exactly the same delay interval, a single-time-point readout will not be valid, since the reactions will be aged differently. The data can still be reduced successfully by performing linear regression on a series of readings, but this requires more effort and sophistication. The point is that one should not assume that all of these considerations have been taken into account by the manufacturer and programmers, especially since few in the industry are familiar with the requirements of CB reactions. The user must verify the timings of injections and readouts to his or her satisfaction.

However, random addressability of the wells, which may have some value in ELISA readouts to compensate for edge and other position effects, is not an essential feature in CB assays. Software that is capable of selecting a rectangular subset of wells and reading them in a consistent and predictable way is adequate.

If the decision is made to take multiple readings and perform linear regression, or if it may become desirable, then ideally the software should have the capability of collecting such data in a single file, in a consistent format, to allow the use of a data reduction macro. It is a simple matter for an experienced mathematical programmer or spreadsheet user to set up an automated routine to perform linear regression on the data from each well, yielding a calculated time-dependent rate of change and a correlation coefficient (discussed further under Section 2.4.1). However, the data should be in a predictable format, and preferably within the same file, to allow this to occur. Of course, it is also essential that the software supply the data in a format that can be used by the regression program (such as Microsoft Excel, using the LINEST feature), or at least in a file type that is readily imported by such a program.

2.2.2.4 Automatic Gain Adjustment   While most instruments take a single series of rapid readings of a given well and integrate the results, there are luminometers that operate in a different way, involving an initial reading followed by a gain adjustment and then the final reading. This gain adjustment is generally accomplished by changing the voltage of the photomultiplier tube (PMT). It would appear that adjusting the gain to the luminance of the individual well might provide additional accuracy, but the author has observed three problems associated with this method. First, the reading delay was no longer constant, at least in some of the original instruments. When this interval is undefined, it becomes difficult or impossible to plan a complex experiment based on CB assays, for reasons mentioned above. Second, the gain adjustment slowed the reading process, again causing potential interference with carefully designed protocols. Finally, linearity over the dynamic range observed with gain adjustment was not as strong as that seen with the other instruments. Although technological improvements may have eliminated some or all of these problems, the assay developer is encouraged to probe these features of a luminometer employing gain adjustment prior to making a purchase decision.

2.2.2.5 Whole-Plate Imaging and CCD Detectors   The modern trend in luminometers is toward reading of the entire plate at once. Although there are various ways of accomplishing this, the CCD is now in the lead. U.S. Patent 6,518,068, filed in 2000 and issued in 2003 (91), provides a full description of a luminometer based on this principle. While it represents a major advance, the use of the CCD does not eliminate the problem of cross talk among samples, and the patent specification makes it clear that rigorous collimation and rejection of stray light are essential aspects of data collection.

The author has not had the good fortune to perform CB measurements using a whole-plate imaging system, but many of the advantages are obvious. With no delay between the readings of individual wells, treatment of the data is greatly simplified. However, the reactions will of course be differently aged unless they are also somehow initiated simultaneously by a specialized automated injection system.

2.2.3   Luminescent Microplates

As of this writing, 96-well luminescent microplates are available in bulk for approximately $3 each. The black ones are sometimes less expensive than the white. There is a small but noticeable difference in performance between the two colors; black plates yield slightly lower signals, but may also generate less cross talk. Both types of plates are usable for most applications, but the user should choose one for a particular assay and stay with it to minimize sources of variability (or at least test the effects of both colors). Other than the choice of color, little variation in quality among vendors or types has been noted by the author.

2.3   COUPLED BIOLUMINESCENT ASSAY PROCEDURES AND PRECAUTIONS

This section addresses the general nature of these assays and presents certain principles with wide applicability. The idiosyncrasies of particular assay methods are left to their individual chapters.

Although some photoproteins are not technically "enzymes" because they do not turn over within the usual reaction timescale, CB assays are still essentially enzyme assays. This is true of many modern procedures, including ELISA, but the philosophy of CB assays is entirely distinct from that of ELISA, leading one to make the further distinction that CB assays are generally enzyme *kinetics* assays. (ELISAs may also be performed as kinetics assays—see, for example, Reference 92—but this done only rarely.) Moreover, they are by definition coupled assays. While in the general sense this means merely that a process of interest is coupled to the light generation phenomenon, in practice the process is frequently the catalytic activity of an enzyme as well, so that the CB assay becomes a coupled enzyme assay. Many coupled enzyme assays have been performed for various reasons, of which the most important is that it is often the simplest way to measure the activity of a particular enzyme, and as a result a number of fundamental principles of these assays are known and can be applied to the development of CB assays.

To draw another obvious tautology, CB assays are also bioluminescent assays, and this aspect of their nature also allows some degree of generalization. Some of this has been presented in Section 2.1.1.1.

CB assays are often very rapid and sensitive. In the early days of the polymerase chain reaction, one's coworkers were often surprised and at times contemptuous of the measures required to ensure cleanliness: hair nets, separate reagents and pipettes, and so on. Similar measures may be indicated for successful performance of certain CB assays.

Finally, CB assays involve either luciferases or photoproteins, which often have special handling and storage requirements.

2.3.1   Coupled Enzyme Assays

Enzymologists have coupled the activities of enzymes for many decades. A few complex kinetics studies have been performed to demonstrate superdiffusional coupling of enzymatic activities, leading to enhanced rates (93, 94), but most coupled enzyme assays are done simply to produce a desired signal from a process of interest (see Reference 95 for an early example of a clinically useful coupled assay). The practical aspects of these methods more than compensate for the added complexity of multiple simultaneous reactions, but kinetic calculations using coupled reactions are, not surprisingly, much more involved than their simpler counterparts and may involve defined corrections based on the relative rates of the independent reactions (96) and other manipulations and assumptions. The choice of coupling isozyme may also influence the results (97).

The usual objective of coupled enzyme assays, that of measuring the catalytic rate of a single enzyme, is accomplished most simply by ensuring that this rate is the sole *limiting* factor in producing the observed signal. The concept of "limiting" in this context is easy to define but profound in its implications. We use the term "limiting" to refer to a single step in a series wherein acceleration or deceleration of that step affects the overall rate. When it is the only limiting step, we are able to draw the important conclusion that variations in the observed rate of the coupled system, due to substrate titration, for example, reflect variations in the phenomenon of interest, without interference from other limiting factors. If the variation is linear, the enzyme response is linear; if exponential, the enzyme is exhibiting some such phenomenon as hysteresis (98, 99) or autoactivation; and so on. If, conversely, another reaction in the series is at least partially limiting, drawing conclusions about the response of the enzyme of interest to the perturbation is far from simple.

Our chosen example involves measurement of the activity of the enzyme phospho-β-glucosidase (PBG), which catalyzes the hydrolysis of cellobiose-6′-phosphate to glucose-6-phosphate and simple glucose (100). Cellobiose-6′-phosphate is a disaccharide, a dimer of two sugar molecules, of which one is phosphorylated. Since neither of the products is readily detected by a direct, simple, and rapid method, the authors chose to couple this hydrolysis to the activity of a second enzyme, glucose-6-phosphate dehydrogenase (G6PDH). This enzyme oxidizes glucose-6-phosphate, one of the products of the first reaction, to produce a phosphorylated aldose, which is important in the *in vivo* reaction, but is merely a by-product from the point of view of the coupled system. Simultaneously, G6PDH *reduces* the oxidized enzymatic cofactor, nicotinamide adenine dinucleotide phosphate ($NADP^+$), to produce the reduced form nicotinamide adenine dinucleotide phosphate (NADPH), which absorbs strongly at 340 nm. Thus, the activity of the enzyme under study, phospho-β-glucosidase, is measured by an indirect process in which one of the products of its reaction is further altered, and the prosthetic group in this further reaction, rather than the direct product, is the species that yields a measurable signal (Fig. 2.1).

We begin by supposing that it is desirable to observe a linear response with respect to the true activity of PBG, the enzyme under study. If the activity doubles, the signal should double. How can this be achieved? Only by ensuring that every other component of the coupled system that participates in the rate is present in *excess*, that is, in such quantities that providing additional reagent would have no effect on the rate. This must be true of every reagent apart from the reagent we wish to be limiting, in this case PBG. For example, the substrate cellobiose-6′-phosphate must be present in large quantities. In practice, the important kinetic parameter is the famous Michaelis constant or K_m, which is the substrate concentration at which the half-maximal rate is observed. If the substrate is present at a concentration of 20 times the K_m or greater, this is generally sufficient excess, since the theoretically observable rate is

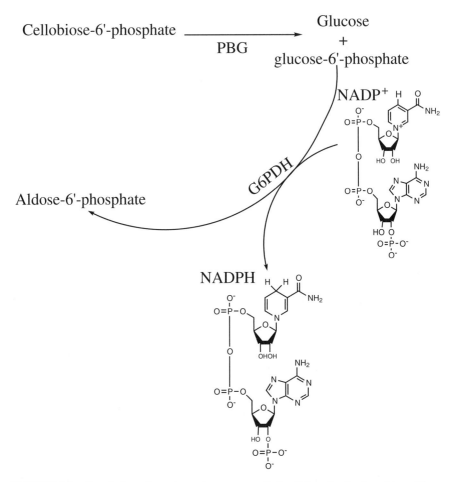

FIGURE 2.1   Coupled reaction scheme to measure activity of phospho-β-glucosidase. Cleavage of the disaccharide carried out by the test enzyme is coupled to oxidation of glucose-6-phosphate, yielding the reduced cofactor NADH and generating an absorbance signal at 340 nm.

some 95% of the maximum possible rate; however, more substrate might be even better.

Moreover, each component of the second reaction must also be present in excess to achieve the desired result, in this case, the G6PDH enzyme and its cofactor NADP$^+$. The latter must be at a concentration suitably higher than its own K_m; but what is the proper concentration of the coupling enzyme, G6PDH? The answer is that its *rate* must not be limiting. In other words, it must be present in sufficient quantity to handle the glucose-6-phosphate produced by the first reaction immediately. If glucose-6-phosphate accumulates in this system, then the signal does not reflect the true rate of the PBG. Hypothetically, it might be possible to find a literature value for the catalytic rate of G6PDH and calculate the needed amount on this basis, but in practice, coupled enzyme assays are developed by titrating the various components until the single reagent or process under study is isolated as limiting. On occasion, exceptions must be made, and other reagents are allowed to be partially limiting (this is often true of the CB cytotoxicity assay described in Chapter 3), but when this occurs, the nature of the partial limitation and the reasons for it should be understood. Some reactions can be made to respond in linear fashion to one component even when another is partially limiting, given adequate time and effort for optimization. Apart from isolating the parameter of interest, another goal of the assay developer is usually to achieve *steady-state* kinetics, that is, a situation in which the change in the signal is linear with time. This type of kinetics is further described below.

A critical feature of coupled enzyme reactions is that they depend on multiple serial events. Thus, a coupled reaction series is effectively inhibited if any of the enzymes in the chain is inhibited. This represents a point of vulnerability of coupled enzyme assays and of CB assays in particular. Obviously, molecules, that block luciferase activity cannot be expected to yield any other information in this kind of assay, no matter how they perturb the other enzymes. Aside from this, enzymes in coupled pathways may be structurally related to each other, or merely similar, especially if they are parts of the same biochemical pathway. Thus, attempts to identify inhibitors of one of the enzymes, perhaps as drug candidates, may inhibit a different enzyme as well, or instead. These issues can be addressed with secondary screens, but it is important to know that they exist and to recognize that the nature of coupled assays implies the existence of multiple failure modes. In particular, firefly luciferase uses ATP to make light; ATP contains phosphate residues, and phosphate is one of the major components of many signaling pathways. Enzymes of great current interest in drug discovery, including kinases, phosphatases, and phosphodiesterases, are closely involved in phosphate metabolism. Molecules that inhibit these may also interfere with luciferase, or with other enzymes in the coupled pathways that give rise to ATP formation. Thus, while it is possible to employ such enzymes in CB assays, desirable inhibition, when observed, should be confirmed through the secondary step of testing the enzyme of interest by an independent or *orthogonal* method.

The blossoming of fluorescence technology in recent decades has brought with it an increasing number of coupled enzyme assays with fluorescent readouts. A few products are naturally fluorescent at useful wavelengths, but most of these assays require fluorogenic substrates in the final step. Thus, some of the points regarding artificial fluorogenic substrates in Chapter 4 also apply to these coupled fluorescent assays; however, other potential interferences should be noted, such as quenching of fluorophores or other critical reagents by enzymes or intermediates and unanticipated sequestration by enzymes. Despite these and other potential drawbacks, coupled fluorescent assays can provide excellent sensitivity and speed, and the ability to excite the fluorophore without strongly affecting other aspects of the signal often provides an added element of specificity. Coupled fluorescent assays have much in common with CB assays, in that the process of interest is being studied by feeding its chemical output into a further step or series of steps designed to produce a large number of detectable output photons. However, since the fluorescent assays often require artificial substrates, in a sense they have more in common with chemiluminescent reactions.

2.3.2   Steady-State Kinetics

Among the many achievements of the immortal J. B. S. Haldane were the mathematical rescue of evolutionary theory and an appreciation of the ability of the human frame to resist an astonishing array of self-administered poisons, ingested purely for the cause of science, but he also made fundamental contributions to the theoretical basis of enzyme kinetics. *Briggs–Haldane* kinetics provides a mathematical approach to enzymes whose reaction rates are observed to be constant over a wide enough interval of time to take useful measurements. (George Edward Briggs was a Cambridge botanist whose best-remembered scientific contributions are the work he did with Haldane on enzymes (101, 102).) This concentration regime and the associated mathematical treatment are also known as steady-state kinetics. Fersht's 1988 volume is useful for those with deeper interest in the subject (103), as is the following Web site as of this writing: http://www.ch.cam.ac.uk/magnus/michmenten.html. Here, we make the simplifying assumption that the rate of product formation, including the generation of light (which is of course a product of the canonical luciferase reaction), directly depends on the concentration of the *enzyme–substrate* or ES complex, that is, the molecular complex formed by the enzyme and its cognate substrate by the association event prior to the transformation to product. We also adopt the simplification of referring to a single ES complex for purposes of our discussion of kinetics, despite the fact that many enzymes exhibit multiple ES complexes, including several types discussed in this volume. We do this because the existence of multiple intermediates has little practical impact on the question of how to achieve a steady state. Given these twin assumptions, then, the rate of observed product formation depends on the concentration of ES, and if this concentration is constant, the reaction is in a steady state. This is desirable, but how is it to be achieved?

Here, we discuss four kinds of phenomena that may prevent a coupled enzyme reaction from exhibiting steady-state kinetics, while acknowledging that there are other, less common conditions that may lead to this result. The first of these effects is *saturation*.

2.3.2.1 Saturation in Enzyme Kinetics   Saturation is often defined phenomenologically as the situation in which the signal responds to the stimulus (usually the analyte concentration) in sublinear fashion, or does not respond at all. The range of potential causes is very broad, from the simplest case, physical saturation of the available enzyme molecules (i.e., there are not enough molecules of enzyme present to process the analyte, so that additional analyte changes the observed rate little or not at all), to product inhibition (in which a product of the reaction slows later iterations of the same reaction; this is very common), to changes in pH or redox conditions, or even perturbations of the physical conformation of the enzyme due to the presence of excess substrate. However, saturation can also be observed as a failure of the equipment to reflect the true signal.

Whatever the cause, if the saturation is undesirable, the analysis procedure essentially involves logical isolation of each component, including the equipment itself, so determine whether it is the source. This is accomplished by perturbing the component under study while holding everything else constant. If added enzyme relieves the saturation, this answers the question. Product inhibition is slightly harder to establish but can be studied by using the products of spent reactions, sometimes with the remnant of enzyme either removed by chromatography or other means, or denatured chemically or by heat treatment. Equipment saturation may be identified by testing the equipment with other reactions yielding comparably high signals. The life scientist is familiar with these principles and needs only to apply them to discover and perhaps ameliorate the cause.

2.3.2.2 Lag Phases

A lag phase is an initial period of slow, nonexistent, or even negative signal prior to establishment of an ordinary steady-state signal. Several phenomena can cause this. Some are of scientific interest, while others are artifacts caused by imperfect technique or reagents. Certain enzymes require a period of activation *in situ*, or even undergo autoactivation (such as calcineurin; see Chapter 5). Hysteresis (99) is a separate phenomenon in which some aspect of the enzyme-catalyzed reaction tends to alter the enzyme itself to a more active form. A similar yet distinct effect is seen in multisubunit enzymes that undergo rearrangements upon activation by substrate encounters, such as dissociation of inhibitory subunits. Any of these natural, well-established effects may manifest itself as a lag phase. However, other potential causes of an accelerating increase in signal are temperature artifacts (i.e., the reaction is warming after initiation, causing higher rates), mixing artifacts (in which the full reaction rate is not realized initially because the reagents have not been mixed properly), and incomplete solubilization of a component. Finally, in CB reactions, there is the important possibility that exposure of the luciferase or photoprotein to light may have caused a backward reaction, leading to the formation of intermediates that must be exhausted before the true reaction rate appears; such processes can sometimes cause an apparent lag phase. These possibilities may be checked by controlling appropriate parameters: heat, light, mixing, and so on. Mixing artifacts involving organic solvents are especially pernicious and can lead to either positive or negative apparent signals, as well as attenuations or perturbations of real signals. See Reference 104 for an analysis of an interesting mixing artifact with benzoyl tyrosine ethyl ester dissolved in isopropanol that was originally described in a prominent journal as a reaction catalyzed by an artificial enzyme (105).

2.3.2.3 Biphasic Reactions Involving a Single Fast Turnover

Our next example, which is observed more rarely, can resemble saturation but is distinct in its kinetic profile. The best-known examples are the reactions of some of the proteases (106), although many hydrolases exhibit biphasic reactions. With these enzymes, the initial hydrolysis of most substrates is rapid. However, the first catalytic step, while giving rise to a product and sometimes a detectable signal, does not represent a complete enzymatic turnover in these cases. Instead, the enzyme has been left in a state in which one of the two products of the hydrolysis has dissociated, while the other remains covalently bound to an active-site group of the enzyme. For the enzyme to catalyze another reaction, this second product must also dissociate. This requires a second hydrolysis, specifically a deacylation, and this final product-off step is often the slowest on the reaction pathway. Therefore, what is observed empirically is a single rapid turnover, resulting in a burst of signal, followed by a slower steady-state phase during which the enzyme must undergo the full multistep turnover to generate each detectable molecule of product. This type of reaction exhibits deceleration, as does a saturated reaction, but, in contrast, eventually reaches steady state and continues to yield a linear increase in signal. The first turnover is often so fast that stopped-flow equipment is required to observe it, but it can be slowed by conditions and/or substrates engineered for the purpose.

2.3.2.4 Accumulation of Intermediates in Coupled Systems

2.3.2.4 Accumulation of Intermediates in Coupled Systems Technically, this is a form of saturation, but it may be difficult to troubleshoot because it is inherently a problem of coupled assays and may not be observed in systems with isolated enzymes. It occurs when the initial enzyme in the coupled reaction series is not saturated, but an enzyme later in the series is. This leads to accumulation of a reaction intermediate. Since the saturated enzyme is unable to respond to increases in the concentration of the intermediate, the overall coupled system appears to be saturated. The problem is identified by titrating the reaction cocktail with each component independently until the observed saturation is relieved. Another approach to this or any saturation phenomenon is of course to use a lower concentration of analyte to begin with.

2.3.3 Coupled Bioluminescent Reactions

Here, we address general features of these reactions. Some of the principles derive from their character as coupled reactions, and others from the luminescent aspect. A few are rather counterintuitive.

2.3.3.1 Bioluminescence in the Assay Context In theory, every photon emitted by a luminescent sample can be detected, and each such photon may represent a component of a useful signal. This principle is realized in practice in the context of applications using immobilized luciferases, in which single photons are routinely counted to detect activity of the enzyme (107). This is not generally the case with other methods. The implications regarding the potential sensitivity of CB methods are striking. However, the use of microplates for bioluminescent assays, while bringing striking improvements in throughput and convenience, obviates or diminishes some of the advantages of luminescent technology by introducing multiple light-emitting samples into a chamber that ideally should be entirely dark apart from the sample-generated photons. Nevertheless, bioluminescence yields manifold advantages, regardless of format.

2.3.3.2 The Ethics of Sensitivity It may be astonishing to an academic scientist that anyone would question the value of sensitivity. There may be an associated cost, but surely sensitivity, assuming one is detecting a true signal, is purely good.

Some time ago, the author proposed an experiment to a principal investigator. There is no need to specify the place or time. The investigator had publicly proposed an unusual model of a particular biomolecule's activity. The author proposed to try the reaction with a different reagent as a critical test of the model, but the response was, "No, don't do that. If it works, the whole model is no good." The author refrained from asking whether the lab was engaged in science or poker.

There are individuals who believe that the rising value of increasing sensitivity comes to a sharp halt when one starts learning things one does not wish to know. This occurs primarily in the regulatory environment, where the presence of certain molecules in certain preparations, even at very low levels, may be impossible or uncomfortable to explain. "That assay asks a question we don't want answered," is a remark that can be heard in some quarters, usually in very small groups.

It is the author's contention that this approach is not only unethical but also poor business practice. "The public will be confused" is an argument commonly heard for suppressing facts about a myriad of scientific issues, such as whether to report arsenic levels in parts per billion or even trillion. But by covering up the facts, we postpone the debate about *relative* risk that must sooner or later occur. Similarly, those who favor less sensitive methods for reasons of business strategy are (probably) unconscious participants in a conspiracy to keep the public and/or regulators *truly* confused about what safety means. What it means, of course, is not a "zero level" of anything, but a series of public and individual decisions based on valid statistics and accurate and abundant information.

On top of this do-gooder argument, there is also a business case for developing and using highly sensitive assays. In fact, at least three strong bottom-line arguments can be put forward in favor of these methods.

1. Using the most sensitive practical method available is a good practice. If the studies are ever revealed to the public, or to a regulatory agency, this point will bear the strictest scrutiny and make a favorable impression.

2. Small signals in one context may become large signals in another. If a drug exhibits low hepatotoxicity in the tested environment, it may become much worse in patients with different proportions of cytochrome P450s (108, 109) or in those who drink grapefruit juice (110). Thus, by collecting appropriately sensitive data, the researchers may be able to understand, control, or even eliminate a major risk by an appropriate modification.

3. Businesses on the whole, especially large ones, often lack mechanisms to evaluate the opportunity cost of failing to raise barriers to competition available through rigorous quality control (QC). Manufacturing is considered to feed sales, which leads to profits. QC is merely a cost center. But what if one can either do something one's competitors cannot, or show that one's product is better in a way other firms cannot match? On occasion, highly sensitive

assays can provide this kind of advantage. For example, in a pharmaceutical context, if one's business can demonstrate a 10-fold lower level of pyrogens or other contaminants through an improved method, a regulatory agency may soon adopt one's method as the standard, and one's competitors will be in the position of having to catch up belatedly. They may even be unable to meet the standard until they retool to run the improved assay. This can be a case of doing well by doing good. Raising competitive hurdles is a substantial part of the case for moving to CB assays and other ultrasensitive methods.

2.3.3.3 Units for Quantification of Luminescence

Luminescence is almost always reported as "relative luminance units" (RLU). The relationship between RLU and total emitted photons is variable and abstruse, and does not need to be understood for successful CB assay performance. It depends on such imponderables as the size and geometry of the detector(s), as well as parameters that may or may not be under the user's direct control, such as the voltage setting of the photomultiplier tube. In fact with some instruments, the relationship is not even truly linear in theory (though it generally is in practice). For example, gamma counters require exquisite sensitivity because many samples emit very little radioactivity, and as a result the counters frequently have problems with false signals. Some engineers address this issue by considering an event to be a "count" only if *two* photons excite two separate detectors within a given time interval. This is a good approach for gamma counting, because a gamma ray produced by the usual isotopes is usually sufficiently energetic to release numerous photons from the liquid scintillant, so that a single disintegration event is detected and interpreted properly. However, when the same instrument is used to measure luminance, this "feature" turns into a difficulty. Now counts will be registered only if two photons are emitted nearly simultaneously and *both* are detected. This can of course give rise to a sensitivity issue, but it also represents a distortion of the very nature of luminescence, in which every photon is signal.

Consider an experiment in which very few photons are produced by a unimolecular reaction within a sample. A reading is taken. Now the sample is diluted with an equal volume of a buffer that does not change the reaction rate (other than by dilution), half of the diluted sample is reintroduced into the detection chamber, and the luminance is measured again. Because the reaction is unimolecular, the dilution does not affect the total number of photons emitted. However, since only one-half of the original luminescent molecules are now in the detection chamber, the signal should be one-half of the original signal. If the detector is a gamma counter, however, the apparent signal may be reduced by as much as *three-fourths*. This is because when the rate of photon emissions per time interval in a slow reaction decreases by one-half, the rate at which two photons are emitted during a single time interval can be as low as one-fourth (one-half squared) of the original value. (If the reaction is fast enough to cause signals in a high proportion of the time intervals, the effects of dilution are more mathematically complex and do not diverge as far from Beer's law.) Fortunately, this effect is not usually a major problem for users of these instruments, because there is rarely a need to dilute the reaction after the luminogenic components have been added.

Moreover, some of these instruments have a "luminescence" setting that circumvents the problem.

In short, the typical user of CB assays takes the RLU reading more or less for granted. It is extremely useful in comparing luminance levels, but only rarely is it possible to relate RLU directly to photon production. The word "relative" in the term must be taken quite seriously.

2.3.3.4 The Implications of Limiting Rates and Reagents for Coupled Bioluminescent Reactions   The introduction of a luciferase or photoprotein into the coupled scheme does not change the principles enunciated above. However, a great practical advantage of these reaction series, especially those involving firefly luciferase, is that it is usually very easy to provide the luciferase and its associated components in excess. As a result, the linear dynamic range of the simple luciferase/ATP assay, in which the concentration of ATP is reflected in the luminescent signal of the enzyme, is extremely broad, covering at least seven useful orders of magnitude. Generally, larger concentrations of enzyme are required to detect very low ATP concentrations, in keeping with the mass action law, which holds that the encounter rate of two molecules is proportional to the concentration of each of the two. At high ATP concentrations, the enzyme can be diluted, yielding a cost saving. There will of course be a point at which a given luciferase concentration is insufficient to process a large amount of ATP immediately so as to yield a linear signal; however, increasing the luciferase concentration may continue to yield satisfactory results until some other factor limits the readout, such as physical saturation of the detector. The detection capabilities of firefly luciferase are remarkable.

However, this does not solve the problems presented by other steps in the reaction. Taking the aCella-TOX^{TM} CB cytotoxicity from Chapter 3 as an example once more, we observe that there are three enzymatic steps in the reaction series, catalyzed, respectively, by the enzymes glyceraldehyde-3-phosphate dehydrogenase (G3PDH), phosphoglycerokinase (PGK), and luciferase. The principle of the assay (see Fig. 3.8) is the detection of the G3PDH enzyme released from cells that have lost membrane integrity. All other components are present merely to facilitate the determination of G3PDH activity. Therefore, ideally, these components would be present in excess, while G3PDH is of course eliminated from the reaction cocktail to the extent possible. Both the PGK and the luciferase reaction must respond rapidly to any flux through the G3PDH portion of the reaction series to reflect the concentration of this enzyme. While in the best case these reagents would be provided in excess, it is found to be impractical to accomplish this with two of the reagents (PGK and glyceraldehyde-3-phosphate, a substrate of G3PDH) for reasons explained in Chapter 3. However, concentration regimes have been identified in which the response of the system to G3PDH activity is highly linear over a dynamic range of at least three orders of magnitude, despite these difficulties. Methods of extending the dynamic range and working around the problem in other ways are presented in Chapter 3.

2.3.3.5 Initiation of Coupled Bioluminescent Reactions   The question of how to initiate a CB reaction pertains not only to what should happen after the reaction starts, but also to what should not happen before it starts. For example, mixing an enzyme with its cognate substrate under normal conditions generally leads to an enzyme-catalyzed reaction, and potentially to exhaustion of the substrate, production of inhibitory reaction products, and side reactions. Apart from this rather obvious mistake, however, there are several more subtle, yet undesirable prereaction states to be avoided.

Protein Concentrations   Most proteins, including enzymes, undergo denaturation in very dilute solutions, an effect sometimes attributed to adherence to the walls of the storage vessel. Generally speaking, proteins should be either stored at concentrations of 1 mg/mL or greater, or protected from this type of denaturation by the inclusion of inert proteins (bovine serum albumin is often used) or small molecules with an equivalent effect. However, some proteins alter their own properties, for example, proteases capable of autodigestion and autoactivating enzymes such as calcineurin. If the reaction is exclusively *trans* in nature, the rates of these reactions can be lowered by dilution; if *cis*, or both *cis* and *trans*, storage at lowered temperatures and/or the use of inhibitors should be considered. These considerations are not unique to the reaction-initiation scenario, but the problems may be exacerbated by the need to thaw reagents and perform some mixing and preparations prior to assay initiation.

Backward Reactions   In principle, any enzyme can catalyze its cognate reaction in either direction. This is a corollary of the general law of microscopic reversibility and is true regardless of the thermodynamic gap between substrates and products. An analogy frequently seen in the pedagogical literature on enzymology is that of the moving vehicle on a road that leads over a hill; the vehicle is the reactants, the road is the reaction pathway, and the hill is the potential energy barrier to the reaction, which determines (by its height) the rate at which the reaction proceeds. An enzyme merely lowers the hill. It does not raise or lower the level of the ground on either side of the hill. If the vehicle has sufficient momentum (in the form of kinetic energy of the molecules involved), it can cross the hill in either direction, and the phenomenon of "catalysis," through which the energy hill is lowered, accelerates this transition in both directions by equivalent factors.

The rather odd implication of this is that while luciferases and photoproteins can promote reactions that generate light, light can also promote the reactions of these proteins—in the opposite direction. Thus, the canonical firefly luciferase reaction, of which the net reaction is the transformation of ATP to AMP, pyrophosphate, and a photon of visible light (luciferin effects are omitted for simplicity), can be reversed: light, in the presence of luciferase, pyrophosphate, and AMP (with appropriate cofactors and dications), can and does lead to production of ATP. Since the ATP concentration is a critical parameter in CB reactions and must be controlled for the results to be interpretable, this backward reaction may well be harmful. It can be avoided most simply by keeping luciferase away from these substrates, protecting it from light, or both. Figure 2.2 shows the practical effect of protecting luciferase from light, thereby minimizing the backward reaction prior to initiation.

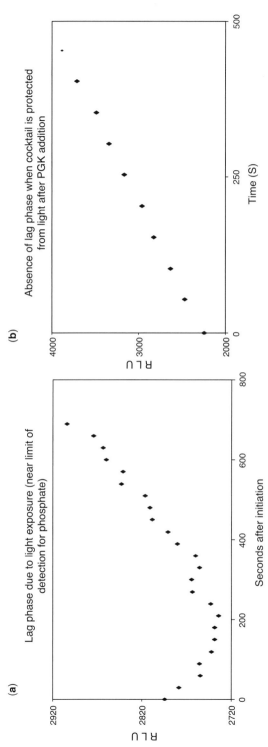

FIGURE 2.2 Potential effect of room light on coupled reactions employing firefly luciferase. (a) Failure to protect reagents after mixing enzymes and substrates before initiation of the full coupled series leads to a prolonged lag phase; the true rate is seen only after several minutes. (b) Protection of the reagents from light, often by initiating the reaction with an injection step in the dark chamber, leads to a linear signal from the outset.

The reversibility of enzyme-catalyzed reactions can be a source of problems in CB systems, but it also represents an opportunity, as we shall see in Chapter 4.

Recommendations for Reaction Initiation   Within the above guidelines, it is generally safe and advisable to combine substrates in one vessel and combine proteins in another, each with appropriate buffer constituents. Obviously, this excludes conditions under which a highly reactive substrate or other molecule is introduced to the presence of a reaction target; pH and redox incompatibilities can also be exceptions. At times it may be necessary or convenient to combine some of the enzymes with small molecules that are not substrates. However, this approach must be considered and tested. We use aCella-TOX once again as our example. It may appear harmless to mix the reactants needed by the test enzyme, G3PDH, with the actual second enzyme, PGK, since no reaction should occur. However, commercial preparations of PGK are virtually always contaminated with G3PDH. As a result, mixing these apparently unrelated reagents (in the presence of ADP) may lead to a G3PDH-catalyzed reaction, followed by the usual PGK reaction and formation of a considerable amount of contaminating ATP prior to reaction initiation. Such unforeseen consequences attend many apparently innocent reagent combinations. All such formulations should be tested before critical runs are planned with new procedures.

2.3.4   Recommended Precautions for Coupled Bioluminescent Reactions

Experienced workers in the life sciences sometimes develop a mind-set according to which there are two types of reactions: those that involve amplification and those that do not. The polymerase chain reaction is the archetype of the latter, and most are aware that a single contaminating molecule in PCR is one too many. However, enzymatic amplification is not limited to the PCR. CB reactions, by using multiple enzymatic steps, may yield strong signals, desirable or undesirable, from very small initial perturbations. During the initial development of aCella-TOX, the author was puzzled to discover that in this very sensitive detection method capable of detecting roughly one-tenth the contents of a single nucleated mammalian cell, at one point about one in every 20 samples yielded an enormous "spike," or a reading differing from the expected by 10 or more standard deviations. The explanation was the shedding of skin cells, and this persisted to some extent even after the use of gloves was rigorously adopted. Cells can be shed from any exposed skin, and gravity comes into play. Adoption of CB assays may require a change in mental approach for operators used to a clear distinction between "DNA work" and "all the rest."

The advent of disposable plastic for nearly every measuring and storage function has dealt an encouraging blow to the problem of cross-contamination and put a great many glasswashers out of work, but complacency remains ill-advised. Aerosols can enter the famous Pipetman^{TM} and its competitors. Special care is needed in dealing with positive controls. In a highly sensitive assay for phosphatase activity, for example (see Chapter 5), it is generally necessary or desirable to include a run with a phosphatase that is known to be working to ensure proper operation of the whole system; this sample should preferably be loaded after all the negative and unknown samples are loaded, and covered, if possible. After it is added, the transfer device should be cleaned, especially if it will ever encounter unknown or negative samples again. Even better is to reserve a device for positive controls, although this may give rise to a need for a separate "negative–positive" control, depending on the degree of care required.

When prepared, initiated, and mixed correctly, CB assays can yield astonishingly precise numerical results, sometimes with coefficients of variation (CVs) of under 1%. In performing linear fits of time-dependent data, correlation coefficients of >0.999 are routine, and four 9s are frequently achieved; in fact, a reaction that yields a coefficient of under 0.999 may be an indication of a problem, which may be a sublinear decay due to consumption of a substrate or some other form of biochemical saturation. (However, lower coefficients of correlation are to be expected, of course, when the actual signal is small relative to the noise, as seen in Fig. 2.2.) The "downside" of this potentially high assay quality is that fairly small variations may represent actually technical errors that require attention, depending on their nature. With sensitive assays, spikes due to contamination and/or cross-contamination will occur, but they should not be accepted as routine.

Firefly luciferase can be diluted in an appropriate diluent (a good one is available from Sigma–Aldrich in bulk) and either frozen or kept on ice for hours without excessive loss of activity. (It is desirable to keep light away, because photooxidation has been reported, although the author has not observed this failure mode.) Other reagents, however, may not be so resilient. In particular, the phosphorylated substrates that are a source of phosphate moieties for transfer to ADP in certain CB formulations can be very labile. Glyceraldehyde-3-phosphate in particular, used in the aCella-TOX cytotoxicity assay as well as in the phosphate detection scheme of Chapter 5, is rapidly degraded at neutral pH even on ice. Acidic conditions enhance the lifetime of this reagent. ADP and AMP are much less labile; the biological reductant NADH, when needed, should be kept away from ambient oxygen until the last possible moment. Finally, some of the prosthetic groups of the luciferases and photoproteins require special care. FMN is so unstable that it must be generated *in situ* in a separate coupled reaction.

A final precaution pertains to the detection instrument. As users of fluorometry may already be aware, it is possible to damage a photomultiplier tube by blasting it with light. This rarely occurs with a properly collimated and interlocked fluorometer, but a luminometer, even if the detector is interlocked against exposure to room light, may offer no such protection from the sample. Extremely bright samples may damage the PMT. The user should therefore remain aware of the maximum possible light output of each reaction and determine that it is within instrument tolerance before proceeding.

2.4 DATA HANDLING FOR COUPLED BIOLUMINESCENT ASSAYS

For glow reactions, a great deal depends on whether a stop reagent is used. Stop reagents for CB assays are a separate field of study in themselves, one which has not yet been extensively investigated by anyone. For most reactions, one can dump in some acid or an inert "washout" compound to terminate the increase in signal, but CB reactions are more complex, because one process (the generation of light) must continue while the reactions that produce the proximate cause of luminescence (e.g., ATP) should be halted.

The kinetics of coupled assays is extremely complex, and an extended mathematical discussion of the subject is outside the scope of this volume; even kinetics with single enzymes can be a subject of extraordinary depth. However, the user of CB assays will be well served by several principles. The most important, enunciated above, is to try to make the phenomenon under study the only limiting step in the entire coupled process. Whether or not this is the case is tested by separately perturbing the concentration of each component that has the potential to contribute to the rate; ideally, no component other than the one under test should do so. However, there are cases in which other components are known to be partially limiting, yet the assay readout responds in linear fashion to the critical analyte. Attaining one of the desirable states in which the system responds to analyte in a robust, linear manner depends mainly on providing as many nonlimiting components as possible in excess, possibly great excess, but there remains an empirical aspect to the process of optimizing these assays.

If a stop reagent is used, the reaction essentially becomes a single-time-point assay. Data reduction of such results is not very different from working with ELISA readouts. However, for various reasons, users may choose to use data from a continuous reaction. These data can still be treated as single-point data, if precautions have been taken to ensure that the reactions are identically aged (or that their ages are accurately known); a further option, leading to greater potential accuracy and the use of a higher proportion of the available information, is to perform linear regression of the readout against time.

2.4.1 Analysis of Coupled Bioluminescent Assay Results by Time-Dependent Linear Regression

There are several cases in which linear regression of repeated readings, yielding a rate of change of the signal, may be preferable to the use of a single-point readout.

- It is impossible or impractical to age the reactions identically; thus, a single-point readout is invalid.
- More precision is desired. This is available through the linear fit because the rate of change is a direct reflection of the enzymatic activity, whereas a single datum is subject to numerous errors.
- The user is unsure of the interval during which the signal increases in linear fashion with time. In such a case, a single-point readout is actually misleading, whereas a careful analysis of multiple readings should allow both identification of the linear range and reduction of these data to yield a correct value. Moreover, the data reduction may reveal the true linear range, enabling the use of single-point readouts in subsequent studies.
- The linearity of the process is one of the features being studied.

- The early phase of the reaction is of special interest, either because the linear phase is known to be brief, or for the opposite reason, that there is a "lag" phase due to mixing or the effects of a backward reaction. Here, the user may obtain both information about the prelinear phase and satisfactory data from the linear phase.

Figure 2.3 presents a data set from a run of the CB phosphatase assay described in Chapter 5. (The data at the upper asymptote have been altered to illustrate the point being made.) The time stamps are on the left, followed by the luminance readings. This dataset was chosen for analysis because the amount of phosphate was near the low limit of detection of the assay and the signal-to-noise ratio is relatively poor for this reason. Our discussion therefore addresses not only the issue of lag phases due to back reactions, but also the question of how to obtain useful information from a noisy data set.

The results of the analysis are presented at the end of the row of data. Several features of CB assays that present potential difficulties are illustrated by this run. First, the data are noisy due to the very small signal in terms of the change in absolute RLU, but as we shall see in a moment, they are nevertheless good enough to use in quantifying phosphate. Also, it is also evident by inspection that the reaction rate is negative for about 3 min. This is a classic lag phase, which may be due to mixing effects, the backward reactions mentioned, or other unknown phenomena. (In this case, a backward reaction is responsible; the excess products of this reaction must be consumed before the underlying rate is seen.) The user therefore faces the decision of which data to include in the linear fit.

A decay phase is also seen in the late phase of the reaction. Ideally, this phase should also be eliminated from the analysis, but at times it must be included because (1) there are too few data without it, or (2) other wells in the same run yielded useful data and it is desired to keep the analysis parameters consistent.

The latter point highlights an important aspect of these analyses. The method chosen should be logically consistent for all the data in the run. This is not always easy to accomplish, and at times it appears that certain experiments are difficult or impossible to carry out within this constraint. A straightforward example is an enzymatic titration over a wide range of concentrations. If the "high" wells contain 1000 or 10,000 times as much enzyme as the "low" wells, the worker may encounter the common situation in which the signal from the low wells develops slowly, but the high wells exhaust their substrate quickly or exhibit some other form of biochemical saturation. The time interval over which both types of wells yield usable linear data may be narrow, or even nonexistent. The user's approach to this problem will depend on the situation and the objectives. For one thing, "consistent" does not mean "identical." It may be acceptable to truncate the "high" data after 1–2 min, but consider the "low" data over a longer interval, perhaps 10 min, especially if the analysis shows that these are clearly the linear ranges in the two cases. To do this, the user should gain confidence through repeated runs and a clear understanding of the response of the assay with respect to enzyme concentration that the conclusions are not being biased by the unusual treatment. Preferably, this assertion will be bolstered by information from the

Time	RLU
0	2794
29	2778
59	2754
89	2755
119	2742
149	2739
179	2738
209	2733
239	2743
269	2763
299	2764
329	2755
359	2760
389	2780
419	2790
449	2808
479	2810
509	2816
539	2842
569	2841
599	2860
629	2863
659	2873
689	2904
719	2906
749	2921
779	2922
809	2925
839	2923
869	2929
899	2928
929	2931
959	2936
989	2933
1019	2930
1049	2930
1079	2938
1109	2935
1139	2933

Range of Linear Fit	Best-Fit Rate
0-1139 seconds	0.213 RLU/sec
239-1139 seconds	0.241 RLU/sec
0-689 seconds	0.194 RLU/sec
239-689 seconds	0.335 RLU/sec

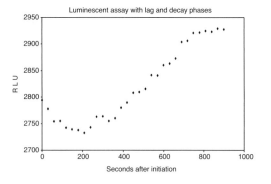

FIGURE 2.3 Linear regression of time-dependent data from a coupled bioluminescent reaction. Both an initial lag phase and a final decay phase are seen in this artificial example. (a) Nonlinear deviations are strongly evident from the graph. (b) Time-dependent data and results of linear regression. Only by selecting an appropriate time interval can an accurate linear rate be calculated by this method.

intermediate data. Alternatively, the user may decide to gather the same information from two independent experiments, perhaps with two separate formulations of the assay cocktail, tailored for the respective concentration regimes.

One way of selecting the linear phase is to do repeated regressions until the highest possible correlation coefficient is obtained. Doing this for all the data in the entire run is rather complex to program (and extremely tedious to perform manually) but is likely to yield a very good choice for the analysis interval. However, if one has any interest in run-to-run and day-to-day comparisons, one must consider whether the next run will be analyzed in such a way as to make the data comparable. It is usually better to establish a particular interval for analysis (e.g., 30–180 s, or perhaps 300–600 s if a lag phase is anticipated) and use it for every run of the same experiment or study, with quality control parameters built into the process to ensure rejection of runs in which the data fail to exhibit adequate linearity, or in which the true linear range is displaced because of technical variances.

2.4.2 Performing the Linear Regression

We now use the above data to demonstrate the process of analysis by least-squares linear regression. This example makes use of Microsoft Excel®, but there are of course many software packages with this capability. For the user with some programming skill, it is not difficult to write a custom macro to perform regressions and correlations of each time-dependent dataset in a run. This may take only a few seconds. However, this automated capability does not eliminate the need to determine which data from the run should be reduced, as explained above.

The LINEST and CORREL functions of Excel are used to obtain the regressed linear rate and the correlation coefficient of the data, respectively. The use of these functions is explained in the Excel Help feature. Briefly, in the simplest form, the LINEST function takes two parameters, for example, a list of data (independent y values) and a list of time stamps (dependent x values), and returns an estimate of the slope m from the equation $y = mx + b$, using the least-squares method. The syntax is

$$<\text{selected cell}> = (\text{LINEST}<\text{selection of y values}>,$$
$$<\text{selection of x values}>) \qquad (2.1)$$

The CORREL function may be used to obtain the correlation coefficient in an exactly analogous way:

$$<\text{selected cell}> = \text{CORREL}(<\text{selection of y values}>,$$
$$<\text{selection of x values}>) \qquad (2.2)$$

Thus, the estimated slope of data and a gauge of the "quality" of the linear fit may be obtained very simply. These functions have other features and capabilities that may be determined by reading the associated "Help" material.

In the analysis of the sample data of Fig. 2.3, we see that the choice of the range to analyze has a significant effect on the results. Analysis of the entire range of the data, from 0 to 1139 s, yields a poor correlation coefficient of ~0.94 and a low calculated rate of increase, as expected. If we decide that the 239 s time point is the first at which we can confidently state that the lag phase has ended, we can analyze the data from this point onward (239–1139 s), but this improves the rate and the correlation coefficient only slightly, because the data are still contaminated by the decay phase at the end of the run. Retaining the decay phase but eliminating the initial lag yields a very poor correlation of ~0.83 but by properly eliminating both nonlinear time intervals, we obtain a rate of 0.335 RLU/s and an acceptable correlation coefficient of over 0.98. All of these analyses except the last should be highly suspect simply because of the correlation coefficients alone. When a low coefficient is observed, a visual inspection will usually reveal the nature of the unexpected variation. Some runs are just very noisy, but in most cases the deviation will be evident. The residuals should also be examined, as explained below. Even a correlation coefficient of 0.98, although it indicates that 98% of the variation in the data is accounted for by the linear model, is still lower than that of an average run and may trigger an investigation into possible causes (in this case, the very low range of the expected signal).

The more advanced and knotty issue of residuals, which constitute one of the best measures of the quality of regression analyses themselves, is discussed primarily in Chapter 13. Here, we note that whether the correlation coefficient is favorable or otherwise, a good linear fit (or nonlinear fit) should exhibit random scatter; that is, the variances of the actual data from the graph of the fit model should not show a pattern. At times, however, linear fits are used to fit data that exhibit some degree of decay in the upper range, and a sublinear deviation is inevitable. The validity of this approach depends on the degree of deviation and the purpose of the assay.

Once the user has obtained the linear rates, these may be used as reduced data for graphing and analysis. Figure 2.4.3 shows the results of an experiment analyzed by single-point methods and linear regression, respectively. In this case, the results and conclusions are almost identical, indicating that single-point data are definitely usable as final readouts when automated injection is available and the timing of the instrument's read function is consistent and well understood. However, even when single-point readouts are known to be useful, the user may decide to compare the two analysis methods on occasion simply as a check on the quality of the data.

2.4.3 Outliers

The problem of outliers is discussed at some length in Chapter 13. Briefly, outliers are data that are so far outside of the expected range that they probably do not reflect the phenomenon under study. Outliers can be caused by technical errors in measurements, contamination or cross-contamination of samples, bubbles, and occasionally (though this is much less common as a cause than human error) instrument malfunctions or "glitches." Unfortunately, one can rarely be certain that an outlier really is an outlier,

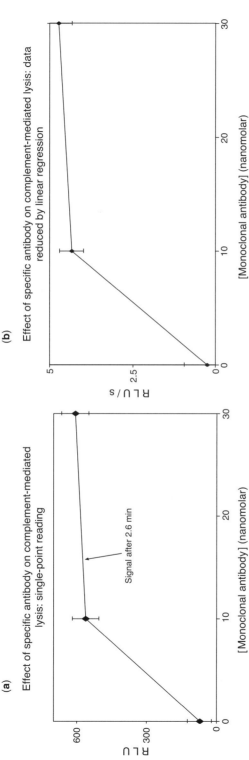

FIGURE 2.4 Single-point reading versus linear regression of time-dependent data. (a) Luminance signal 2.6 min after initiation. (b) The same data, reduced by the linear regression method described in the text. The *y*-axis label is "RLU/s," indicating that the reported data represent a rate of change in the luminance signal.

in the sense that it does not add useful information to the data set. The appropriate procedures for handling outliers depend entirely on what one is doing and how critical the result is. If one is in a discovery phase and one merely wishes to know whether an effect is present and roughly how big it is, a run with a single or even multiple outliers may be enough to answer the question, despite the problems. If one wishes to publish, outliers should rarely be completely ignored, unless there is a highly compelling rationale for doing so. Finally, as explained in Chapter 13, outliers must be dealt with more rigorously in a regulated environment such as drug development or clinical testing.

2.4.4 Special Considerations Relating to High-Throughput Screening

Many features of CB assays make them nearly ideal for high-throughput screening (HTS) procedures, including the extreme sensitivity and rapidity possible, their homogeneous nature, and the relatively low cost of the reagents. It is therefore worthwhile to consider these assays in terms familiar to the practitioner of HTS, especially the Z value (111).

The Z value is sometimes regarded as a metric that represents the "quality" of an assay. It is defined by the following equation:

$$Z = \frac{\mu_{s-b} - 3(\sigma_s + \sigma_b)}{\mu_{s-b}} \qquad (2.3)$$

where μ_{s-b} is the average difference between the positive signal and the negative-control background, σ_s is the standard deviation of the signal, and σ_b is the standard deviation of the background. The absolute level of the background signal is irrelevant to the calculation of the Z factor; only the variances and the separations between the positive and negative signals are of importance. Therefore, in CB systems, a modest level of contamination of some of the assay constituents by reagents under test (such as ATP in the aCella-TOX^{TM} system, or ATP or inorganic phosphate in the phosphate detection of Chapter 5) has minimal effect on the Z value, since the contamination introduces a detectable but *constant* background signal. In some cases, this quirk may reflect reality, in that the statistical distinction between signal and background is what is of true significance, rather than their absolute levels. However, this principle should not be overgeneralized: some sources of noise scale up with the level of the background, and a very high background level can cause various modes of saturation, or even undesirable side reactions.

The Z value may provide correct information as to whether a given degree of change in a phenomenon of interest can be detected reliably; however, reliance on this derived number can lead to problems. The Z value of an assay depends on the actual formulation and, therefore, does not always reflect the fundamental nature and quality of the assay system. Thus, one can often improve one's Z value simply by spending more on reagents. This and other points about the Z value in the context of CB assays are addressed in more detail in Reference 112.

2.5 COMPARISON OF COUPLED BIOLUMINESCENT ASSAYS WITH OTHER METHODS

Under favorable conditions, CB assays are capable of surpassing all other forms of liquid-phase assays in terms of speed, sensitivity, and convenience. This is true of these three parameters in aggregate, and it is also true of each individually, with the possible exception of chemiluminescent assays, which can also yield extreme speed and sensitivity. However, they are not always available for the same systems and can be inconvenient. In any case, these claims for CB assays are not meant to be a pure and straightforward recommendation. CB assays come with important considerations and caveats that should be understood before a decision as to assay type is made.

Most of the comparisons among assay types are made in the individual chapters in which those assays are discussed. What is presented here is primarily the generalizations that can be drawn about the relative merits of the kinds of assays, although even these are subject to numerous exceptions.

2.5.1 Coupled Bioluminescent Assays Versus Spectrophotometric Assays

Here, we include both visible and ultraviolet spectrophotometry, which differ little in practice. The most important advantages enjoyed by CB assays in this comparison are speed and sensitivity. Initially, it would appear that there are other significant advantages, since luminescent measurements require no lamp, all photons produced are useful, and the chamber can be made almost perfectly dark to eliminate stray light. However, in the modern world of microplates, these apparent advantages become less important, or even disappear. It is true that no collimation is required if a single luminescent reaction is being read by a single detector, but a microplate filled with shining reagents does not meet this criterion. Modern plate-reading luminometers must provide excellent collimation and physical coupling of the reading device to the well being read. Alternatively, a high-resolution charge-coupled device may be used to image the plate if the user has the funds to buy it and software sophisticated enough to deconvolute the well images, possibly with the additional step of Fourier transforms to disregard mathematically stray light that cannot be eliminated physically. In short, dedicated luminometers may no longer be called "simple," although at least one can claim the advantage that luminometers never require lamp replacement.

The speed and sensitivity performance rely on the nature of the measurements and the processes being measured. Each molecule of an absorbing material prevents some small portion of the photon stream from reaching the detector; however, many molecules are required to have much effect on the bright lamp emission. If even more of the absorbing species is present (which would seem to present the ideal condition for detection), then little of the light is transmitted to the detector at all, and precision falls off dramatically. Moreover, light is not entirely inert; photoinduced reactions can occur, "bleaching" or otherwise changing the sample as it is being studied, although this is typically more of a problem with fluorescence. In a CB reaction, on the contrary, the luminescent molecules "announce" their own presence. Light, rather than darkness, is the signal, and each emitting molecule is a direct source of that signal. It is not unusual to see a sensitivity advantage of four to six orders of magnitude for CB reactions in comparison with UV/spectrophotometric methods.

The reader will pardon an anecdote. On one occasion the author was presenting CB methods to a contract research organization, one of the many firms that perform specialized research in the life sciences, such as answering questions about drug candidates. Antibiotic screening was the topic, and the presentation of CB methods centered on the extreme sensitivity to small effects and the mechanistic information that could be obtained by distinguishing live and dead cells, killing versus. bacteriostasis, and so on. The competing method was simply measuring the absorbance of the growing bacteria at 650 nm in an armed flask (which does not require removal of an aliquot or breaking of the sterile seal). Very understandably, the firm chose to stick with their extremely simple and convenient method. They wanted to know how fast the bacteria were growing in bulk culture, and that was all. Sensitivity, mechanism, life, death, and bacteriostasis were all unnecessary luxuries from their point of view. However, organizations with specific interest in antibiotic mechanisms and structure–function relationships might reach different conclusions.

2.5.2   Coupled Bioluminescent Assays Versus Fluorescent Assays

One hesitates to place any limits on where fluorescence technology will go. The reader will encounter any number of fluorescent assays in perusing this volume, many of which are ideal or nearly so for their intended applications. Current developments in the field are dizzying, with the regular appearance of phenomena such as surface plasmon-coupled emission, an outgrowth of surface plasmon resonance used to study extremely small systems such as individual muscle fibers (113), and the advent of new, longer lived fluorescent species for improved resolution and imaging of biological samples (this example published, ironically, in the journal *Luminescence* (114)). Thus, it is dangerous to make assertions about what fluorescence technology cannot or will not do.

Within these constraints, however, it is fair to point out that fluorophores still do not emit their own light. That light ordinarily comes from a lamp (although it may be perturbed and/or reemitted by any number of fluorophores on the way), and phenomena such as photobleaching, stray light from the lamp, wavelength issues, and so on will always have to be considered in performing fluorometric assays. Dust particles can scatter light and appear to be fluorescing. Molecules that do not absorb light themselves can exhibit "Stern–Vollmer quenching," a phenomenon in which electronic excitation energy is bled away via collisional mechanisms and is no longer available to contribute to the signal; thus, some fluorescent measurements are strongly dependent on the presence or absence of small molecules that appear irrelevant to the chemistry under study.

Luminescent assays are generally held to be more sensitive than fluorescent assays by a factor of 100–1000. This is an overgeneralization, however, and so many parameters enter into the comparison that the sensitivity ratio alone is almost meaningless without the rest of the information. One of the many comparisons that may be drawn is that between aCella-TOX and Promega's various fluorescent means of measuring cytotoxicity, both of which are described in Chapter 3. The sensitivity advantage of aCella-TOX is real, but the fluorescent formulations also have clear advantages, and there are numerous other considerations relating to convenience, expense, analysis, and the degree to which the assays are established in the scientific community.

2.5.3   Coupled Bioluminescent Assays Versus Chemiluminescent Assays

Chemiluminescent methods are available for some assays and have achieved dominance in a substantial portion of these markets; for example, electrochemiluminescence employing artificial substrates of horseradish peroxidase or β-galactosidase is a superb way of detecting the results of Western blots and other experiments involving associating or annealing with immobilized reagents. However, chemiluminescent substrates for many reactions are not yet available, and when they are available, the luminance is often of the "flash" type, since the dioxetanes on which so much of the field is based decompose quickly. As a single example, a patent providing chemiluminescent protease substrates was issued to Tropix, Inc. (now part of PerkinElmer) in 1999 (115), but these methods have not gained traction in the marketplace. The competitive CB methods developed by Promega (see Chapter 10) are very new and largely untried, but promising and flexible, although one must add that their creation has involved a great deal of skillful chemistry and molecular biology.

PART II

BIOMEDICAL APPLICATIONS OF COUPLED BIOLUMINESCENCE

3

COUPLED BIOLUMINESCENT CYTOTOXICITY ASSAYS

3.1 INTRODUCTION

"Add the compound and see if the cells die." This could be an instruction from a principal investigator to a graduate student or from a manager of a drug discovery enterprise to a subordinate. Unfortunately, there is no single definition of the word "die," especially when cells are concerned. Scientists investigating apoptosis may be satisfied with a demonstration that caspases have been activated (116), whereas a complement researcher may not be convinced that death has occurred or will occur even if the cells are leaking massive amounts of cytoplasm (117). In cases of drug safety testing, of course, looking for death is not good enough. Toxicity that stops short of killing the cells may well signal undesirable properties in the drug candidate. In short, the concept of "cell death" carries its own weight of scientific and philosophical baggage.

Since even "death" is not well defined, the researcher must make a choice as to which aspect or correlate of death is of interest (perhaps more than one). A worker studying apoptosis will be interested primarily in many known phenomena that accompany this extraordinarily complex process, although even in this case, the use of other methods may be desirable to assess the degree of necrosis or mechanical lysis in the samples under study. Other parameters associated in various ways with cell death include at least loss of membrane integrity, loss of mitochondrial membrane potential, disruption of other biochemical pathways characteristic of actively metabolizing cells, markers of activation of death pathways, and (especially for microbial cells) failure to form colonies. In some cases, what is measured in the course of a cytotoxicity

Coupled Bioluminescent Assays: Methods and Applications, Michael J. Corey
Copyright © 2009 by John Wiley & Sons Inc.

assay is the absence of a characteristic of living cells; these methods are actually viability assays and are discussed in Section 3.4. In fact, it is a common practice to market viability assays as cytotoxicity assays, but there are statistical difficulties attending this approach, as explained below.

A number of vendors, notably Promega, have systematically addressed the issue of "type of death" in their technical literature. The following document, available online as of this writing, is a general treatment of the issue of selecting a cytotoxicity assay, and it includes a useful summary of the time-dependent correlates of apoptosis, necrosis, and necrosis secondary to apoptosis: http://www.promega.com/paguide/chap4a4.pdf. Understandably, the document does not address coupled bioluminescent (CB) assays, since Promega does not currently market any such assays for cytotoxicity measurement.

The CB cytotoxicity assays discussed here fall into two categories: membrane-integrity assays and viability assays. These approaches are very different in both their practice and their implications. However, it is possible to use a single CB assay method to obtain both pieces of information from a single sample. To minimize confusion, the two varieties of CB assay will be described separately, whereupon the "dual mode" will be described as an independent method.

CB methods can address many, but not all, of the needs of researchers performing cytotoxicity assays. For many critical applications, the use of multiple orthogonal methods of measuring identical or parallel phenomena can yield much greater confidence in one's results. This is good science, and it may also be required by regulatory agencies. Moreover, nearly all methods are subject to interference by one or more classes of chemicals. The concept of "orthogonal" therefore involves partly the assurance that the independent assays in use are not likely to be subject to the same artifacts. This assurance requires an understanding of the nature of the assays.

To assist the researcher in making a choice of one or more cytotoxicity assays, it may be useful to list the types of assays available and describe their intended use. Table 3.1 is meant to provide a very general overview, and it distinguishes among assay philosophies, as opposed to detection methods; therefore, considerable variability is to be expected in some of the parameters.

Cytotoxicity assays deserve their own book, but for our present purposes, the methods apart from CB assays can be described in only modest detail.

3.2 MEMBRANE INTEGRITY ASSAYS

3.2.1 Radioactive Isotope Methods

Membrane integrity assays rely on diffusion of a compound through a ruptured or weakened cellular membrane. However, many subtypes of assays fall within this category, varying by the type of molecule released, its origin, and the scheme for detecting it.

These radioactive isotope assays have all the familiar advantages and disadvantages of many assays employing radioactivity. The ^{51}Cr-release assay (CRA) is prominent, but other isotopes are also employed to obtain different types of information.

TABLE 3.1 Comparison of Broadly Defined Strategies for Cytotoxicity Measurement

Type	Applications	Sensitivity	Speed	Interferences	Cost (estimated)
Isotope release	Broad	High	Very slow	Few	High
Other label release	Broad	High	Slow	Few	Low to moderate
Natural enzyme release	Broad	High	Fast to very fast	Various	Low to moderate
Direct growth measurement	Mainly bacteria	Low to moderate	Moderate	None	Lowest
Mitochondrial potential	Eukaryotes	Moderate	Moderate	Many	Low

3.2.1.1 The ^{51}*Cr-Release Assay* The dominant method in the class of radioactive membrane integrity assays is measurement of the release of ^{51}Cr, which is loaded into the target cells prior to the experiment (118). This assay has been used very widely in cytolytic measurements of various phenomena, from complement attack to pore-forming detergents to cell-mediated killing. Like many approaches employing radioactivity, it provides excellent sensitivity, but the most important reason for its success is the autocatalyzing process of acceptance by virtue of prior acceptance. Because there is such a large database and such profound and widespread understanding of the assay and what its results mean, it is tempting to continue doing the same thing.

As an example, we consider the challenging problem of analyzing the cytolytic behavior of cytotoxic T cells (CTLs; generally these are CD8$^+$ cells, although the coincidence of the two sets is imperfect). This example is chosen primarily because the ^{51}Cr-release assay is used almost exclusively at present for assessing CTL activities and its results are widely accepted by regulatory agencies for this purpose. The primary function of these leukocytes is to recognize and eliminate cells of the body that have been infected by viruses, although other pathogens and some cancers may generate signals that invite CTL attack as well. The recognition process involves the major histocompatibility I (MHC I) complex, for which the word "complex" is a serious understatement in terms of both structure and function. Briefly, this complex displays short peptides derived from viral or other antigens for recognition by T-cell receptors, which are fascinating and highly diverse proteins integral to the membranes of CTLs with many features of antibodies. This system is capable of recognizing a cognate antigen and transducing the association event into activation of the CTL, but *only* if the antigen is "presented" in the context of the cell-surface MHC I molecule, and only if other stimulatory signals are also present and inhibitory signals are absent (119). The whole system thus resembles a complex gating device within an electronic computer, in which a procedural pathway is activated if and only if all the appropriate conditions are met. If so, the CTL initiates killing of the "guilty" target cell by various mechanisms, including at least the following: the release of perforin, a powerful pore-forming agent, in the specific direction of the condemned cell; release of granzymes, enzymes that stimulate apoptotic pathways and are especially effective when the target cell has been "softened up" by perforin; and release of interferon-γ, which rings general antiviral alarm bells throughout the body.

Most methods of assessing CTL activity fall into two categories: means of measuring CTL secretion and techniques for quantifying lysis of the target cells. The former include granzyme assays carried out with cell-culture supernatant (120) and assays, especially immunoassays, for interferon-γ (121). Interferon-γ assays are especially useful when one wishes to study the activities of individual CTLs, since modern methods of carrying out the Elispot assay enable this degree of sensitivity (122). Granzyme assays can also employ all the latest techniques of detecting protease activities, including specific fluorogenic substrates (123). While these methods are useful in screening procedures and are excellent ways of answering specific questions about CTL activation in defined systems, they are second best in terms of answering the most important question about CTL function. No extent of analysis of the CTL itself can ever address the central issue of whether the intended target cells are actually being killed. This issue is critical, because it is entirely possible to observe secretion of interferon-γ, for example, in the absence of lysis (124). Thus, for the growing field of cell therapy in which CTLs are employed directly as a treatment for viral diseases and cancer (125), demonstration of the lytic properties of the CTLs is an obvious and essential element of quality control.

The CRA is described briefly here to give the reader a sense of the amount and types of labor involved. Preparation of the CTLs and recombinant (or other) expression of the cognate antigen(s) within the target cells are not described; this is typically accomplished by means of expansion of a particular cell type of interest or introduction of the antigen via a viral vector, such as vaccinia, considered to have low immunogenicity. Once available, the target cells must be loaded with the isotope by incubation with labeled sodium chromate for some 2 h. They are then washed extensively to remove label by repeated centrifugation and resuspension, and at this point, of course, leakage leading to a background signal begins to occur. Finally, the cells are counted and concentrated to the desired density by another centrifugation and the experiment can begin. Humans or robots aliquot the target cells (usually in roughly identical numbers) into microplate wells. CTLs and any modulating agents are added and mixed, and the lytic process takes place. The readout step follows, but is highly sensitive to technique, because typically the upper fifth to half of the supernatant is removed without, it is hoped, disturbing the settled cell layer. If the suspension is mixed to any significant degree, a significant outlier will almost surely result. The method also depends on free diffusion of the label without mixing, but the label is a small molecule and, absent significant charge effects, this is usually not an issue. Scintillation counting is usually accomplished by air drying of the plates, addition of a fixed amount of scintillant, and reading in a gamma-counter or other suitable luminometer. It is worth recalling that all of the equipment used, including the water bath, the pipettes, the bench, the cell-culture hood, the centrifuges, and the counter itself, must be approved for radioactive use, monitored, and cleaned according to rigidly defined and usually state-mandated protocols. This affects cost, as well as employee morale on occasion.

If all goes well, the results may resemble those shown in Fig. 3.1. Here we see the plot beloved of T-cell immunologists, in which "effector:target (E:T) ratio" on the X-axis is plotted against "% Lysis," a value obtained from the release of radioactivity by a simple calculation based on a no-effector background and a "Max Lysis" accomplished by detergents. ("Effector" is a synonym for "CTL" in this case.) Max Lysis is sometimes considered as an important parameter of CTL activity, but it frequently has more to do with the targets and their preparation method than it does with the effectors, especially if a clear asymptote is seen at the upper range (e.g., approximately 52% in Fig. 3.1). The failure of the CTLs to lyse more than 52% of the targets, despite the provision of very large numbers of CTLs, implies that the targets are somehow heterogeneous. However, this effect can also be due to transient target effects, which in turn may be related to a limited ability of the CTLs to lyse targets with low levels of the cognate antigen. These effects can often be seen and distinguished from true target heterogeneity if multiple time points are taken.

Although the E:T ratio is what immunologists are used to seeing, it is quite misleading. The use of the term implies that the results of the experiment are somehow independent of the *absolute* numbers of cells and depend only on this ratio. This, of

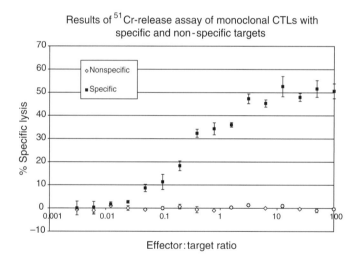

FIGURE 3.1 Sample ^{51}Cr-release assay of cytotoxic T lymphocyte (CTL) lytic function against specific and nonspecific target cells, infected with vaccinia vectors containing the cognate target protein and an unrelated protein, respectively. CTLs were grown by rapid-expansion method [126]. Following labeling of the target cells for 2 h, they were washed and transferred to V-bottom microplates. CTLs were added in an equal volume and the mixtures were incubated for 4 h at 37 °C. After incubation, plates were spun for 2 min in a tabletop centrifuge to pellet the cells. Aliquots amounting to 20% of the total volume were removed by careful pipetting from the upper half of the liquid and transferred to a scintillation plate, followed by gamma-counting in a Top-Count instrument (PerkinElmer) after drying overnight. The results of specific target lysis show the characteristic sigmoidal response to CTL number, due to the log-linear nature of the graph. The asymptote is approximately 52% lysis, indicating that nearly half of the specific target cells were not subject to lysis by the CTLs.

course, is incorrect, since the situation is almost precisely analogous to Michaelis–Menten enzyme kinetics, in which the three important parameters are (1) the concentration of the "substrate" (target cells) *relative to its* K_m (which is difficult to define in CTL reactions, but not impossible), (2) the turnover rate of the "enzyme" (CTLs), and (3) the issue of whether the reaction is in the linear range or not with respect to both types of cells. Briefly, if many of both cell types are present *and* the number of target cells is sufficient for partial depletion of the pool of active CTLs (this depends on dwell time as well as the number of lytic sites per CTL), the kinetics are complex. If not, then the reaction is in the linear regime with respect to targets and there is at least a chance of learning something about the inherent kinetics and affinity of the interaction. However, if few cells of each type are present, saturation is unlikely and both cell types are probably in their linear regimes. Thus, an E:T ratio of 1:5 can be associated with either nearly complete saturation or near-zero saturation; the ratio alone conveys almost no information. Complex data sets, whether reported by E:T ratio, effector number, or target number, may also be reduced to "apparent" kinetic parameters, using, for example, a four-parameter method (minimum signal, maximum signal, half-maximal concentration, and cooperativity or slope are the possible choices of disposable parameters), but this requires more mathematical sophistication.

Despite the density of this discussion, the assay can be used quite effectively for intraday comparisons of CTL and/or target effects. The method is tedious, expensive, lengthy and slow, subject to artifacts, and demanding of personnel, but it has the advantages of excellent sensitivity and widespread acceptance.

3.2.1.2 The $^{86}Ru^+$-Release Assay $^{86}Ru^+$ has also been used as a label for assessing cytotoxicity, but it was found in work of the 1970s and early 1980s to indicate a different type of event, identified rather generally as the inability of the cell membrane to control ion exchange, since the ion was used in the 1950s as an analog of the potassium ion (127). In addition to its role as a potassium analog, $^{86}Ru^+$ has been used in concert with CRAs to study differential effects on membranes (128). Its use as the sole label in a cytotoxicity assay is now rare.

3.2.2 Other Labeling Methods

Apart from radioactive isotopes, many other molecules that can be detected directly may serve as labels that are released upon loss of cellular membrane integrity. Here we encounter difficulties in categorization, because the set of such labeling methods overlaps with several of our other classes. For example, some of the fluorogenic dyes discussed below in Section 3.2.3 are cleverly designed to be processed by intracellular enzymes, yielding a fluorescent signal only after crossing the cellular membrane; this is therefore an enzymatic method as well. Enzymes themselves can also be labels. Because of the potential confusion, we address only two classes of methods in this section: dye-exclusion methods, in which vital dyes are introduced into cells, whereupon their leakage from the cells indicates failure of membrane integrity; and the special class of enzyme-release assays that involve recombinant expression of the enzymatic label.

3.2.2.1 Dye-Exclusion Methods   Trypan Blue (129) and Crystal Violet (130) are two of a range of vital dyes that have been used to assess membrane integrity by microscopic examination of the cells. These methods are not competitive in speed or accuracy with the automated methods listed here, although the author has compared Trypan-Blue exclusion with an early version of the G3PDH-release assay described in Section 3.3 and found them to give indistinguishable results (131). However, it should be noted that some of these dyes affect the processes under study (e.g., Trypan Blue accelerates cell collapse due to complement attack). This is a general problem with any labeling method—the label can perturb the experiment—but some labels are worse than others. It is of course possible that further developments in both dye chemistry and imaging of individual cells will resurrect these methods as players in high-throughput assay development.

3.2.2.2 Cytotoxicity Assays Employing Release of Recombinant Enzymes   An early assay of this kind was reported in 1992 (132). The β-galactosidase enzyme from *Escherichia coli* was stably transfected into a myeloma cell line, thus establishing a resource that could be used indefinitely for assays of the cell-killing activity of cytotoxic T lymphocytes (see Section 3.3.2 below). Later, the advent of a chemiluminescent substrate enabled a similar luminescent assay (133), although in this case transient transfection of the enzyme was employed. Firefly luciferase itself has also been transfected into immortal cell lines to create luminescent cytotoxicity assays (134), although the half-life of luciferase in cell-culture medium was found to be short, perturbing the results. Although the transfection step renders this approach inconvenient and labor intensive and limits its applicability, the introduction into the target cells of a signaling enzyme that is not normally present does provide a solution to the problem of leakage of enzymes such as glyceraldehyde-3-phosphate (G3P) dehydrogenase from effector cells in the CB method discussed below (see Section 3.3).

3.2.3   General Enzyme-Release Methods

Enzyme-release assays are indirect means of measuring membrane rupture or leakiness, and their value as cytotoxicity assays depends directly on the relationship between such leakiness and the toxic effects under study. The choice of enzyme has important implications for success, not only because of variable stability and enzymatic properties, but also because of differential release.

Another issue is how to measure enzyme activity. Until the advent of the CB cytotoxicity assay (83), these assays typically employed UV spectrometry, with the disadvantages mentioned in Chapters 1 and 2. More recently, fluorometric assays of lactate dehydrogenase (LDH) activity have been employed, alleviating some of the concerns.

3.2.3.1 Spectrophotometric Methods   Traditionally, lactate dehydrogenase (135) has been used most frequently as a marker of the loss of membrane activity associated with cell death. The reaction scheme of the enzyme is depicted in Fig. 3.2, with

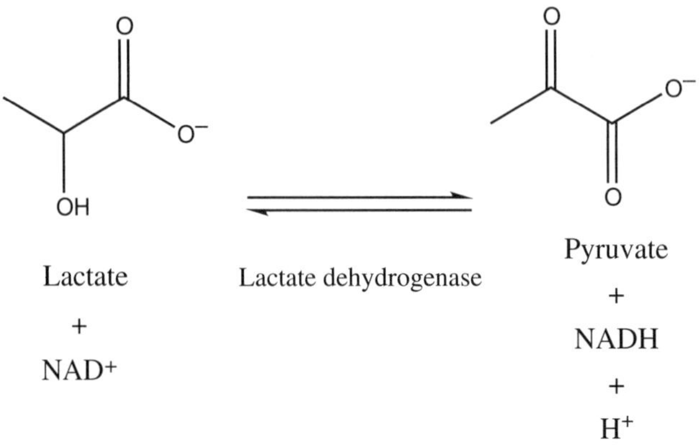

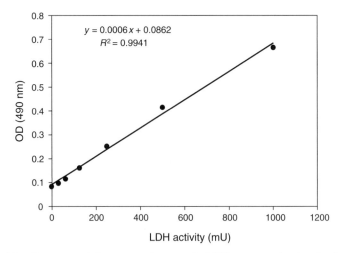

FIGURE 3.2 Reaction scheme of lactate dehydrogenase. The formation of NADH can be followed by spectrophotometry at 340 nm, as can its exhaustion by the reverse reaction of the enzyme.

the usual forward metabolic direction to the right, although the reaction is readily reversible. NAD^+ is consumed in the course of the forward reaction, yielding the reduced cofactor NADH. The long-established method of measuring the activity of LDH is to follow the creation or destruction of NADH by UV spectroscopy at 340 nm. Figure 3.3 presents data obtained by the spectrophotometric LDH method. The enzymatic reaction was carried out for 30 min at room temperature with gentle shaking. The assay detected a fairly small quantity of LDH, which, however, represents the output of thousands of cells, as shown by Fig. 3.4, which depicts the results of a cell dose–response titration. These cells were lysed with the detergent Triton X, and the

FIGURE 3.3 Response of the spectrophotometric LDH assay to titration by the enzyme. A slight sublinear deviation is seen at the highest concentration. Used by permission of Cayman Chemical Company.

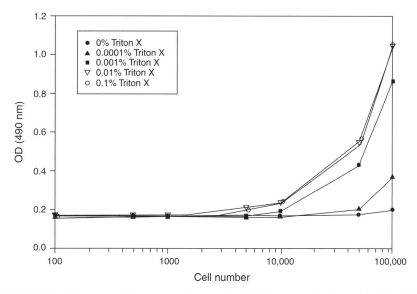

FIGURE 3.4   Response of the spectrophotometric LDH assay to titration with Triton-X, a non-ionic detergent. The effects of both increasing amounts of detergent and titration with cells are shown. The limit of detection may be near 1000 cells. Used by permission of Cayman Chemical Company.

figure shows both the effects of the lytic agent and the response of the assay to large cell numbers. Quantification of numbers of cells below 1000 may be problematic with this method. Another significant problem associated with the use of the LDH enzyme for cytotoxicity determinations is that it is present in rather large quantities in serum, which is frequently used in cell culture of many types. Heat-inactivated serum or serum-free media may be used instead, or the background determined and subtracted. (Similar problems exist with G3PDH, although the G3PDH activity in serum is lower.) Soon after its development, the LDH assay was shown to be statistically comparable to the ^{51}Cr-release assay (118), but spontaneous release of LDH was significantly lower (136), a significant advantage, since this release can give rise to a variable background that may depend on cell type and conditions.

A refinement of this approach is to include a reagent such as the tetrazolium dye WST-1, the sodium salt of 4-[3-(4-iodophenyl)-2-(4-nitrophenyl)-2H-5-tetrazolio]-1,3-benzene disulfonate (137), which is useful as a reagent for cytotoxicity detection in its own right. The dye then reacts with NADH, a product of the LDH reaction, to produce a yellow formazan, which is detected at 450 nm. Alternatively, the tetrazolium salt 2-(4-iodophenyl)-3-(4-nitrophenyl)-5-phenyl tetrazolium chloride (known as INT) reacts with pyruvate, a product of the LDH reaction, to yield a formazan that absorbs at ∼500 nm. These improvements may yield a limit of detection closer to 100 cells. Other major vendors marketing comparable kits include Cayman Chemical, MBL, Roche, Promega, BioAssay Systems, Biovision, and Takara Bio; list costs are as low as $0.18 per well for bulk purchases (assuming 96-well plates).

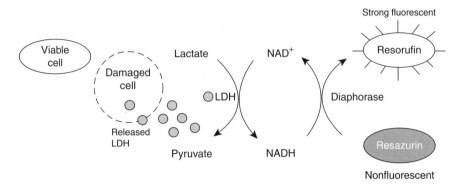

FIGURE 3.5 Schematic view of resazurin/resorufin-based fluorometric LDH assay. The enzymatic activity of LDH released from dying cells is coupled to reduction of NAD$^+$ to NADH, which in turn is utilized by diaphorase to reduce resazurin to resorufin, with strong fluorescence at 590 nm upon excitation at 530–560 nm. Used by permission of Anaspec, Inc.

3.2.3.2 Fluorometric Methods   If one's microplate reader is capable of fluorometry, it is hard to justify the use of a colorimetric assay for LDH, since fluorometric methods are much more sensitive, faster, and of comparable or lower expense. In the most commonly used formulation (Fig. 3.5), the NADH produced by LDH enzymatic activity is coupled to the activity of diaphorase, which reduces resazurin to the fluorescent resorufin, simultaneously oxidizing NADH to NAD$^+$. The strong fluorescence of resorufin results in excellent sensitivity for LDH activity.

The SensoLyte^{TM} kit from Anaspec, for example, enables detection of fewer than 100 lysed cells (see Fig. 3.6), although some unexplained sublinearity is seen,

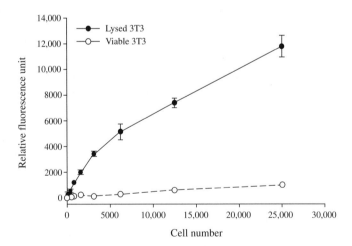

FIGURE 3.6 Dose–response curve of fluorescence signal to quantity of lysed cells using fluorometric LDH assay. Lysis buffer or growth medium was added in a small volume to 3T3 cells. Assay reagents were added and fluorescence read after 10 min. The limit of detection is slightly under 100 cells. Used by permission of Anaspec, Inc.

probably due to exhaustion of a reagent in the coupling scheme. The readout was obtained only 10 min after initiation, qualifying the method as a true high-throughput procedure. Errors are small. Comparable assay kits are sold by many other vendors, including Molecular Probes (Invitrogen), Sigma-Aldrich, Promega, and Wako Chemicals.

Molecular Probes (Invitrogen) also took an entirely different approach to fluorescent assessment of cell viability and cytotoxicity, beginning in the early 1990s. In their LIVE/DEAD Viability/Cytotoxicity kits, they provide two separate dyes for quantifying live and dead cells, respectively. The integration of the two methods has yielded useful results in studies of adherent cells (138) and certain tissues (139), establishing its general applicability even with difficult sample types. In this approach, the calcein AM dye we have seen elsewhere is used as the viability marker, since the intracellular esterases that process this dye to a highly fluorescent product decay upon cell death. The other wing of the method uses Ethidium homodimer-1, which fluoresces strongly upon association with nucleic acids. The latter dye penetrates only compromised membranes, and, under the reasonable assumption that free nucleic acids are not abundantly present in the cell-culture supernatant, its fluorescence can be attributed to a failure of membrane integrity in the affected cells. These dyes are useful not only in liquid-phase assays, but in techniques such as flow cytometry and fluorescence microscopy, since the dye molecules generally remain with the cell. (Figure 3.7 shows the superb separation achieved in flow cytometry, implying that

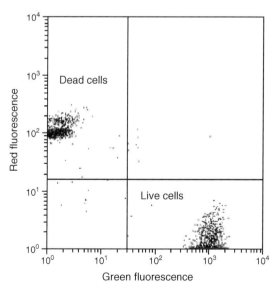

FIGURE 3.7 Strong separation of live and dead cells by flow cytometry using the LIVE/DEAD Viability/Cytotoxicity Kit (Invitrogen/Molecular Probes). A 1:1 mixture of live and ethanol-fixed human B cells was stained with calcein AM (live stain) and ethidium homodimer-1 (dead stain). Flow cytometry was performed after 5 min. Courtesy of Invitrogen, Inc.

liquid-phase assays would also yield highly independent readouts for live and dead cells.) Separate kits are sold for analysis of animal cells, work with bacterial cells, sperm viability, and other specialized purposes.

The fluorescent methods in general must be considered strong competitors for use in high-throughput environments, offering sensitivity and speed, along with the confidence associated with established methods. The data scatter seen in these experiments is usually low, especially considering the imprecise nature of cell-based assays, and statistical considerations therefore support the use of these methods as well. However, for extremely sensitive and very rapid measurements, it appears that CB assays are in the lead, as we shall see in the following section.

3.3 A COUPLED BIOLUMINESCENT ASSAY FOR ENZYME RELEASE: THE G3PDH-RELEASE ASSAY

CB methods that test membrane integrity by enzyme release fall into two categories: assays employing enzymes naturally present in the cells under study, and those involving expression of the signal-generating enzyme by the target cells. The latter set of methods has the obvious disadvantage that an expression system must be worked out and tested beforehand in the target cells of interest, followed by transfection or transformation of these cells and incubation for expression. Perturbation of the metabolism of the target cells is quite possible, complicating interpretation and rendering additional controls necessary. Nevertheless, some success has been achieved by the use of luciferase and β-galactosidase (140) expression plasmids (the latter requiring the use of a luminogenic substrate of the enzyme, although alternative fluorescent methods are possible (141)).

Herein, the focus will be on CB methods that employ naturally present enzymes. The prototype for these assay methods is the G3PDH-release assay reported in 1997 by the author and colleagues, which operates according to the scheme shown in Fig. 3.8 (83). The basis of the assay is the generation of light in the presence of G3PDH released from dying cells via a three-step process: (1) the synthesis of 1,3-diphosphoglycerate (1,3-DPG) by G3PDH; (2) the transfer of a phosphate group from 1,3-DPG to adenosine diphosphate (ADP) by phosphoglycerokinase (PGK), yielding ATP; and (3) the reaction of ATP with luciferase to produce light. Since G3PDH is not supplied in the reaction vessel, the only source of this enzyme is the ruptured cell, making light generation dependent on the lytic process.

The development and use of this idea over an 8-year period exemplify many of the principles discussed in Chapter 2, while the capabilities and characteristics of the technique illustrate the potential value of CB methods. The story of the G3PDH-release assay is therefore provided below in some detail as a "case study" of the evolution of a successful CB assay.

3.3.1 Development and Features

In 1996, the author was working on enhancement of complement-mediated lysis of tumor cells by antibodies directed against complement Factor H. These studies

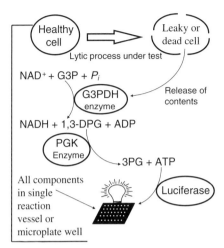

FIGURE 3.8 Reaction scheme underlying the aCella-TOX™ CB cytotoxicity assay. Cells with damaged membranes leak G3PDH enzyme, which oxidatively phosphorylates glyceraldehyde-3-phosphate, yielding 1,3-diphosphoglycerate. The second enzyme in the series, phosphoglycerokinase, then transfers a phosphate group from 1,3-DPG to adenosine diphosphate, yielding ATP and a by-product. ATP provides the chemical energy to power light production by luciferase, which is therefore dependent on the quantity of G3PDH present.

eventually led to the identification of a murine monoclonal antibody capable of accelerating the attack of the alternate pathway of complement on cancer cells *in vitro* by over 10-fold (131). Soon after the inception of the project, the author grew dissatisfied with the cytotoxicity assays available at the time. In particular, lysis of mammalian cells was frequently measured by dye exclusion, a tedious, insensitive, and inaccurate method that is subject to artifacts, especially in complement studies. In the course of the project, a coupled bioluminescent G3PDH-release assay was developed and used to measure complement attack on human cells, while the well-established hemoglobin-release assay was employed in parallel for work with rabbit erythrocytes (142). Later it was discovered that the CB assay is also capable of measuring erythrocyte lysis, with the unprecedented sensitivity of approximately one cell. The original CB method, called the "GPL" assay at the time (for G3PDH, phosphoglycerokinase, and luciferase, the three enzymes coupled to yield the light reaction), involved three separate steps, as well as separations of cells from supernatant. Although it was somewhat inconvenient, the sensitivity of the assay (<0.1 mammalian nucleated cell equivalent) justified continued development. Eventually, the improved one-step homogeneous CB assay (143) shown in Fig. 3.8 replaced the GPL technique and is currently on the market (trade name: aCella-TOX™) (144).

While the GPL scheme as originally conceived surprised the author by yielding a strong signal the first time it was attempted, it is hoped that this book, and this section in particular, will spare the reader part of the long and painful educational process that lay ahead.

3.3.1.1 Storage Conditions of G3PDH-Release Assay Components The G3PDH-release assay has many components (143). Since it is highly inconvenient to assemble the reaction mixtures on each use, it becomes necessary in practice to mix some of the components and store them in this state. But which ones should be mixed, and how should they be stored? Of course, this issue is even more problematic during an optimization phase, when it is necessary to vary component concentrations.

One of the first lessons to be learned is that enzymes should be kept away from substrates when possible—not only their own cognate substrates, but others as well. The reason is that enzyme preparations, especially commercially available preparations, are rarely of high purity. In other words, by mixing nominal substrate and enzyme preparations, one may be unintentionally causing a reaction of contaminating enzymes with the actual substrate, or even with other contaminating molecules. Often these reactions generate ATP or intermediates on the pathway to ATP formation, thereby increasing the static background signal, sometimes severely. It should be evident that even when the mixtures have been tested and the rate of generation of undesired products has been shown to be minimal, it is still preferable to store such component mixtures in very cold conditions. As a straightforward example, it might appear that one could mix the PGK enzyme with NAD^+, inorganic phosphate, G3P, and ADP in the presence of appropriate buffer constituents, perhaps with the luciferase reaction components present as well, and be ready to perform the G3PDH assay simply by adding the cells and test reagents. However, preparations of PGK are almost always contaminated with small amounts of G3PDH. (This is hardly surprising, since these two enzymes are sequential elements of the glycolysis pathway and are found in the same cellular compartments; G3PDH is also a highly abundant enzyme.) Since G3PDH is present, the reaction proceeds during storage, yielding a high ATP background or, worse, short circuiting the desired biochemical scheme and exhausting the reaction substrates before the experiment even commences. This is only a single example, and the reader is urged to consider which enzymes can be safely stored with which small molecules, keeping in mind that the label on the bottle does not necessarily reveal all of its contents. Another potential source of trouble with any CB assay employing beetle luciferases is the class of enzymes that exchange phosphate groups among nucleotides, known as terminal transferases. These can also be present in nominally high-purity preparations of other enzymes, and can readily convert ADP to ATP (generating AMP as well in the process). Again, this raises background and exhausts a critical substrate.

Of course, individual components must also be protected from inappropriate storage conditions. Here again, an understanding of the nature of the assay as well as the chemistry of each component is essential. The two components of the G3PDH-release assay that cause the greatest stability problems are G3P, which is highly labile to hydrolysis, especially at neutral or (worse) basic pH, and the luciferase reagent itself, which should be protected from light and heat. Luciferase also undergoes chemical degradation, although steps can be taken to minimize this problem, especially provision of adequate reducing power. For the G3PDH-release assay, component decay is not a serious issue, because assuming that the enzymes mentioned in the above paragraph are kept away from the substrates, there are no components that yield a serious background or inhibition problem upon degradation. However, refer to Chapter 5 for an example of exactly such an issue in the case of the CB phosphate assay.

3.3.1.2 Backward Reactions A basic principle of enzyme catalysis is that any enzyme reaction can run backward. This is a direct consequence of microscopic reversibility. An enzyme is merely a catalyst that lowers the energy barrier to the reaction and does not "favor" one side over the other (145). Backward reactions can give unexpected results, some of which may affect the quality of an assay.

As a simple example, recall the reaction scheme of phosphoglycerokinase, which is also, of course, embedded in Fig. 3.9.

Note the double-headed arrow. This enzyme can convert 1,3-DPG and ADP to 3PG and ATP, but it is equally adept at catalyzing the reverse transformation. If one supplies 3PG and ATP to the PGK enzyme under appropriate conditions, ADP and 1,3-DPG will be generated. The ADP would cause no problem with the CB G3PDH-release assay, but 1,3-DPG is in the direct reaction series leading to ATP. If one performed this backward reaction and then removed the purines ADP and ATP, thinking that one had eliminated sources of background, one would be incorrect, since 1,3-DPG will give rise to a signal as soon as ADP is added. This mistake is unlikely to occur in practice, since there is rarely a reason to add ATP to this reaction and never 3PG.

Now consider the luciferase reaction. Light is a product of this reaction. Therefore, For the process to run backward, light must be supplied. Unfortunately, light of the appropriate wavelengths is only too abundant in life science laboratories. Therefore, if luciferase, its chemical reaction products (i.e., AMP and pyrophosphate), and luciferase cofactors are left under illumination in appropriate buffering conditions, the reaction will be driven in the direction of formation of ATP. Light can also photooxidize luciferase (146). Mixtures containing luciferase should therefore be protected from light when possible, but even this precaution may be insufficient. Ideally, the

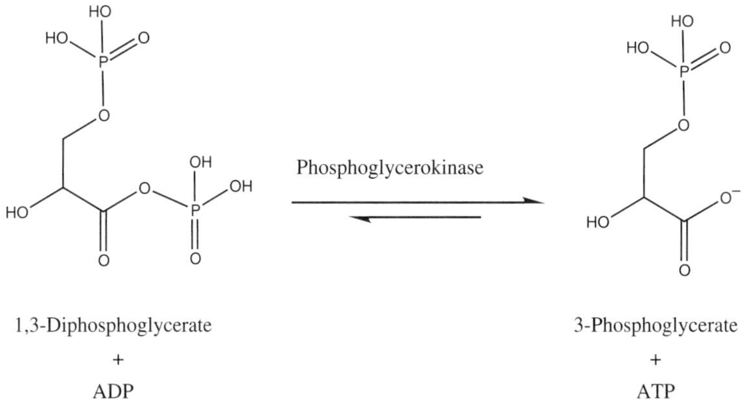

1,3-Diphosphoglycerate
+
ADP

Phosphoglycerokinase

3-Phosphoglycerate
+
ATP

FIGURE 3.9 Reaction scheme of phosphoglycerokinase, which transfers a phosphate group from 1,3-diphosphoglycerate to ADP, yielding ATP.

luciferase should be kept separate from ADP, ATP, and components that may generate either of these molecules until it is desirable to initiate the light reaction.

Figure 2.2 illustrates the importance of protecting the reaction from light. Here it is clear that the problem is not irreversible photooxidation. If the reaction is exposed to the light prior to measurement, a long lag phase is observed, during which the signal decreases for several minutes, presumably reflecting accumulated ATP due to a backward reaction, although other interpretations are possible. In any case, partial protection of the reaction mixture from exposure to light ameliorates the problem (Fig. 2.2). Although comparative data in the complete absence of light are not available from this experiment, subsequent work has shown that the lag phase can be eliminated by protection from light. If an automated injector is available, the researcher should design the experiment so as to mix the enzymes with the substrates at the last possible moment. This generally resolves the issue.

3.3.1.3 Sensitivity to Contamination   The developments of the past three decades in the life sciences have enhanced the sensitivity of our detection tools by several orders of magnitude. PCR can now detect one plasmid molecule (147), as opposed to a limit of detection of thousands or millions of molecules for earlier techniques. CB reactions can detect one cell death. With these improvements, however, comes a drastically increased need for cleanliness and care. As in clinical and forensic PCR, CB reactions must be protected from contamination of a nature that can lead to false conclusions. Shed skin cells contain enzymes that can yield a signal. Although gowning is generally not necessary for most procedures, the worker should consider whether a laminar-flow hood will improve results when critical experiments at the limit of sensitivity are being performed. The author has observed "spiky" data, in which about 5% of the data consisted of outliers with luminance values far higher than the replicates. These results were clearly due to shed skin cells, but of course other sources of contamination may be present, including aerosols and other experiments in the same lab.

This issue is mentioned here specifically because it arose in development of the G3PDH-release assay, since G3PDH is highly abundant in human cells. Clearly, other CB assay types necessitate attention to different sources of contamination—for example, a phosphate assay requires punctiliously phosphate-free buffers, reagents, and reaction vessels, while acetylcholinesterase assays mixtures must be kept from human sources of the enzyme, including saliva droplets. Some of these points are discussed at greater length in the various chapters covering their respective assay topics.

3.3.1.4 Reaction Timing and Substrate Exhaustion   Workers experienced in other biochemical systems often observe "nothing" the first time they try a CB assay. The absence of time-dependent or reagent-dependent signal can occur for either of two reasons, both rather obvious: (1) Something is keeping the reaction from proceeding; or (2) The reaction has run to completion, exhausting one or more of the substrates. Of these possibilities, (2) is surprisingly common, and was frequently seen in the early days of development of the G3PDH-release assay, simply because even a few cells contain so much of the enzyme that the reaction can run its course in seconds or milliseconds. Of course, the experienced biochemist will immediately distinguish between (1) and (2) by the appropriate use of controls and standards, but the point is that like every newly developed assay, CB assays require suitable "ranging" to identify the useful quantitative regime. It may not be intuitively obvious that reagent purity is a major player in this issue. The reason is that in CB assays, the background signal is due almost entirely to impurities in the input reagents (such as G3PDH contamination of the PGK preparation, or ATP in the ADP, in the case of the G3PDH-release assay). These impurities and their associated background effects limit the concentrations achievable, which in turn limit the dynamic range of the assay.

3.3.1.5 Reaction Initiation Rapid reactions are a boon to the high-throughput screening enterprise. They are also a challenge to the individual investigator without an automated injector, such as the author during the period he was developing this assay. Clinicians, inspectors, management, vendors, and technicians frequently demand simple time point readouts that are easily interpreted. However, loading a full 96-well microplate (to say nothing of the denser formats) rapidly and with sufficiently accurate staggering delays to enable a single-time-point readout of a typical CB assay is virtually impossible.

Fortunately, there is a perfectly valid alternative for workers not blessed with injectors. Instead of reported a single time point, they can take luminance data over a short interval (a 3-min read is usually sufficient), perform linear fitting of the results, and report the rate of increase of luminance as the readout. This method was used in almost all the development work. Statistically, it is actually superior to the single-time-point method, but the data reduction required is either tedious or, if automated, somewhat less trustworthy. It can also become yet another issue for regulatory scrutiny. However, as shown in Chapter 2, the linear-fit and single-time-point methods give virtually identical results when carried out correctly.

3.3.1.6 G3PDH in Serum Serum preparations used as supplements for cell growth are often contaminated with unpredictable amounts of active G3PDH, leading to a potential background problem. The best tissue practices generally call for the use of heat-treated serum, which ameliorates or resolves this issue. In the author's hands, an hour at 60°C proved superior to 30 min at 56°C in killing residual G3PDH activity, without significantly affecting the properties of the serum as a supplement. Alternatives are to wash the cells, use serum-free media (clearly preferable for a product stream from a regulatory point of view, although not always achievable), or accept the presence of a moderate background level, which should be fairly constant. Note that serum contains much higher levels of other active enzymes, such as lactate dehydrogenase, which in particular may become further elevated in some pathologies (148, 149).

3.3.2 Performance Characteristics

The G3PDH-release assay has been tested in cytotoxicity mode with a large range of mammalian cell types and with a more modest panel of bacteria. There seems to be little doubt that the method will perform well with any type of cell that can be lysed by processes compatible with the activities of the enzymes involved, but this in itself is a significant limitation in dealing with very tough cells such as *Cryptosporidium* and certain environmental strains of bacteria (discussed in Section 3.4).

The CB G3PDH-release assay was invented as a means of assessing complement attack on cell lines derived from bladder cancers (83). The effectiveness of the methods of enhancing complement lysis varied with the cell line, but all cell lines tested yielded strong and consistent signals when killed with detergents. In fact, the variation in G3PDH content among the three lines used in the initial work was only about 20% from least to most; however, the reader should not conclude that all mammalian (or human) cell lines will behave in this way. For one thing, G3PDH content likely depends on parameters such as life cycle, metabolic activity, and senescence characteristics, and definitely varies during apoptosis (150). In fact, the coupling scheme of this assay may be useful in measurements of the flux of G3PDH into the nucleus during apoptosis, although this has not been tested.

The assay has been studied with mammalian cells undergoing various types of lytic challenge, including at least complement attack, cytotoxic T-cell attack, antibody-dependent cellular cytotoxicity, and detergent lysis. In some cases, comparative data are available from spectrometric, fluorescent, and/or other methods, enabling competitive evaluations of performance along with other parameters.

3.3.2.1 Complement-Mediated Lysis

The experiments reported here employed the alternative pathway of complement (151), but results should be similar with the classical (152) or lectin (153) pathways. Briefly, the alternative pathway is generally regarded as the antibody-independent means of complement activation, whereas stimulation of the classical pathway involves association of particular structural features of antibodies with activating proteins. However, *specific* antibodies are capable of dramatic activation of the alternate pathway (131). The antibodies employed in these experiments were created by the use of human complement regulators as immunogens.

Irrespective of whether complement activation occurs by the classical or alternate pathway, the membrane attack complex (154) is formed in either case, causing serious ruptures in the cell membrane. However, certain especially tough cancer-derived cell lines appear to have evolved rapid membrane-repair mechanisms (155), rendering the connection between membrane leakiness and cell death more tenuous.

The early experiment whose results are shown in Fig. 3.10 indicated both the power of the concept of using antibodies against complement regulators and the value of the CB G3PDH-release assay in these experiments (83). A 10-fold level of complement acceleration had not been reported by the use of any reagent up to that time. These cells had been weakened with puromycin several hours prior to the readings; however, a similar acceleration factor is seen with the same antibody in the absence of puromycin treatment, although the overall levels of lysis both with and without antibody are some five-fold lower. The smaller signal is readily distinguished by the CB method.

Although the actual limit of detection of the assay is discussed below in the context of viability assays (Section 3.4), the possibility of detecting very small amounts of lysis

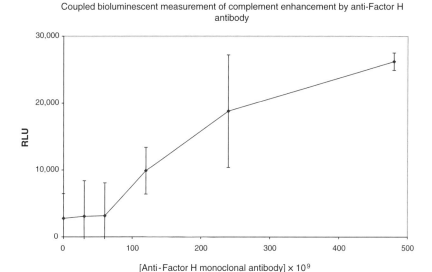

FIGURE 3.10   Measurement of enhancement of complement-mediated lysis of the Raji cell by an anti-Factor H antibody, using the coupled bioluminescent method of Fig. 3.8. No-complement blank of 3420 has been subtracted from all data. Reproduced from Reference 83 and used by permission of Elsevier, Inc.

can have a qualitative impact on research, enabling one, for example, to detect modest rates of attack by cytotoxic cells (T cells, natural killer cells, antibody-dependent cellular cytotoxicity, etc.), or lysis due to small amounts of chemical or mechanical stress. However, this degree of sensitivity can complicate or invalidate established experimental designs. Living cells leak small amounts of G3PDH (156), gradually raising the background over time and potentially confusing interpretation.

The issue of single-time-point read versus linear fits of time-dependent data is addressed by the experiments whose results are depicted in Fig. 2.3. Here a rabbit polyclonal antibody directed to the complement regulator Factor I (157) is employed at two concentrations to accelerate complement attack on the prostate-cancer cell line PC-3. Time-dependent data were captured for 20 min. The automatic-injection feature of the Berthold LB96V luminometer was used to initiate the reactions. Linear fits were calculated for each replicate and the resultant slopes averaged to yield the reported data. However, Panel B of Fig. 2.3 shows the results of the same experiment analyzed in a different way. The data gathered at 2.6 min were used for this graph; no other data were considered in the analysis. A visual comparison of the graphs shows that they provide almost identical information. Thus, when using automated injection, the researcher may consider a readout at a single time point to be just as informative as the much more complex linear-fit analysis. Of course, one must be highly confident that one is in the linear range with respect to the time and concentration variables for this approach to be valid. This must be established by prior studies.

3.3.2.2 Cytotoxic T-Cell-Mediated Lysis   Cytotoxic T lymphocytes are both one of the most important and one of the most complex aspects of the body's defenses against viruses and cancer (for reviews see References 158 and 159). Circulating CTLs recognize non-self-antigens presented as short peptides derived from the antigens as the cleavage products of specific proteases. The lytic process of CTLs is described above in Section 3.2.1, along with various means of assessing this activity. Because of the need for a direct demonstration of killing, most studies of CTL activities that have proceeded well beyond the academic stage employ at least one means of measuring lysis of the target cells directly. To date, nearly all experiments of this type have incorporated a specific labeling step, in the course of which either a radioactive molecule or a fluorophore is introduced into the target cells. Unfortunately, this process has a number of inherent drawbacks. The cells must be prepared and labeled in a step, which is often laborious and lengthy. Excess label must be washed away. Finally, once they are labeled, one is never sure quite what one has. Does the label or labeling process change the phenotype of the target cells? Do their responses to CTLs resemble the responses to be expected *in vivo*? Does the label change the behavior of the CTLs? These issues must be addressed by complex equivalency experiments or left unanswered. Moreover, beyond the inherent difficulties, the specific choices of label present additional difficulties, such as the inconveniences associated with radioactive waste and the potential unknown perturbations of cellular metabolism.

Methods that depend on release of intrinsic molecules avoid these problems (though, ironically, at the cost of introducing a different issue—that of distinguishing target-cell lysis from spontaneous release of the test molecule by the CTLs themselves). As an abundant and essential enzyme, G3PDH is readily released in large quantities by cells under cytolytic attack. Figure 3.11 depicts the results of an experiment in which 440 clonal CTLs grown by the patented REM technique of Riddell and Greenberg (160) were mixed with immortalized target cells obtained from the same patient, transfected to express either the specific cognate antigen of the CTLs or an unrelated antigen from the same virus (Hepatitis B). These are smaller numbers of CTLs than are typically used in such experiments, and, in particular, much smaller numbers of target cells. After subtraction of background effects, including leakage, the specific CTLs yield a robust signal with only modest scatter, compared with an essentially flat signal from the nonspecific targets.

3.3.2.3 A Note About Data Analysis   For the sake of immunologists, several points of distinction between this assay and the traditional ^{51}Cr-release assay (described under Section 3.2.1) should be noted. CRA data are typically reported as a sigmoidal plot of "percent of maximum release" of label on the *Y*-axis against the "effector-to-target ratio" on the *X*-axis, which is usually log-transformed. ("Effectors" is a synonym for "CTLs" in this case.)

Although scientists get comfortable with what they are used to, it is hard to imagine a more misleading way of presenting data. As pointed out above, the mysterious ratio of effectors to target cells present has no magical properties. If there are a hundred effectors present and one target cell, the E:T ratio is 100, but it is unlikely that any lysis will be observed, even if it occurs. Moreover, although some scientists

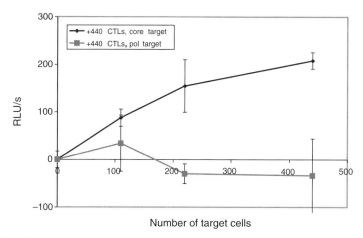

FIGURE 3.11   CB assay of cytotoxic T lymphocyte lytic function against specific and non-specific target cells. Cells were treated as described in legend to Fig. 3.1, except that since careful turbulence-free sampling was unnecessary, cells were incubated at 10 × higher concentration. After incubation, 1/10 of the reaction volume was transferred to a ninefold greater volume of CB reaction cocktail and counted in a TopCount (PerkinElmer). Reported data are singlepoint reads 2.33 min after reaction initiation.

are comfortable with the appearance of the characteristic sigmoid shape of the plot, the shape is misleading. Typically, no positive cooperativity is seen in *in vitro* lytic experiments with CTLs; indeed, negative co-operativity is more commonly observed, although in the author's hands it has generally proven to be artifactual. This curve form, appropriately used to represent truly cooperative processes such as oxygen binding by hemoglobin (161), has been co-opted for a system in which such effects have not been shown to exist.

Instead of the E:T ratio, the important parameters in these experiments are, first, the absolute number of effectors and, second, the relationship between the E:T ratio and the true half-maximal E:T ratio. In enzymatic terms, the absolute number of effectors is analogous to the V_{max} and the half-maximal E:T ratio is similar to the K_m. In cases in which the number of target cells present is substantially below the "K_m" and effectors are not limiting, the reaction is in the linear range with respect to targets: addition of n-fold more targets will accelerate the reaction by n-fold, assuming one does not exceed the linear range. In this regime, the reaction rate is insensitive to addition of effectors (even though the addition will perturb the E:T ratio) and highly sensitive to addition of target cells (though this perturbs the same ratio). In the opposite case, the asymptote representing V_{max}, all the effectors are saturated with targets. (Ironically, since most scientists performing this experiment titrate a fixed number of targets with variable numbers of effectors, this V_{max} region is usually achieved only at very low E:T ratios; in other words, the data points that are the most informative as to the CTLs' true rate of action are also the least accurate, because few CTLs are present.) In this regime, perturbations of the target number have no effect on the observed rate; only changing the number of effectors will affect it (yet either manipulation alters the E:T ratio). Finally, if one wishes to perform true CTL kinetics, the number of target cells should be reduced to the point where it has a negligible effect on the active CTL pool (analogous to reducing complex quadratic kinetics to the simple Michaelis–Menten case (92)). It is then possible, at least hypothetically, to titrate a fixed number of targets with CTLs as desired and determine the true half-maximal E:T ratio using principles derived from Michaelis–Menten kinetics (162).

To accomplish this by means of the CB G3PDH-release assay, one measures the leakage rates of the targets and effectors separately at various concentrations, performs a linear fit, and obtains a normalized rate per cell for each type. This can be used to calculate the anticipated background leakage rate for any of the reactions. Of course, as target cells disappear their leakage rates require adjustment for true accuracy, but the same problem exists with labeling methods, although it is usually ignored.

Although the above theoretical discussion is rather complicated, running and interpreting the CB G3PDH assay for CTL activity are not. No loading of the cells is necessary. The CTL reaction is run, and the reaction product can usually be assessed directly, if appropriate controls and standards are included, in a procedure requiring a single-reagent addition and a reading time of 3–5 min. Analysis of the replicate data points is simple, but the overall process of understanding how effectively the CTLs are killing the target cells is qualitatively similar in difficulty to the same problem using the CRA, except that the sensitivity of the CB procedure is greater and the number of effectors needed is accordingly smaller. These characteristics should allow more informative kinetics determinations.

3.3.2.4 Detergent Lysis   Nonionic detergents are frequently used to lyse cells rapidly and completely. However, both cells and detergents have special features that affect this process, and both have been investigated from this point of view (163, 164). A detergent was sought as the lytic agent in the dual-mode assay described below. The detergents tested initially and qualitative results of testing led to selection of Nonidet P-40 and NP-40 for further work; however, commercial providers of viability assays involving a lysis step often use their own proprietary formulations. (*Note*: NP-40 is distinct from Nonidet P-40.)

Figure 3.12 shows the dose–response curve of NP-40 concentration versus lysis of the 841CON cell line as detected by CB G3PDH-release assay, along with viability data from application of the dual mode (see below). The half-maximal luminance value is seen at ∼0.03%, slightly greater than but still close to the CMC value of ∼0.02%. When using a detergent in an enzyme assay, however, especially a coupled assay, one should determine the effect of the detergent on the rate of the coupled enzymes independently.

Figure 3.13 shows the concordance between a G3PDH-release assay of detergent-killed HL60 cells and visual estimates of the viability of the same samples by a worker who was not aware of the CB data. The agreement would be even closer if it were not for the dilution effects of the CB reagent cocktail, which were not corrected.

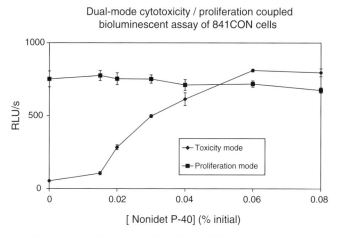

FIGURE 3.12 Dual-mode CB assay of 841CON cell line. Cytotoxicity due to increasing amounts of Nonidet P-40 detergent was first read by linear regression of data from the first 3 min after initiation. All cells were then killed by addition of 0.2% detergent (final) and a further (viability) reading was taken, indicating even cell plating and a fairly constant response of the assay, although a slight effect with increasing detergent concentration is seen.

In the cytotoxicity/viability dual mode, the cytotoxicity measurement is taken as always, followed by addition of a lethal concentration of lytic agent in a small aliquot and a further reading. The nature of the lytic agent depends on the cell type; 0.2% Nonidet P-40 is usually enough to lyse any mammalian cell, while bacteria may

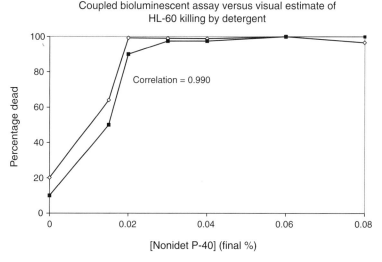

FIGURE 3.13 Comparison of CB assay results with "blinded" visual estimate of cell death by expert technician. Cells were inspected by visual microscopy and viable/dead cells estimated without knowledge of the assay results. Open diamonds: CB assay.

require more exotic combinations of lysozyme, detergents, and pore-forming agents, and very tough cells such as *Cryptosporidium* may present nearly intractable problems in this step. In any case, the lytic agent either must not interfere with the CB assay or must interfere in a defined way (obviously a dilution effect will result from addition of a reagent as well). In Fig. 3.12, the "viability" readout was achieved simply by adding 0.2% Nonidet P-40 to each well and reading the plate again. Equal numbers of cells were added to each well, so the nearly flat readout is consistent with this, although the slight decrease in signal is consistent with a small incremental effect of additional Nonidet P-40 on the efficiency of the CB reactions.

The commercial version of the G3PDH-release coupled bioluminescent assay, aCella-TOX^{TM}, provides a kit for running the same reaction described (165).

3.4 VIABILITY ASSAYS

Many methods described as cytotoxicity assays by vendors are actually viability assays, in that they measure quantities of living cells, while dead cells yield no signal (at least in theory). Among these many methods, we shall discuss the ATP-release assay, dye-exclusion techniques, total enzyme-release assays, and means of assessing mitochondrial membrane potential.

3.4.1 ATP-Release Assay

The ATP-release assay is the most widely used viability assay (166). It is conceptually simple, involving merely the rupture of all cells present and subsequent measurement of ATP released into the supernatant by the methods described in Chapter 2. The sensitivity of the method is good, and very broad dynamic ranges have been achieved. However, the method suffers from a number of practical and theoretical drawbacks, especially as a cytotoxicity assay.

On the practical side, most vendors of this assay recommend the use of a step to separate the supernatant to be tested from the lysed cells. Thus, the method is not homogeneous, adding to the expense and complexity of its adaptation to high-throughput operations. Like the other total-release methods described below (including CB methods), the ATP-release assay results in loss of the cells. However, the major theoretical shortcoming of the ATP-release assay as a means of assessing cytotoxicity is a statistical problem that is shared by all the viability assays used in this way, unless a separate cytotoxicity measurement is taken as well.

3.4.2 Assays of Vital Cellular Functions

Ironically, many "viability" assays kill the cells being studied. This category is an exception. Dyes such as MTT (167), XTT (168), and resazurin (169) can be used as indicators of the "health" of the target cells, since their response depends on the integrity not only of the plasma membrane, but also of mitochondrial function. The blue dye resazurin, for example, is reduced by mitochondria to red resorufin in a reaction that strictly depends on the metabolism of living cells, since the energy-intensive redox properties of intact mitochondria are involved. These methods have enjoyed very widespread acceptance for cell quantification and assessment of cell proliferation. They are also frequently used in cytotoxicity assays, despite the caveats discussed below. The author has spoken with many researchers who use viability assays in this way and concluded that they are either unaware of the drawbacks or locked by long habit, SOPs, or institutional inertia into the use of procedures that are far from optimal for the purpose.

The primary concerns relating to the use of cell-viability assays in measuring cytotoxicity are statistical. Suppose the researcher wishes to determine whether a drug candidate has desirable killing properties against a given cell line. The researcher sets up duplicate wells with appropriate numbers of replicates, one control set, and a number of "treatment" sets depending on the number of concentrations to be tested. The drug candidate and vehicle control are added. Large numbers of cells are killed at high concentrations, more modest percentages at lower; the experiment is a success. But the next step is to test *undesirable* toxicity. Now the researcher encounters a statistical difficulty. The process of aliquoting the cells onto the plate has an associated error. The treatment may cause a reduction in the number of viable cells, but if the undesired effect on the treated cells is close in magnitude to the plating error, an important mode of toxicity may be missed. This is an inevitable result of the difficulty of attempting to measure a small quantity by subtracting two large numbers, each with its characteristic uncertainty. The answer is to use a true cytotoxicity assay, or, if possible, both a viability assay and a cytotoxicity assay.

Another issue relates to the nature of the process. The fluorescent dye accumulates over time in a manner associated with the integral of metabolism. Thus, it is not well suited to detection of *sudden* phenomena, such as the death of large numbers of cells over a short period of time, which can readily occur as a result, for example, of complement attack. If all the cells die due to a highly cooperative process after 2 h, a release assay will reveal the situation, but the change is essentially invisible to this form of viability assay if the time point is taken soon after the event.

More recently, multiple publications have reported fairly serious problems with regard to these assays, including interference by such substances as serum albumins (170) and differential effects of artificial surfaces, such as carbon nanotubes (171). Mitochondrion-based viability assays are useful for quantification of live cells, but the reader is cautioned not to depend solely on this method in measuring cytotoxic effects.

3.4.3   Simple Cell-Growth Assays

The simplest possible assay of viability is measurement of the turbidity of a culture. Although this method is not generally used with mammalian cells, which adhere, clump, and otherwise refuse to behave as discrete light-scattering entities, it is quite useful for bacteria. It does not have the technological appeal of the complex series of enzymatic reactions described elsewhere in this chapter, but this factor does not trouble the many researchers who rely on the absorbance of a culture at (usually) 650 nm to inform them of the presence not only of cytotoxic agents, but also of cytostatic agents that halt growth without killing, which may be missed by other methods. The major disadvantage, of course, is that one must wait for the culture to grow; however, doubling times of many bacteria are such (as little as 18 min under optimal conditions) that growth may often be measured in the time it takes to run a fluorometric assay. No reagents are added, no separations or special devices are needed, and the vessel does not even have to be opened if it is designed for the purpose. In considering cytotoxicity assays of bacterial cells, the assay developer is encouraged to include this minimalist approach in the evaluation, despite the potentially devastating effects on the bottom lines of the reagent suppliers.

3.4.4 Total Enzyme-Release Methods

In practice, this is the same as an enzyme-release cytotoxicity assay with 100% killing. However, the lysis method depends critically on the cell type, and sometimes even the *strain* of a species (see Chapter 15 for an example). Lysis of cells is a topic worthy of a volume in itself. NP-40 was found to be a satisfactory lytic agent with aCella-TOX for many cell types, but bacterial cells may require additional substances such as polymyxins (172, 173) (strong pore-forming agents) or lysozyme (an enzyme that cleaves peptidoglycan, a major constituent of prokaryotic cell walls). Some microscopic eukaryotes, such as *Cryptosporidium*, have evolved to resist a wide range of physiological and environmental lytic substances, and are difficult to lyse without denaturing the proteins; even chlorine cannot be counted on to kill "crypto." This is an area in which an understanding of the cell type, a review of the literature, and considerable empirical effort are essential.

Another caveat pertains to the enzyme-release step. Just because a cell wall is destroyed does not automatically imply that all molecules in the cytosol immediately diffuse into bulk solvent. Enzymes in particular may have evolved to associate with internal cellular structures (174); this association often has an ionic or polar component, and altering the concentration of salt in the reaction buffer is one thing to try if difficulties are encountered.

If the problems of lysis and enzyme release are solved, assessment of total enzyme activity can be a good, though destructive, way of measuring the number of viable cells. The use of standards is essential, since even different cell lines from the same type of the same cancer can harbor varying quantities of household enzymes.

4

THE ROLE OF COUPLED BIOLUMINESCENT ASSAYS IN KINASE SCREENING AND STUDY

4.1 THE MANY ROLES OF KINASES IN BIOLOGY

Kinases, enzymes that transfer a single phosphate group from one molecule (usually a nucleoside triphosphate) to another, are among the most celebrated and studied proteins in all of the life sciences. It is difficult to overstate the importance of kinases or to overestimate their diversity. Some are involved strictly in "blue collar" energy metabolism, while many others are the "executives" of biochemistry, managing the functions of proteins and functional units much larger and more numerous than themselves.

That phosphorylation and dephosphorylation of proteins can profoundly affect their activity has been clear since the 1950s with the immortal contributions of Fischer and Krebs to our understanding of the regulation of glycogen phosphorylase. Nevertheless, new kinase activities and roles for kinases are still being discovered at a high rate. The BCR-ABL kinase is the target of what appears to be the first "miracle" cancer drug, Gleevec (175), which causes prolonged relapses in a surprisingly high percentage of patients suffering from chronic myelogenous leukemia and is beneficial in other cancers as well, but that, alas, has proven to be cardiotoxic, almost certainly because of the same kinase-inhibitory properties that make it useful. It takes nothing away from the brilliant work that led to the development of this drug to point out that improved assay methods may facilitate the development of even more specific kinase inhibitors with wider therapeutic windows.

Kinases are classified into various ways. Possibly the most important distinction is drawn between *protein* kinases, a class that constitutes a vast array of enzymes

Coupled Bioluminescent Assays: Methods and Applications, Michael J. Corey
Copyright © 2009 by John Wiley & Sons Inc.

that phosphorylate surface residues of proteins, usually targeting hydroxyl groups, although histidine kinases are known (176); and all other kinases, most of which act upon smaller molecular substrates. The major classes of protein kinases are tyrosine kinases and serine/threonine kinases. These enzymes function almost exclusively in signaling roles. The mitogen-activated protein kinases or MAPKs, for example, are the subject of intense current research interest because of their participation in growth regulation, as well as their involvement in both suppression and promotion of tumor growth (177). Kinases with small-molecule substrates range from essential players in signal transduction, such as phosphatidyl inositol-3-kinase (PI3K) (178), to enzymes central to energy production and phosphate metabolism (e.g., phosphoglycerokinase and creatine kinase (ck)) or biosynthesis.

4.2   CURRENT STANDARD KINASE ASSAYS

The dominant approach to measuring kinase activity has long been detection of the phosphorylated product. This type of kinase assay is so well entrenched that it is unlikely to be superseded in the foreseeable future. The great advantage inherent in these methods is specificity: the fact that a specific product is detected constitutes strong evidence that one has measured the intended activity. However, the disadvantages are also significant. Since the product itself is rarely able to generate a specific detectable signal on its own, other reagents that interact with the product must be introduced and, moreover, must be engineered to yield luminance, fluorescence, ionizing radiation, or another usable readout. These secondary reagents can be expensive and difficult to handle or use. When they are antibodies, slow diffusion due to large molecule size is another concern.

An alternate strategy that has appeared fairly recently is the measurement of the *disappearance* of the kinase substrate ATP. This CB method circumvents most of the objections cited above: ATP is readily detected by the luciferase reaction, no molecular engineering is needed, and the reaction occurs very rapidly. However, what is lost is the assurance of specificity: anything that degrades ATP will yield a signal in this system. Thus purified or well-characterized assay components are more amenable to this strategy. Moreover, this method yields a negative signal, which is a potential source of difficulties from a statistical point of view.

A third strategy, also a CB approach, relies on the ATP substrate, but in a different sense. In the "reverse" kinase assay, ATP is actually synthesized from supplied ADP and a phosphorylated substrate. This method yields a positive signal with kinase activity and also avoids some of the objections to the first method; however, thermodynamic and other issues arise. These are discussed below in Section 4.3.3.

Finally, the second product of kinases is ADP, and there is now yet another approach involving direct detection of this product. BellBrook Labs markets an assay kit that includes a labeled antibody against ADP and an ADP tracer, enabling time-resolved fluorescence resonance energy transfer (TR-FRET) signals to be modulated by the presence of ADP. This novel method is discussed below in Section 4.2.3. BellBrook has been a strong advocate of the ADP detection approach in kinase assays and also markets a fluorescence-polarization (FP) kit for this purpose.

The two CB methods mentioned above presently play a minor role in the world of kinase screening, which is dominated by fluorometric methods that are relatively slow, complex, and expensive, yet many have the advantage that they are familiar to the enormous fluorescence community, with their depth of experience and sizable investment in equipment. Techniques incorporating radioactive labeling are also still important. A more comprehensive description of these methods is needed to understand the potential advantages of CB assays.

4.2.1 Fluorometric Kinase Assays

4.2.1.1 Antibody-Based Assays   The largest class of these assays by far involves detection of the phosphorylated product (usually a protein) by use of an antibody specific for the modified amino acid residue, which is often phosphotyrosine but can also be phosphoserine or phosphothreonine. These methods are made possible by the exquisite specificity of the antibodies and the quality of modern detection reagents. For example, the DELFIA-enhanced immunofluorescence platform from PerkinElmer combines antibodies with high specificity for the substrate with a signal-amplification system incorporating biotin–streptavidin capture, wash steps, and detection *via* time-resolved fluorometry of the Eu^{3+}-labeled antibody. DELFIA is not restricted to kinase assays but represents a general set of reagents and procedures applicable to cytotoxicity, detection of fusion proteins, and other applications for which antibodies are available. Though sensitive, with a limit of detection in the low nanomolar range, the kinase assay as described is slow, with two incubations of 1–2 h, and expensive, since it requires several complex reagents and conjugates. It is also operationally complex, with a wash step and three separate incubations. These characteristics are typical of antibody-based detection platforms. Other approaches include screening with phosphorylation-sensitive antibodies against peptide arrays on chips (179) and fluorescence-polarization methods using tracers (e.g., TKXtra Explorer[TM] from Molecular Devices).

All assays based on these principles face the same constraints: the slow diffusion characteristics of antibodies, as well as the necessity of introducing a fluorescence detection scheme. A few companies are developing "minibodies" and other alternatives to antibodies with faster diffusion rates (180), but so far the expense of these research efforts has limited their objectives primarily to therapeutic targets. The formulation of the fluorescent-signal-generating apparatus can be a significant source of expense as well.

4.2.1.2 Other Fluorometric Assays   CalBiochem's TruLight[TM] method exemplifies a class of alternatives to antibody-based techniques. This strategy employs microspheres coated with both fluorescent polymers and clusters of gallium ions on linkers; the latter capture the phosphate groups, enabling quenching of the fluorophores by the peptide products (which have been labeled with a quencher prior to the experiment). The maker claims high sensitivity due to superior quenching of the polymers in comparison to monomeric fluorophores. However, the assay time is still 2 h, and the list price as of this writing is an eye-popping ~\$5 per well for 96-well plates or ~\$2.50 per well for the 384-well format.

The Molecular Devices unit of MDS Analytics Technologies markets IMAP® or the immobilized metal affinity platform, which enables antibody-free detection of phosphate groups. Trivalent metal ions fixed on particles are able to chelate the target phosphates, and the proprietary technology then yields an FP signal. This interesting technology is described in more detail in the context of phosphatase assays in Chapter 5 but should be considered a competitor in the kinase arena as well. The IQ^{TM} Technology from the Thermo Fisher Scientific unit of Pierce is also antibody-free but employs the traditional concept of a fluorophore-labeled peptide together with a proprietary iron-containing reagent that quenches phosphorylated substrate molecules, while dephosphorylated ones fluoresce. This technology employs an ordinary fluorometric readout and is described further in Chapter 5.

4.2.2 Radiolabeling Kinase Assays

Methods of measuring kinase assay with radioactivity employ γ-labeled ATP. Labeling of phosphorus is relatively inexpensive, and the isotopes ^{32}P and ^{33}P are both used for these purposes. They present minimal health risks unless ingested or inhaled. Unfortunately, the real expense of these techniques does not appear in the reagent prices. Disposal is of course very costly, but less so than the intangible and untraceable "overhead" costs in workers' productivity, and sometimes in morale.

Traditional kinase assays employing radioactive labels involved (1) mixing label, kinase, and target; (2) separating unincorporated label from labeled target; and (3) measuring the radioactivity of the target (181). Aside from the tedium and complexity of this method, an important parameter of success lay in the specificity of the kinase. Proteins other than the target are necessarily present in such an assay, including the kinase itself. For the many kinases that are not entirely specific, it was often necessary to separate the labeled target from other phosphorylated proteins by means of capture by an appropriate antibody, assuming one was available.

The development of the scintillation proximity assay or SPA has eliminated some of these difficulties. Briefly, a chemical "scintillant" or molecule that emits visible or UV light in response to the radioactive decay of the isotope is added to the reaction, allowing direct photon detection by an ordinary photomultiplier tube or CCD camera. Early SPAs required a separation step, which is one of the most costly and difficult procedures to automate, but over the past two decades improved methods employing immobilized ligands have been developed, and these, while slow relative to CB techniques, can loosely be termed "high-throughput" assays (182–184). However, development and utilization of SPA have slowed down in comparison to other methods, and most of the recent citations of SPA kinase assays describe side-by-side runs with competing approaches, usually to demonstrate equivalence or superiority of the nonradioactive techniques (185, 186).

4.2.3 Fluorometric Assays Involving Detection of ADP

A novel approach to measuring kinase activities that deserves separate treatment is the specific detection of ADP. While slightly unconventional, this idea is a straightforward and logical means of assessing kinase-catalyzed ATP hydrolysis. It enjoys an advantage shared by the negative-signal CB assay we shall examine below (see Section 4.3.2), in that the method is independent of the nature of the kinase and its substrate; this of course also renders it less useful for distinguishing among activities if multiple kinases are present, but such is rarely the case in the high-throughput screening scenario. Moreover, by yielding a positive signal, ADP detection methods avoid the statistical problems of "subtracting two large numbers" that can attend negative-signal assays and may require them to be run past the linear range to generate an adequate signal.

At present ADP detection can be accomplished by TR-FRET assays using anti-ADP antibodies, as in the Transcreener™ product from BellBrook Labs (187), illustrated in Fig. 4.1. Despite the use of several synthetic compounds and conjugates, the list price is ~$0.50 per well, somewhat lower than average for antibody-based assays. The stated time required for the assay is 1 h; while hardly rapid, this is somewhat shorter than the time required for many antibody-based assays, possibly because in Transcreener the rate of association depends on diffusion of the small molecule ADP, rather than that of the antibody. Of course, the method requires equipment and expertise in measuring time-resolved fluorescence; the instrument in particular is not yet commonplace.

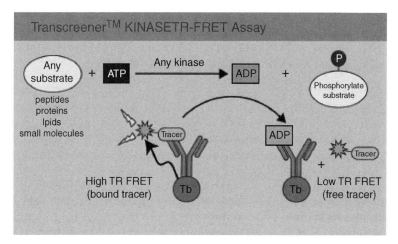

FIGURE 4.1   Time-resolved fluorescence resonance energy transfer is observed when the Tb^{3+} (terbium) ion covalently bound to an anti-ADP antibody is close to the ADP-fluorescein tracer. The TR-FRET signal is diminished when free ADP produced as a result of kinase activity displaces the labeled ADP from the antibody. Used by permission of BellBrook Labs.

4.3 COUPLED BIOLUMINESCENT KINASE ASSAYS

Most kinases employ ATP as the energy source (although GTP-dependent kinases are not rare). Since beetle luciferase is dependent on ATP, this suggests an obvious "negative-signal" assay for ATP-dependent kinases: assess the activity by measuring the reduction in ATP concentration and the associated decrease in luminance. This is indeed the major mode of CB kinase assay in current use, but there are still two surprises in CB kinase assay technology: first, the versatility of the described method, and second, the *reverse* assay, which yields a normal, positive signal. We begin by giving an example of the second for historical reasons, proceed to the negative-signal assay, and conclude with the modern developments in the reverse technique.

4.3.1 A Historical Note: The Coupled Bioluminescent Creatine Kinase Assay

The first CB kinase assay, and, indeed, the first CB assay ever patented, was a logical choice. Creatine kinase is the enzyme responsible for the first-line "energy buffer," continuously providing phosphate groups for transfer to ADP to re-create ATP for use. Unlike most other kinase reactions, this one is thermodynamically favorable in the direction of ATP creation, and therefore makes a logical choice for Arne Lundin's development of a coupled assay (188). Dr. Lundin also speculated about the eventual development of many forms of CB assays. The creatine kinase assay was later patented (189).

Lundin's formulation is conceptually very simple (see Fig. 1.11). By supplying creatine phosphate, ADP, and an active beetle luciferase system, it is possible to make the appearance of light dependent on the presence of creatine kinase. Alternatively, one could supply the enzyme instead of creatine phosphate, resulting in an assay for the latter.

Lundin's CK assay is exergonic in the forward direction, and it might be thought that this characteristic is critical for success. Such is not the case, however. Reactions that are endergonic can still be rendered exothermic by appropriate manipulation of the law of mass action (real thermodynamicists prefer to talk in terms of "chemical potential"). This is discussed below in Section 4.3.3.

4.3.2 ATP Depletion Kinase Assays

We turn now to the more obvious case of ATP depletion by the kinase. Any ATP-dependent kinase activity can be measured in this way (although as in other coupled systems, some unusual enzymes may require conditions that are incompatible with luciferase activity, necessitating a two-step assay). Here is a case in which there is a huge gap between the utility of the assay and the extent of its use. In 2006, a large group at Boehringer Ingelheim published a comparison of three highly orthogonal methods of identifying kinase inhibitors: a DELFIA technique, a competitive binding assay for a fluorescent probe of the ATP-binding site, and a CB ATP-depletion assay (190). While the results of the three methods were in agreement over 57% of the hit set, the

CB assay was the clear winner in terms of accurate lead generation, producing "the largest number of unique hits, the fewest unique misses, and the most comprehensive hit set, missing only 2.7% of the confirmed inhibitors identified by the other two assays combined." When one considers the speed, sensitivity, and ease of use of this method, one must conclude that the primary reasons it has not enjoyed widespread adoption are ignorance and simple inertia.

A further indirect reason, however, is probably the intellectual property position. Evidently, the method is in the public domain, which lessens the enthusiasm of many firms to expend scarce funds and bear the opportunity cost of developing a technology that cannot receive broad protection. However, Promega Corporation enjoys an advantage in this area, as in others, because of its patented thermostable luciferase (Ultra-Glo Recombinant Luciferase (38)) with a half-life of over 5 h. Promega offers Kinase-Glo® and Kinase-Glo Plus® as its versions of this assay (191–193). Although the individual researcher can certainly develop such an assay independently, the availability of a first-tier commercial supplier with an established kit selling at approximately $0.15 per standard well (list price; less for 384-well and denser plates) generally shifts the cost–benefit analysis in favor of spending R&D dollars elsewhere. The research and biotechnology communities have responded with numerous publications reporting the use of the CB kinase assay from Promega for diverse purposes, including high-throughput screening (194, 195), basic research (196–199), and even a structural study of protein kinase A (PKA), using the CB assay in combination with NMR (200).

PerkinElmer has also entered this arena with the easylite-Kinase™ Luminescent Assay System. One of its claimed advantages is the elimination of dithiothreitol (DTT) from the reagents as an "odor hazard." This appears to be a broadside at Promega, since one of Promega's fundamental luciferase patents (33) relates to the use of high concentrations of DTT and other pungent sulfurous reducing agents to stabilize luciferase.

While negative-signal assays generally have a poor reputation among enzymologists for statistical reasons (discussed in Chapter 6), in this case a great deal of the necessary homework has been done to insure against the more important failure modes. Like the other CB assays discussed herein, kinase assays can and should be run at room temperature to avoid the inevitable differential heating effects in an incubator or plate warmer. Because of the nature of the reagents, contamination of the reaction is a much less serious problem than in other CB assays, but the reaction is somewhat sensitive to initial ATP concentration. Finally, without intermediate coupled enzymes, the only major interference problem pertains to compounds that inhibit luciferase itself.

Still, the worker using a negative-signal assay must learn to think in a different way. The data in Fig. 4.2 from Promega Technical Bulletin #372 may be startling at first to those accustomed to the "positive" way of doing things. Here, the decreasing signal represents *increasing* concentrations of protein kinase A, the enzyme under test. In panel A, we see an approach to the asymptote at the quantity of enzyme, although clearly the ATP has not been exhausted in 5 min. However, in panel B, a luminance reading of zero is reached (it is not clear from the information provided whether

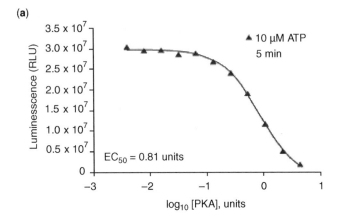

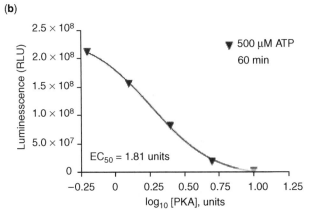

FIGURE 4.2 Negative-signal CB kinase assay. Consumption of ATP by the kinase lowers the concentration of the energy source for luciferase, diminishing the luminance signal. (a) Low range of ATP concentration. The fit is excellent and the scatter is slight. (b) High range with longer incubation. Used by permission of Promega Corporation.

the data have been corrected for background). As for the high absolute numbers, the term "RLU" or "relative luminance units" covers many sins, since it depends on many malleable parameters, such as the amount of luciferase used, the voltage setting of the instrument's photomultiplier tube, and the integration time. An RLU reading is rarely useful as an absolute measurement of anything; it must be analyzed in relation to other readings under the same conditions (see Chapter 2 for a more detailed discussion).

The word "relative" that appears in "RLU" also carries other implications in the world of inhibitor assays. There are fundamental differences between the enterprise broadly referred to as high-throughput screening and what an academic would call "real" research. The goal of HTS is simply to identify leads, often as defined in *relative* terms. A lead is a lead simply because it is an outlier, according to rules defined a priori. The project director may decide to query the chemical library in a more complex way than usual—for example, one can envision a first-round screening in which the elements of the library must pass a two-step test by inhibiting this kinase and *not* that one—but the goal of the process is still to identify candidates, not to answer a medical or biochemical question. The director looks at the Z values as representative of the quality of the assay run and notes a value of 0.72 (see Chapter 2 for a discussion of Z values). At this point, some of the most promising candidates may be subjected to an evaluation of EC_{50}. Once they are identified, they undergo secondary screening by an orthogonal method (one hopes), and the survivors are subjected to further characterization, ADME–tox studies, and so on. Eventually the K_i of the inhibitor is measured, probably by a low-level technician poring over a protocol written decades earlier by a traditional biochemist.

4.3.2.1 Determination of K_m by a Negative-Kinase CB Assay

An academic enzymologist forced to watch such a procedure would squirm in frustration. What is the K_m? The K_i? The competitive mode? Fortunately, CB assays provide means of answering these questions in "absolute" biochemistry as well, but in the case of negative-signal kinase assays, it requires some work. Unfortunately, Promega no longer provides the K_m data that were present in an earlier version of this bulletin, but the K_m is determined by titrating with substrate over a wide range. When there are two substrates, as there are here (ATP and peptide), the concentration of one should be held constant in a regime at least 10 times greater than the K_m (assuming substrate inhibition is not present; this must be tested separately). If the data exhibit a sigmoidal response to the substrate under test, fitting software such as GraphPad Prism® may be used to determine the phenomenological K_m. If the response is not sigmoidal, they usually provide at least a clue as to whether the correct concentration regime is higher or lower than the tested range. The experiment is then repeated in the indicated regime and the K_m is calculated.

Although Promega no longer provides the substrate titration, they have given us a comparative wealth of inhibitor data. Fig. 4.3 must be considered to be as useful to the assay developer involved in a high-throughput screening project as it is frustrating to the academic enzymologist. Here, the H-89 inhibitor has been titrated against fixed concentrations of both the enzyme and the two substrates (although the concentrations of the substrates are varied panel to panel). The increasing RLU reading in each panel represents *inhibition* of the enzyme, due to the negative signal.

The HTS manager may well be satisfied, or even delighted, with these data. They are clean, with small errors, and the EC_{50} values provide information for later comparisons with the effects of other inhibitors or lead compounds. However, the academic researcher, or the manager with an academic background, is likely to be slightly less enthusiastic. Although inhibitor performance parameters are reported, and are probably accurate, useful information is missing, in the form of *absolute* kinetic parameters. In other words, valuable further steps in data reduction could and should have been performed. These steps would depend on two changes in the study strategy: (1) determination of the K_m of each substrate and (2) repetition of the titrations over a more appropriate concentration regime.

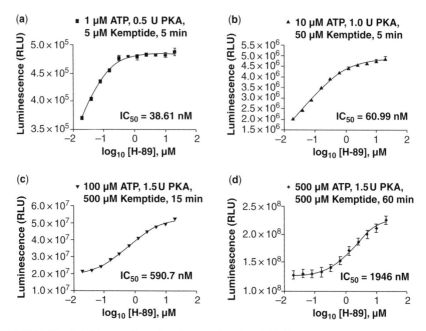

FIGURE 4.3   Inhibitor studies using the negative-signal CB kinase assay. The concentrations of the substrates, the amount of enzyme, and the reaction time are varied in tandem, probably to demonstrate the assay performance in a wide range of conditions. All the data are good, but those taken at the highest substrate concentrations exhibit noticeably more scatter for unknown reasons. Used by permission of Promega Corporation.

4.3.2.2 Use of the CB Negative-Kinase Assay to Measure Inhibition Constants

Enzyme inhibitors range from simple analogs that compete for a substrate binding site (*competitive* inhibitors) to molecules that perturb the enzyme's catalytic rate at a location independent of substrate association (*noncompetitive* inhibitors) to molecules that inactivate the enzyme in time-dependent ways, or perhaps do all of these things at once. In the case of competitive inhibition, the kinetic effect is observed as an apparent increase in K_m, described by the equation

$$K_{m(app)} = \frac{K_m(1 + [I])}{K_i} \tag{4.1}$$

where [I] is the inhibitor concentration and K_i is the characteristic competitive inhibition constant. When the inhibitor is present at the concentration represented by its inhibition constant, the apparent K_m is exactly double the true K_m. This does *not* mean that the reaction necessarily proceeds at half its uninhibited rate. In a low-concentration substrate regime, this will be true, that is, the enzyme will behave as though its K_m had doubled, meaning that twice as much substrate will be needed to achieve the same rate. However, this effect gradually vanishes as the substrate concentration is raised. If the substrate is present at 10 times the true K_m, then it is also present at five times the inhibited K_m, and the effect of the inhibitor is only

$$1 - \frac{5K_{m(app)}/(5K_{m(app)} + K_{m(app)})}{10K_m/(10K_m + K_m)} \tag{4.2}$$

or approximately 8% inhibition. This tendency of the inhibition to be suppressed by increasing concentrations of substrate is diagnostic of competitive inhibition.

In the opposite case of pure noncompetitive inhibition, no change in apparent K_m is observed; instead, the k_{cat} or maximum turnover rate of the enzyme is affected, and no amount of substrate can overcome the effect. The classic example is binding of a noncompetitive inhibitor at an allosteric site, causing a conformational change in the enzyme that prevents catalysis (201). The simplest form of this effect is described kinetically as

$$k_{cat(app)} = \frac{k_{cat}}{1 + [I]/K_i} \tag{4.3}$$

Thus k_{cat} is apparently reduced by the degree of inhibition, with $[I] = K_i$ at half of the maximal effect.

It is also possible, and in fact quite common, to see *mixed* inhibition, with both competitive and noncompetitive components. In such cases, effects on both K_m and k_{cat} are seen. The catalytic rate is reduced at low substrate concentrations, and the inhibition is only partially relieved by raising the substrate concentration to saturating levels. One of numerous examples is tacrine, an inhibitor of acetylcholinesterase that was considered as a treatment for Alzheimer's disease (202). By careful titration with both substrate and inhibitor and appropriate curve fitting, it is possible to deconvolute the two inhibitory modes, and such results can provide mechanistic information as well (203).

Returning to the data of Fig. 4.3: it is possible to determine both a K_m and a K_i from a single experiment, but only if the relevant concentration regimes are probed. These two parameters are by definition half-maximal values of something (enzyme catalysis and inhibition, respectively). To determine a half-maximal value with accuracy, one should titrate *through* that value. It is true that it is theoretically possible to fit an equation that contains, say, three disposable parameters with only three data points, but the confidence level in the result would be too low to express. The presented data, while demonstrating both the utility of the assay method and the technical skill of the operator(s), are not entirely satisfactory to the academic scientist, simply because they do not provide information about the entire concentration range surrounding the kinetic parameters of interest. In Panel A, the lowest concentration is not low enough, making it impossible to place the lower asymptote. Since the range of the data covers 1.3×10^5 RLU and the graph is cut off at 1.3×10^5 RLU, it is impossible to conclude whether or not the asymptote is zero, or whether there may be a high static background. It is true that the fitting software can provide a good guess, but we cannot rule out the possibility of an unexpected phenomenon at a low concentration of the inhibitor from these data. Since we do not know the K_m, we also cannot calculate a true inhibition constant. The legend of the figure points out the good agreement with literature at the low ATP concentration, but since the reported IC_{50} values vary over a range of 200, it is not entirely clear which IC_{50} number should be reported. Calculation of the true K_m and K_i would solve this problem. In Panel B, the problem is similar to that in Panel A, but in this case neither asymptote is known with confidence, while in Panel C, the upper asymptote is the target of guesswork. However, it should be noted that unanticipated deviations from hyperbolic saturation behavior are seen only rarely, and very rarely indeed when the known data are as clean as these. Another set of data in the same bulletin, taken with a noncompetitive inhibitor, again shows the excellent performance of the assay and the operator. However, because all of the parameters were varied in tandem, including the concentrations of both substrates and the enzyme as well as the length of incubation, a conclusion that the inhibition is noncompetitive in nature cannot be drawn with complete confidence. To measure noncompetitive inhibition, one should vary the concentration of the substrate (or of each substrate) independently, holding other parameters constant; if the apparent K_i remains fixed, the inhibition is noncompetitive.

4.3.2.3 Z Values and Overall Quality of the Negative-Signal CB Kinase Assay
Before leaving this interesting and useful assay, we will briefly examine the Promega Z values (Fig. 4.4) and other overall indications of quality. The Z value is discussed in Chapter 2. Briefly, it is an indication of the statistical power of an assay to distinguish

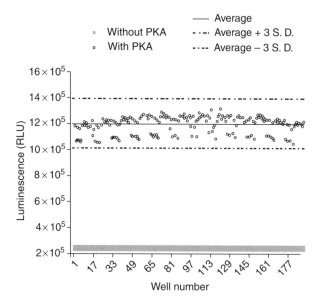

FIGURE 4.4 Z-value determination for the negative-signal CB kinase assay. Used by permission of Promega Corporation.

hits from noise in screening applications. The drawbacks of using the Z value to conclude that a method is *inherently* sound or superior are mentioned in the referenced section, but it can be used successfully to establish confidence in the value of results.

The Z values reported for this assay are about 0.8, well above the general industry cutoff of 0.5. However, two caveats are noted. First, the scatter appears to be highly nonrandom, especially in the no-enzyme controls (the upper band of points). Numerous clusters of points are seen at regular intervals, while in the 100-μM data, individual points appear above the mean, also at regular intervals. These patterns remain unexplained, but probably represent small systematic biases (likely plate-position effects). Since some of the points in the lower (plus-enzyme) band approach the 3-SD line, these biases could be of significance. Second, the plus-enzyme data, which exhibit lower scatter, are slightly misleading. Running a reaction to virtual completion, that is, forcing the data to the asymptote, has the effect of damping any noise present. That the assay has been run very near to completion is shown by the fact that many data points lie significantly above the mean line, but none significantly below it. In other words, this is the lowest luminance that can be achieved. This yields an artificially small standard deviation, since at the asymptote there is very little freedom for the data to deviate. A better indicator of assay quality would have been a run in which a smaller amount of enzyme was used and the reaction stopped during the time-linear phase. Users of the negative-signal CB kinase assay are encouraged to investigate the time dependence of the signal and address the technical issues of enzyme quantities and run length in accordance with their screening or research strategies.

Finally, it should be noted that this assay as commercially available is not a continuous assay, but a time-point assay. It is not clear at this point whether reagents can be developed that will allow mixing and monitoring in real time. Many of the statistical questions would be answered by the availability of time-dependent data, whether obtained by the rather tedious process of repeated sampling in a two-step assay, or provided by a continuous process. Nonetheless, time-point assays are an acceptable alternative when incompatibilities of reagents do not allow for the continuous method, and some managers in drug discovery programs actually prefer to have single readouts rather than deal with the complexities of time dependence. Of course, single readouts are available from either kind of assay.

4.3.2.4 Negative-Signal CB Kinase Assay: Conclusions

Despite a number of objections and potential drawbacks, this method, like many CB methods, represents a significant advance in terms of speed, convenience, and sensitivity over all other available assays, including fluorescent, SPA, and other techniques. Unlike many other CB assays described in this volume, the kinase assay is simple and direct, with no coupling enzymes needed other than luciferase, simplifying interpretation. Market acceptance of the Promega assay has been gradual, in keeping with both the advantages and the conservative nature of life scientists. One insurmountable objection, of course, is that consumption of ATP is not specific to any given kinases; thus, antibody-based methods are considered superior in regard to this point, since they are driven by formation of a specific product, rather than exhaustion of a nonspecific substrate. However, it is hard to see how this will create difficulties in a purified system. Reference 190 above should provide all the confidence anyone needs that the CB method performs at least as well over a range of molecules as the much more cumbersome alternatives.

4.3.3   Reverse CB Kinase Assays

At this point, it is critical to distinguish between the *negative-signal* assays we have been discussing, in which the quantitative readout varies inversely with enzyme activity, and the *reverse* assays, in which the enzymatic reaction is made to run in a direction that is the opposite of the presumed physiological or "intuitive" direction. We now turn to the latter approach as applied to measurement of kinase activities.

The principle of microscopic reversibility holds that any chemical reaction can be run in reverse. Virtually all enzymatic reactions (except those in which outgassing, product instability, or inactivation of the enzyme occur) take place in both directions at once, and the flux in the direction of a particular product is determined by thermodynamics, rather than vectorial selectivity of the enzyme (which enzymes in general do not possess). The principle holds for kinases as well, and the double-headed arrow in Fig. 1.11 represents the fact that creatine kinase can transfer inorganic phosphate from creatine to ADP, but it can also do the opposite.

But creatine kinase is a special case, since the formation of ATP is energetically favored. Most kinases use the so-called "high-energy" bonds of ATP as an energy source, driving their cognate reactions toward phosphorylation of an enormous variety of target molecules. Rephosphorylation of the ADP product using the phosphate groups from these same molecules seems almost wrongheaded. Nevertheless, it can be done, and what is more, the principle of microscopic reversibility further tells us that if a molecule inhibits the reaction in one direction, it must inhibit it in the other, and to the same extent. The author has encountered a surprising amount of skepticism regarding these points, especially from life scientists without specialized training in enzymology. The best way to convince oneself is to consult any of the excellent enzymology texts available (145), but a simple analogy may help. A frictionless car traveling at a randomly chosen speed coasts over a hill with its engine off. The flat ground on one side of the hill is higher (in potential energy) than the other. A bump is encountered at the top of the hill, representing the inhibitor. The bump lessens the probability that the car's energy is sufficient to surmount the hill. The point is that this is true regardless of the direction the car is traveling. Although in the case of molecules the reaction is more likely to proceed from the region of high potential energy to the other side, the inhibitor reduces the probability of travel in both directions, and to the same extent.

4.3.3.1   The Welch Approach   The reversibility of the reactions of virtually all enzymes, including kinases, was exploited by Anthony Welch in his reverse kinase CB assay, for which a Patent Cooperation Treaty (PCT) application was submitted in 1999 (204). (All enzymatic reactions are reversible in principle, but in practice those that generate molecules that outgas or destroy the enzyme itself are difficult to run in reverse.) The simple scheme that underlies Welch's method is precisely the reverse of the Promega negative-signal method: ADP and a phosphorylated substrate are provided, and the kinase catalyzes the transfer of the phosphate group to ADP, forming ATP and enabling the luciferase-dependent signal. Welch begins by making the central point in the background section—that enzymes that hydrolyze ATP can also make it—in an elegant way: "When the reporter enzyme is saturated with all substrates other than ATP, the enzyme reaction catalyzed by the reporter enzyme will be thermodynamically favored such that the reaction catalyzed by the target enzyme will be pulled in the metabolic reverse direction to synthesize, rather than hydrolyze NTPs." In fact, the patented material is not simply a method of measuring kinase activity, but a new way of looking at measuring the activity of all enzymes that hydrolyze nucleoside triphosphates in the course of ordinary metabolism.

The Welch idea was tested successfully with protein kinase A (this was done by Welch, and also independently by the author, who tested protein kinase C as well before he became aware of the Welch patent application), as well as ADP-heptose synthase, a uridylyl transferase, and CT236, an activated form of CTP:phosphocholine cytidylyltransferase (205). The reader may notice that the concept is also identical to Lundin's creatine kinase assay described in Chapter 1, with the sole exception that in the case of the Welch enzymes, the kinase-catalyzed formation of ATP is endothermic (i.e., heat absorbing). The most important point to remember in developing these assays is that while the thermodynamic challenge can often be overcome, it does not simply vanish when the reaction scheme is laid out on paper.

One philosophy of kinase screening holds that when one is screening for an inhibitor, the target of the enzymatic activity is of no more than moderate relevance. In keeping with this idea, commercial preparations of the very inexpensive protein casein were used by Welch as the source of phosphorylated substrate. In the author's hands, technical-grade casein was insufficiently pure for this work, and in fact it appeared to exhibit kinase activities of its own. Purified α-casein was purchased from Sigma and dissolved in water for use without further treatment. Casein exhibits a considerable degree of natural phosphorylation. There are in fact several enzymes known as casein kinases, with general serine/threonine kinase activities, but a preference for specific substrates, including casein (206). Thus a phosphorylated substrate suitable for the testing of the broad class of general protein kinases is readily available. The PKA substrate titration data allowed estimation of a pseudo-K_m for this substrate of 5–10 μg/mL. While the poorly defined molecular nature of the substrate may be an impediment to mechanistic studies, this knowledge of a quantity roughly corresponding to a K_m enables the user to treat the casein as a proper substrate in screening for competitive and noncompetitive inhibitors, depending on whether the pseudo-K_m is perturbed or not.

Although these CB assays have not received the attention they deserve from either academics or industrial workers, their development demonstrates the wide range of possibilities remaining to be explored, especially to the assay developer who understands that thermodynamics can be a tool, rather than simply a barrier. In these assays, the activities of enzymes that hydrolyze nucleoside triphosphates other than ATP are coupled into the luciferase system by nucleoside diphosphate kinase, which transfers terminal phosphate groups among the various nucleoside triphosphates. Welch also points out the range of kinases whose activities may be assessed by this general method: not only protein kinases, but also carbohydrate kinases such as hexokinase and many other enzymes that phosphorylate glucose derivatives, amino-acid kinases, and even phosphoglycerokinase (207), the enzyme that plays a central role in the CB cytotoxicity assay described in Chapter 3.

The Welch application was evidently not pursued at the either national or international level. To the author's knowledge, the methods it describes are in the public domain, but legal counsel should be sought before attempting to develop or use a method of which the intellectual property history and status are unclear.

4.3.3.2 Thermodynamic Principles Underlying the Reverse Kinase CB Assay

The fact that synthesis of ATP by a kinase from ADP and a phosphorylated protein substrate is energetically unfavorable by 5 kcal/mol or more cannot simply be ignored in formulating these assays. Instead, the principle enunciated by Welch of providing all substrates favoring the reverse reaction at saturating levels is no more than a restatement of the principle of mass action, knowledge of which dates back at least to the work of Le Chatelier, whose statement of it is translated roughly as "If a dynamic equilibrium is disturbed by changing the conditions, the position of equilibrium moves to counteract the change," or, more simply, as "Systems tend to act so as to relieve applied stress." In this case, the stress is applied in the form of high concentrations of the reactants of the reverse reaction, in the absence (initially) of ATP.

Consider the simple potential energy diagram of Fig. 4.5. The equilibrium of the reaction lies strongly to the left. However, and this is the point, if only the reagents on the left side of the reaction scheme are supplied, the reaction has "no choice" but to proceed to the right. A chemical reaction does not require the *average* molecule to have enough energy to cross the potential barrier; if any of the appropriate molecular complexes have enough energy, there will be flux in the direction of ATP formation.

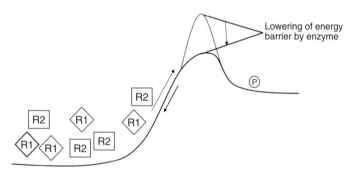

FIGURE 4.5   Potential energy diagram illustrating the requirements for an endothermic CB assay. R1 and R2 are reactants; P is a product. Although the energy barrier is higher from the side of the reactants, these molecules are present in great excess, so that at any given point in time, the probability that a molecule will cross the barrier, that is, undergo a kinase-catalyzed reaction, is greater from the left (the reverse reaction) than from the right (the usual reaction). The net flux of the pathway is therefore toward the right. The kinase itself lowers the barrier and increases the probability and therefore the rate of travel in both directions.

Because of the energy barrier, this will occur at a slower rate than the "forward" reaction, but the rate can be enhanced in three ways: (1) by supplying the reactants in excess, (2) by supplying a relatively large amount of enzyme, and (3) by raising the temperature. This latter point is entirely in keeping with Le Chatelier's Principle. One of the products of the forward reaction is heat. Therefore, by supplying additional heat, one tends to drive the reaction in the endothermic (heat-absorbing) direction.

Reverse kinase reactions have received significant attention in the academic literature, and for some kinases many of the kinetic parameters of the reverse reactions are known (208–210). Creatine kinase is of course the outlier, in the sense that formation of ATP is actually the "forward" direction in metabolism, and this reaction has received considerable study (211). Ikebe and Hartshorne (212) studied the myosin light-chain kinase and its activity in the reverse direction, concluding that k_{cat} was in the range of $0.5 \, s^{-1}$ (compared with $\sim 30 \, s^{-1}$ in the forward direction) and that the K_m of ADP for the reverse reaction is actually lower (i.e., stronger in terms of enzymatic rate) than that of ATP for the forward reaction, consistent with the observation that dissociation of ADP is the rate-limiting step in the forward reaction.

If we accept these values as representative of protein kinases generally, how realistic is it to expect to measure the reverse activity of this enzyme with a CB assay? In fact there should be no difficulty. The main consideration is that to obtain a robust signal, one should supply a considerable amount of enzyme, together with the phosphorylated substrate and ADP at saturating levels. The apparent K_m of ADP is 30 μM; thus, the maximum rate is approached at 0.5–1 mM, which is easily achieved. Note, however, that commercial preparations of ADP are frequently contaminated with ATP, which in this case will yield a static background. The choice of ADP concentration may be settled as a trade-off between signal and noise, or ADP can be purified from ATP (213). The phosphorylated substrate is potentially a larger issue, but, as argued above, the exact identity of the substrate is often of no more than moderate concern in high-throughput screening for inhibitors. Thus, in many cases, casein or another widely available phosphorylated protein may be substituted for the natural substrate. In general, it should be possible to develop a reverse kinase CB assay, yielding a positive signal, for most kinases of interest, limited by such considerations as (1) the expense of the kinase, (2) excessive difficulty in identifying and procuring an appropriate phosphorylated substrate, and (3) the possibility that in isolated cases, the ΔG of the reaction is so unfavorable that creating the proper conditions to generate a robust signal is impractical or impossible, due to expense, solubility limits, and/or substrate inhibition.

4.3.3.3 Comparison of Negative-Signal and Reverse (Positive-Signal) Kinase Assays

In comparing the alternative strategies of measuring ATP depletion and ATP synthesis by kinases, the following points arise. Both approaches have been established with major classes of protein kinases, although the negative-signal assay has undoubtedly seen more widespread use. The negative-signal assay has also been performed with a much larger range of enzymes, and since there is no thermodynamic barrier to the process (apart from very unusual enzymes such as creatine kinase), the confidence level that the assay can be successfully applied to a specific new kinase is high. On the other hand, assay methods yielding negative signals can lead to the statistical difficulties mentioned above, including the "clustering" of plus-enzyme standards at the lower asymptote, leading to artificially low scatter and convoluting the question of where a particular reading lies on the progress curve or at the end point. Ideally, all data should be taken in the time-linear range of the assay, but several iterations may be required with a negative-signal assay before the proper starting amount of substrate is determined. This is much less of a concern with assays that yield a positive signal.

4.3.4   Kinases as Participants in Other CB Assays

Finally, there are at least two cases of CB assays on the market in which kinases play subsidiary roles in reaction schemes involving other enzymes or measurements of other species. These are aCella AChE^{TM}, a CB assay for acetylcholine and acetylcholinesterase activity, and cAMP-Glo^{TM}, a CB assay for quantification of cyclic AMP from Promega. These methods are discussed in Chapters 6 and 9, respectively.

4.4   CONCLUSIONS

For the hordes of enterprises performing high-throughput screening of kinases, the CB assays seem to be a superior choice to any of the fluorescent or radiolabeling methods. All the advantages mentioned in Chapter 2 are pertinent, while some of the disadvantages are not applicable. For instance, in this case there already is an established system that is commercially available at reasonable cost, and the coupling series is so simple that there are few difficulties in interpretation, apart from those inherent to a negative-signal system. The choice between this and the reverse-kinase, positive-signal assay comes down to the balance between a solid method that is in growing use in both research and drug discovery, and a theoretically valid approach that may require considerable effort to bring online, with the promise of a positive signal and its associated benefits. The thermodynamic consideration also cannot be ignored. For many managers involved in lead discovery, this may well tip the decision toward the established method.

5

COUPLED BIOLUMINESCENT PHOSPHATASE ASSAYS

5.1 INTRODUCTION

It would be difficult to overstate the importance of the phosphate group in biology. Phosphate is a central feature of structural, energetic, and signaling activities in all known organisms. Indeed, it is the availability of the free energy of hydrolysis of the phosphodiester bond that makes life possible and, a point of at least equal importance, enables most of today's coupled bioluminescent assays. Means of measuring the concentration of free phosphate have a rich history, incorporating figures such as the venerable Bruce Ames, whose method (214) was the first ever used by the author. Here, we present and compare some of the many available methods of assessing phosphatase activities and examine the potential role of CB assays in these systems.

5.2 PHOSPHATASES

The chapter on kinases has already demonstrated something of the range of those enzymes' activities, virtually any of which is susceptible to measurement by CB methods. The phosphatases, enzymes that hydrolyze phosphate–ester bonds, are nearly as diverse a group as the kinases. Among them are numerous proteins involved in a wide array of signaling pathways, regulating activities ranging from muscle contraction (215) to memory (216). As with the kinases, phosphatases acting on protein targets represent the major, but by no means the only, significant set of signaling molecules in the class. As a molecular target of drug-discovery programs, phosphatases likely

Coupled Bioluminescent Assays: Methods and Applications, Michael J. Corey
Copyright © 2009 by John Wiley & Sons Inc.

rank behind only GPCRs and kinases in importance. Although their biochemical role is nearly always reversing the effects of a kinase, and in this sense the kinases and phosphatases act as regulatory partners, the phosphatases are entirely distinct from the kinases in structure and catalytic properties. Protein tyrosine phosphatases (PTPs) (discussed further in Section 5.2.2) are a hot topic in cancer research, since they contribute to dysregulation of many signaling processes, for example, by promoting activation of oncogenes (217) and participating in carcinogen-induced neoplastic transformation (218). The prominent tumor suppressor PTEN (phosphatase and tensin homolog deleted on chromosome 10) has both lipid phosphatase and protein phosphatase activities, each of which may contribute to its role in preventing cancer (for reviews see Reference 219 and 220).

Because of their robust properties, phosphatases have also frequently been used as tools in molecular biology. For example, alkaline phosphatases conjugated to antibodies are frequently employed as detection reagents (221, 222). These applications are probably inaccessible to currently existing CB techniques, which require rigorously phosphate-free buffers and reagents. Today's detection methods employed in ELISA, Westerns, and similar assays are generally adequate in any case, especially with the advent of fluorescent and electrochemiluminescent readouts, often employing enzymes other than phosphatases (223).

5.2.1 Catalytic Properties of Phosphatases

Unlike kinases, which are actually phosphotransferases, phosphatases require no acceptor molecule other than water. The enzymes simply hydrolyze phosphate–ester bonds, yielding free inorganic phosphate and (typically) a free hydroxyl group. The fact that the phosphatase reaction is not simply the reverse of the kinase reaction is of supreme biological importance, since both of these steps, phosphorylation of a target from ATP and dephosphorylation to water, are nearly always exergonic, that is, they are thermodynamically favorable. Thus nature provides means for signaling interactions to occur readily and reversibly, virtually without energetic difficulties in either direction. These processes expend the energy of the cell, of course, but the profound consequences of the signaling behaviors more than compensate for this by modulating aspects of metabolism that far outweigh the energy expenditure of hydrolyzing individual molecules. We have already seen the consequences of the thermodynamic properties of kinase reactions in regard to the possibility of developing CB assays of various types. In a sense, the challenge of phosphatases is even more daunting, since these enzymes have nothing to do with ATP at all; fortunately, nature has been generous in providing enzymes and reagents that enable us to harness thermodynamics in our favor (see Section 5.4.2).

5.2.2 Classification of Phosphatases

Two important ways of classifying phosphatases are by structure and by activity. Among the protein phosphatases, the two superclasses are serine/threonine phosphatases (see Reference 224 for fundamental aspects of these enzymes) and tyrosine phosphatases (225–227). The former fall into four major structural classes: types 1, 2A, 2B, and 2C; although the first three classes exhibit sufficient homology to imply a relatively recent common origin. These phosphatases generally consist of complexes of catalytic and regulatory subunits and catalyze their cognate reactions with the assistance of two metal ions. The protein tyrosine phosphatases, in contrast, contain the motif HCX5R (see Appendix A for amino acid abbreviations), in which the essential cysteine residue acts as the active-site nucleophile (226). These enzymes are the largest family of phosphatases, and the importance and diversity of their biological roles reflect this fact. PTPs are further classified into "classical" PTPs, enzymes of about 280 residues with specific phosphate-binding motifs, and "dual-specificity" PTPs, a more heterogeneous group that nevertheless retains the catalytic motif. Detailed descriptions of these groups and a great deal of information about PTP biology are provided in the excellent above-cited review (226).

Phosphatases that act against substrates other than proteins represent a rich group in themselves, including enzymes involved in such activities as modulation of second-messenger signaling, such as the inositol (228) and phosphatidate (229) signal transduction systems. Other phosphatases participate in biosynthesis (230). These enzymes are often less specific than protein phosphatases in their substrate requirements, and a growing class of synthetic chromogenic, fluorogenic, and chemiluminescent molecules has addressed the problem of how to measure phosphatase activity for many workers studying these enzymes. These substrates have proven to be useful with a subset of the protein phosphatases as well.

5.3 CONTEMPORARY PHOSPHATASE ASSAY TECHNOLOGIES

Excluding CB assays, which are discussed in a separate section, there are four primary detection modes in high-throughput phosphatase assay technologies: fluorescent, colorimetric, chemiluminescent, and radiolabeling. A further important classification parameter regards the entity being detected. For protein phosphatases, most of the fluorescent methods involve detection of the state of the target peptide or protein (i.e., phosphorylated or not), although there is at least one method that follows free phosphate by fluorescent means (231). Radiolabeling, by contrast, is a conceptually simple approach, in that the phosphate group itself contains either ^{32}P or ^{33}P that is incorporated in the target in some manner; after the enzymatic reaction, the hydrolyzed phosphate moiety must be separated from the remaining phosphorylated target so that scintillation-counting methods may be employed. These methods are generally not homogeneous and have the well-known disadvantages of radioactive assays, but, as seen with the kinases, the use of radiolabeled phosphate groups enables certain types of studies that appear virtually impossible with other methods. Colorimetric assays exist for detection of both the free phosphate moiety and the dephosphorylated product, with the phosphate detection reagent malachite green dominant in this area (see Section 5.3.2), while chemiluminescent assays involve the generation of light associated with hydrolysis of a synthetic substrate.

The concepts underlying these two strategies, that is, probing the state of the substrate and tracking the free phosphate after hydrolysis, are quite distinct. An advocate of the former methods might argue that simply measuring the appearance of inorganic phosphate is not sufficiently rigorous, since ATPases and other enzymes can generate this species, and it can be difficult to obtain phosphate-free reagents in any case. However, the fluorescent methods currently used to detect substrate dephosphorylation tend to be complex, dependent on highly unnatural reagents, often slow, and nearly always breathtakingly expensive. For our purposes, the fact that detection of the inorganic phosphate produced by phosphatase-catalyzed hydrolysis is a recognized and established way of measuring the enzyme's activity is of significance in the context of current CB methods.

5.3.1 Fluorometric Phosphatase Assays

In judging whether to try a CB method, one should be aware of the range of competing methods available, especially of the astonishing variety of fluorometric methods. If this section merely helps the reader to select the best fluorometric assay, rather than a CB alternative, it will have served its purpose.

One sign of the widespread recognition of the biological importance of phosphatases is the sheer amount of corporate energy that has been invested in developing ever more exotic fluorogenic sensors to distinguish the phosphorylation states of proteins and peptides. Many of these reagents are useful in both kinase and phosphatase assays, and as a result some listed assay methods also appear in Chapter 4. The reader will pardon the fact that in a field enjoying such a degree of inventive flux and such an extraordinary variety of offerings, it is inevitable that some valuable technologies will be overlooked or will appear between writing and publication of this book. It is hoped that this section of the chapter will provide a helpful overview of what is possible with fluorescence.

Numerous phosphatase assay methods employing fluorescence are briefly described below.

5.3.1.1 Proprietary Reagents Binding Specifically to Phosphorylated or Dephosphorylated Protein or Peptide   This category includes nearly all of the antibody-based methods, but recently small-molecule phosphate sensors have also appeared, in several varieties. Most of the antibodies require conjugation chemistry as well as (frequently) special probes. The small molecules also require sophisticated chemistry, but may enjoy the advantage of more rapid diffusion.

The major thrust within this category is the development of antibodies to phosphoserine/phosphothreonine, phosphotyrosine, or free (dephosphorylated) counterparts of these amino acid residues. These antibodies can be either nonspecific (i.e., associating with any residue of the given type) or highly specific for the particular phosphorylated or dephosphorylated protein (see Reference 232 for a recent example of the use of both types of reagents in the same study). One problem with antibody-based assays is high background fluorescence from the free antibody. The antibody is usually conjugated with a fluorophore, but this in itself is insufficient to solve the problem. Washing or physical separation on the basis of size is expensive in terms of time and reagents. Means of distinguishing the fluorescence signals in homogeneous formats include PerkinElmer's LANCE and DELFIA, both detect europium-labeled antibodies by means of time-dependent interactions with other fluorophores (233). LANCE is a homogeneous method, whereas DELFIA involves wash steps. These methods generally involved competition with a free form of the ligand, followed by measurement of the time dependence of fluorescence resonance energy transfer (FRET) between the associating reagents. They have the advantage of great generality, in that nearly anything that is the target of an antibody can be measured, albeit with considerable assay-development expense in some cases. However, antibodies diffuse slowly, and for this reason all methods based on association of antibodies with phosphorylated targets must allow incubation periods measured in hours, followed by additional detection steps. Moreover, these methods require multiple conjugated labels and tend to be very expensive as a result. Evidently, few researchers are currently using these antibody-based time-resolved FRET (TRF) methods in phosphatase studies. A PubMed search produced no hits correlating "phosphatase" with the LANCE technology, but a study using DELFIA to measure the activity of prostatic acid phosphatase as a marker of prostate cancer has been published (231), and more recently another group employed DELFIA to assess tyrosine phosphorylation of receptors as an indirect measurement of phosphatase activity (234).

An important alternative to LANCE and DELFIA is to measure fluorescence polarization (FP) (235–237), which was called fluorescence anisotropy for years after it was developed. FP is described in Chapter 1. In brief, this technique allows one indirectly to measure the size of a molecular complex by following the loss of directionality of fluorescent emissions, initially in a fixed plane, as the target molecules rotate in solution. Larger complexes rotate more slowly, yielding greater levels of polarization. The phenomenon has been exploited to develop a phosphatase assay based on the association between an antiphosphotyrosine antibody and a phosphorylated peptide, which was inhibited by the enzymatic action of T-cell protein tyrosine phosphatase (237). This approach may have advantages in terms of convenience over methods that require antibodies to associate with protein or peptide targets in the course of the assay, since the antibody can be incubated with the phosphopeptide in batches prior to the assay run. FP can be exploited in other ways for phosphatase assays that do not rely on antibodies; these methods are discussed in Section 5.3.1.

5.3.1.2 Superquenching

The advent of biosensors employing superquenching is a relatively recent event (238). The reader is strongly encouraged to read the original article, along with the companion commentary (239), which are currently available free on-line through the generous policy of the Proceedings of the National Academy of Sciences of the USA.

Briefly, superquenching is a phenomenon associated with conjugated polymers, typically of fairly large molecular weight ($\sim 10^6$ Da), that fluoresce by mechanisms that have been employed in such technologies as OLEDs, all-polymer electronic circuits, and solar cells. (Some prefer the term "luminesce" to "fluoresce" in this context, arguing that the emission is not true fluorescence. Here, we use "fluoresce" because we reserve the term "luminescence" for those entities that transduce the chemical energy of discrete molecules to light.) These polymers have also been found to be extraordinarily sensitive to quenching by small cations (238). The rate of the quenching event, that is, the photoinduced electron transfer (240), is less than a picosecond (241), an interval that is unlikely to present any significant barrier to the development of rapid assays.

In their seminal study, Chen *et al.* (242) showed that biosensors utilizing superquenching were possible, but the incorporation of these methods into kinase and phosphatase assays was a later development. This system employs Ga^{3+}-associated poly(*p*-phenylene-ethynylene) (PPE) as the fluorescent polymer, which is superquenched upon association with a dye-conjugated phosphopeptide, of which the phosphate moiety forms a complex with the metal cation. Both the photophysics and the actual synthesis of these materials are hardly trivial, but the result is a biosensor capable of responding to exquisitely small concentrations of analyte at extreme speed.

Superquenching is yet another technique equally applicable to kinase and phosphatase assays. The TruLight[TM] "superquenching" assay kit available from Calbiochem uses a phosphorylated tracer that binds to the sensor and is displaced by phosphorylated protein. The vendor claims much greater sensitivity than is observed in FRET assays. Phosphatase and kinase assays with this method evidently require about 2 h and the list price as of this writing is a hefty $2.40 per well in a 400-test quantity. Users must supply the quencher-labeled peptide, another significant source of expense. However, great developments are to be expected from the phenomenon of superquenching in the near future and the prices are likely to drop.

Compounds other than antibodies can be made to associate with protein and peptide substrates or products of phosphatase action. Small iron-containing molecules have been developed that associate with phosphorylated peptides and simultaneously quench the fluorescence of fluorophores attached to the same peptides. Methods employing such molecules include the IQ[TM] Technology from Pierce. This phenomenon is distinct from superquenching, in which the polymer fluorophore is quenched by the association event; here the fluorophore is small and is attached covalently to the peptide substrate. The quenching can be measured directly to assess kinase activity, and of course the reverse phosphatase-catalyzed process destroys the complex and relieves the quenching. Figure 5.1 is a schematic depiction of the IQ technology. An important advantage of IQ technology over the antibody-dependent methods is the much greater diffusion coefficient of a smaller species; however, the manufacturers' recommended assay periods for time-point mode are 30–120 min, and the assay cannot therefore be considered a rapid process by CB standards. These methods also require the synthesis of two specialized molecules. Different binding agents are required for serine/threonine and tyrosine residues. These issues render the methods quite costly, with list prices as of this writing of roughly $2 per well, although the ability to distinguish between the phosphorylated residues at least brings the small molecules level with their highly specific antibody competitors. Figure 5.2 shows the results of an assay employing the IQ technology in a titration of the phosphatase PP2A with the inhibitor okadaic acid, using the phospholabeled peptide LRRA-pSLG as substrate.

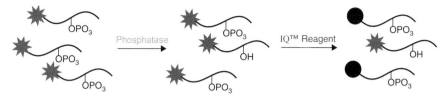

FIGURE 5.1 Schematic representation of the Thermo Scientific IQ phosphatase assay. Labeled peptides are dephosphorylated by the test enzyme. Subsequently, the antibody-free IQ reagent quenches the fluorescence of only those peptides with remaining phosphates, yielding a positive fluorometric signal corresponding to phosphatase activity. Used by permission of Thermo Fisher Scientific.

5.3.1.3 Fluorogenic Substrates The use of fluorogenic substrates, which yield fluorescent compounds upon hydrolysis by phosphatases, has a great deal to recommend it, whether one is performing high-throughput screening or basic enzymology. Most of these substrates are not designed and synthesized with particular phosphatases in mind, but instead are general reagents much like *p*-nitrophenylphosphate (pNPP). Before the details are presented, we will consider one caveat, which relates to empirical observations of the quality of results obtained with fluorogenic substrates. Because these molecules are not natural, methods employing them may yield information of modest biological relevance, because the reaction that is being studied or modulated is different from the *in vivo* phenomenon, perhaps with different kinetic parameters and qualitative characteristics. Moreover, this skewing, if it occurs, will be systematic throughout the study, assuming that the work employs a single substrate and the bias is due to its specific characteristics. That such "failure modes" (which seems too strong a term) can occur is suggested in published work comparing a screening procedure using a fluorogenic substrate with a rather complex chip-based mobility shift assay of protein tyrosine phosphatase 1B (PTP1B). In Reference 243, the authors claim that the chip assay, which employs natural substrates, yielded a broader range

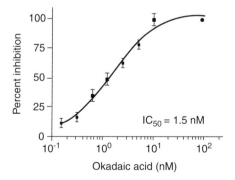

FIGURE 5.2 IC_{50} determination of okadaic acid inhibition versus protein phosphatase 2A using the Thermo Scientific IQ technology. The substrate was LRRApSLG (see Appendix A for one-letter codes for amino acid residues). Used by permission of Thermo Fisher Scientific.

of useful molecules with more drug-like properties, using a narrowly targeted library of potential inhibitors. It does not appear that their assay was compared with other methods using natural substrates, however. In any case, other groups, with other targets, procedures, and libraries, may observe something very different using unnatural fluorogenic substrates.

Fluorescein diphosphate (FDP) was the earliest described fluorogenic phosphatase substrate (244). The use of this substrate yields a roughly 50-fold improvement in sensitivity over the chromogenic substrate pNPP (245). Since FDP exhibits almost no fluorescence, while the monophosphate fluoresces weakly and free fluorescein very strongly, with a quantum yield of ~0.92, hydrolysis of the phosphoester bonds by phosphatases yields a powerful signal. However, the strength of the signal depends on maintaining alkaline conditions, since the fluorescence yield of fluorescein declines upon protonation. FDP is still actively marketed by a number of firms, including Invitrogen/Molecular Probes and Anaspec. It has played a role in published investigations not only of alkaline phosphatases (246, 247) but also of phosphatases involved in signaling networks, such as protein tyrosine phosphatases (248, 249).

The more recently developed fluorogenic substrate 4-methylumbelliferyl phosphate (MUP) is an alternative to FDP, with simpler kinetics, since only a single phosphoester hydrolysis is involved. Among other applications, MUP was widely employed in immunoassays developed to detect *Listeria* in food products in the 1990s as an improvement over the time-consuming culture method (250, 251). The critical nature of this application drove efforts using several approaches, and it is interesting to note as an aside that an unusual coupled bioluminescent assay was developed and used for this purpose (252). In this fascinating system, the *Listeria*-specific bacteriophage A511 was engineered to express the prokaryotic luciferase from *V. harveyi*. The food product could be treated with the modified phage and any *Listeria* present would emit light a few hours later. Unfortunately, this clever idea was superseded by fluorescence-based ELISAs that were more flexible and technically less demanding, but bioluminescent strains of *Listeria* were later used to trace contamination and survival in food products (253).

In a 1999 study that compared methods of detecting the important marine toxin okadaic acid (a protein-phosphatase inhibitor) (254), fluorometric assays employing FDP and MUP were tested in parallel with assays using the colorimetric substrate pNPP and the luminogenic luciferin-phosphate (see Section 5.4.1). (The chemiluminescent substrate CDP-star, discussed in Section 5.3.3, was not hydrolyzed by the particular enzyme used.) Both fluorometric and the luminometric methods were superior in accuracy and sensitivity to the colorimetric method, although all yielded some false negatives in comparison with the "gold-standard" murine bioassay, and even ELISA scored better than direct colorimetry.

The Molecular Probes Division of Invitrogen markets the patented 6,8-difluoro-4-methylumbelliferyl phosphate (DiFMUP) substrate. This molecule is held to produce a 100-fold more sensitive readout than MUP in assays of prostatic acid phosphatase, a widely used marker in prostate cancer research (255). These assays are run at the mildly acidic pH of 5.5 (see Fig. 5.3). DiFMUP is a general substrate for any phosphatase traditionally assayed with pNPP or MUP (256, 257). At ~15 min in

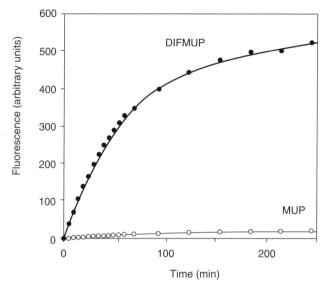

FIGURE 5.3   Assay results for prostatic acid phosphatase with the two generic phosphatase substrates DiFMUP and MUP (see text). The patented DiFMUP substrate yields a far stronger response to the phosphatase activity, with no loss of data quality. Conditions: 0.002 units of enzyme were assayed with 100 µM substrates in 100 mM sodium acetate buffer at pH 5.5. Courtesy of Invitrogen Corporation.

length, the assay protocol is relatively rapid, with a single mixing step. The sensitivity advantage of DiFMUP does not apply, however, to alkaline conditions; MUP, with a pK_a of 7.8, is well suited to assays of alkaline phosphatases. The cost of DiFMUP and other hydrolyzable fluorescent substrates is a pleasant surprise for workers used to antibody-based assays; the list price of DiFMUP, for example, is some $0.16 per well.

5.3.1.4 Antibody-Free FP and FRET Assays of Phosphatase Activity   These promising assay methods are essentially the reverse of the corresponding procedures for kinases (see Chapter 4). MDS/Molecular Devices's IMAP®, or Immobilized Metal Affinity Platform, depends on two events: (1) phosphorylation of a molecule containing a fluorophore and (2) a signaling process that depends on the association of the phosphorylated species with a nanoparticle. The surface of the nanoparticle is decorated with trivalent metal ions that strongly associate with phosphate groups. To generate an FP signal in this system, it is sufficient for the phosphorylated, fluorescent molecule to associate with an entity of large molecular weight. The rotational correlation time of the complex is much greater than that of the free peptide, leading to a large increase in observed polarization. In phosphatase assays, of course, the phosphate group is removed, causing a failure to associate with the nanoparticle and a *decrease* in the FP signal.

A similar strategy can generate a FRET (258) signal, but the FRET procedure requires an additional functionality of the nanoparticle. The fluorescent emission of the captured fluorophore is reemitted and captured by the proximate fluorophores of the associated particle. In principle, this interaction can lead to either a change in the observed wavelength of the readout or a quenching of the readout (the latter if the photon-absorbing species does not reemit at all). However, the IMAP approach invokes yet another possibility, known as TRF, in which the long lifetime of the accepting fluorophore is exploited by following the time dependence of fluorescence emission subsequent to excitation. Light emission due to the specific interaction of interest (between the phosphorylated peptide and the nanoparticle) can be resolved from background signals by these means. Of course, in measuring phosphatase activity this interaction is abolished by the catalytic process, leading to a *decrease* in the time-resolved FRET signal.

The IMAP system can be used with a wide range of protein phosphatases, including those with both general and specific substrate requirements, as long as the phosphorylated substrate is synthetically accessible (and the phosphate group is accessible to the nanoparticle). Figure 5.4 shows a dose–response curve from an experiment employing 100 nM of a specific phosphopeptide substrate with low concentrations of PP2A in a 1-h incubation, as measured by the IMAP FP system. The listed cost of this assay is $0.89 per well at the 8000 quantity.

5.3.1.5 Fluorogenic Phosphate Sensors
The *E. coli* phosphate-binding protein in a conjugate with the fluorophore N-[2-(1-maleimidyl)ethyl]-7-(diethylamino) coumarin-3-carboxamide (MDCC) was originally used as a real-time phosphate

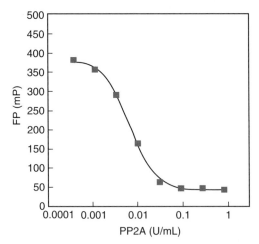

FIGURE 5.4 Substrate titration results for protein phosphatase 2A obtained with the IMAP FP technology. The substrate was 5FAM-GRPRTS-pS-FAEG-OH. The IMAP binding reagent of high molecular weight associates with the labeled peptide if it is phosphorylated, slowing its rotation and giving rise to an FP signal. Used by permission of MDS Analytical Technologies.

sensor by the Webb group in London (259) and was later patented, along with other related formulations (260). The fluorescence output of this conjugate is enhanced over fivefold with a shift in spectrum by association with free phosphate under the low ionic strength conditions used. This is the converse of the IMAP concept: an antibody-free reagent that yields a fluorescent signal upon association with free phosphate, rather than with the phosphorylated substrate. Invitrogen currently markets this conjugated protein as the "Phosphate Sensor." (Although the phosphate-binding protein used in the original work incorporated a site-directed mutation of an alanine to a cyteine, to allow conjugation of the fluorophore by sulfhydryl chemistry, the Invitrogen product literature does not mention this; however, it is implicit in the material, since a "cysteine" residue is mentioned and the native protein has no such residue.)

The Phosphate Sensor is claimed to be 100-fold more sensitive than malachite green, with a limit of detection for free phosphate in the high-nanomolar to low-micromolar range and a correspondingly strong K_d for phosphate of about 100 nM. Figure 5.5 is a schematic representation of the detection mechanism of this sensor, which appears simple but involves sophisticated chemistry. The spectral difference upon phosphate binding is strong, though not in the 100-fold range of some fluorescent sensors (Fig. 5.6). Figure 5.7 displays the results of an experiment utilizing the Phosphate Sensor to assess the activity of PTP1B, with a limit of detection in the subnanomolar range for the enzyme and a sensitivity advantage relative to malachite green of greater than two orders of magnitude. The list price of the Phosphate Sensor in large quantities is a very attractive $0.13 per well.

5.3.1.6 Coupled Fluorometric Approaches   Yet another unusual approach to measuring protein–phosphatase activity is coupling it to the action of a protease. This concept was pioneered by a Japanese group and published in 1999 (261), based on earlier work demonstrating the possibility of fluorogenic substrates for protease assays using the same concept (262). A peptide is synthesized with three covalent modifications: two fluorophores that engage in FRET when in proximity (an aminomethylcoumarin

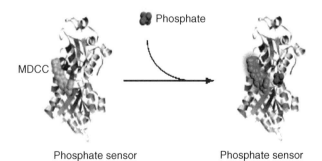

Phosphate sensor   Phosphate sensor

FIGURE 5.5   Space-filling representation of operation of the Invitrogen Phosphate Sensor. The sensor is a phosphate-binding protein from *E. coli* modified with a fluorophore adjacent to the site of phosphate association. The blue fluorescence increases dramatically upon binding. Courtesy of Invitrogen Corporation. (See the color version of this figure in the Color Plates section.)

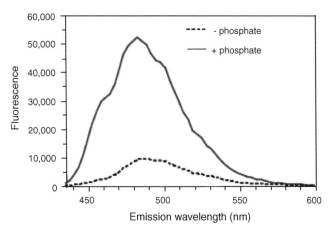

FIGURE 5.6 Emission spectra of the Phosphate Sensor in the presence and absence of phosphate. The sensor is present at 0.4 mM, with or without 1.7 mM phosphate. There is surprisingly little shift in the wavelength of maximum emission, but the fluorescence enhancement appears to be greater than 400% at that point. Courtesy of Invitrogen Corporation.

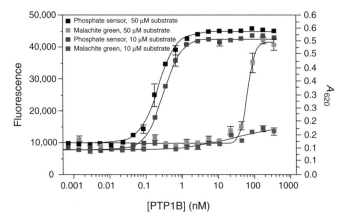

FIGURE 5.7 Comparison of phosphate sensor with malachite green under phosphatase-assay conditions. The fluorescence scale on the left represents the Phosphate-Sensor signal; malachite-green absorbance is recorded against the scale on the right. Protein tyrosine phosphatase 1B was titrated into solutions with 10 and 50 μM DADE-pY-LIPQQG. Both of the lower curves are malachitegreen data. The Phosphate Sensor appears to be more sensitive to the enzymatic activity by about two orders of magnitude. Courtesy of Invitrogen Corporation.

is frequently used in combination with a dinitrophenyl group, resulting in strong quenching of the fluorescence of the coumarin derivative) and a phosphate group on an appropriately located serine, threonine, or tyrosine. Removal of the phosphate group has virtually no effect on the fluorescence of the peptide, but subsequent digestion by a protease results in a large (>100-fold) increase in fluorescence as the FRET quenching effect is relieved (other FRET phenomena, such as a shift in emission, could also be observed with different fluorophores). The activity of the protease strictly depends on the removal of the phosphate group, making the assay highly specific for hydrolysis of the specific bond of interest. Substrates such as "Protein Tyrosine Phosphatase Substrate III" (MCA-GDAEpYAAK(DNP)R-NH$_2$, where MCA stands for 7-methoxycoumarin-4-yl, DNP is dinitrophenyl, "p" is the phosphate moiety, and the single-letter amino acid code is given in Appendix A) and the ProFluor® assays from Promega employ this concept. In the Promega product, the substrate is coupled to bisamide rhodamine 110, a fluorophore that does not require FRET to yield a signal; its natural fluorescence is quenched by its covalent bond with the peptide, which is then cleaved by the supplied protease. There are also corresponding kinase assays, PKA Profluor and Src Profluor. Kinases yield a negative fluorometric signal using this method, while phosphatases yield a positive signal.

The phosphatase assay process using this technique requires an initial incubation for phosphatase activity, followed by addition of a separate reagent package containing the protease and a digestion step. The fluorescence depends on both phosphatase activity and successful digestion of the dephosphorylated peptide. One of Promega's kits provides a substrate that has been tested with several serine/threonine phosphatases, including PP1, PP2A, PP2B, and PP2C, while another is directed to protein tyrosine phosphatases. It is likely that the principle could be extended to virtually any protein phosphatase if the market for such a product is sufficiently robust to support development of further conjugated peptides and testing with appropriate proteases. The recommended incubations are 60 min for the phosphatase activity, followed by addition of the protease/phosphatase stop reagent, a further 30-min incubation for digestion (according to the vendor, this step should be carried out at room temperature), addition of a "stabilizer," and reading. Thus, the procedure is moderate in terms of rapidity and is also of moderate complexity.

In developing such an assay, one must of course deal with each step independently; however, Promega's data show that the protease step is quite robust. Figure 5.8 displays the high degree of linearity achieved in this process, indicating both high achievable precision and good performance over the dynamic range. Figure 5.9 illustrates Promega's approach to the issue of converting peptide hydrolysis to actual phosphatase activity. Panel A shows the raw data from a hydrolysis experiment using PTP1B against the ProFluor substrate. The curve is convincing and the scatter is small. (One attractive feature of this assay is the very low background signal; evidently, the rates of both spontaneous hydrolysis of the phosphopeptide and adventitious digestion of the peptide are very low.) Panel B shows the normalized data, allowing the worker to select an appropriate enzyme concentration, perhaps for inhibitor screening. Figure 5.10 shows an inhibitor titration with sodium vanadate against PTP1B, yielding a clean result of an IC$_{50}$ of 17 nM with low and consistent

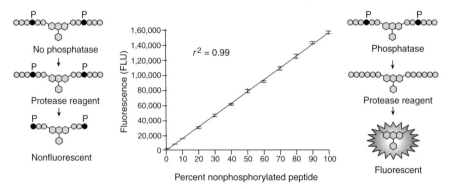

FIGURE 5.8 Data and schematics of Promega's protease-based phosphatase assay. The protease target site in the peptide varies, depending on whether the peptide is phosphorylated. Cleavage of the dephosphorylated peptide liberates the fluorophore. Conditions: Rhodamine-110-labeled serine/threonine phosphatase substrates in various combinations of phosphorylated and nonphosphorylated varieties were incubated for 90 min with the protease, followed by fluorometric readout. No phosphatase was used in this assay. Used by permission of Promega Corporation.

scatter (though higher at very low inhibitor concentrations, which is not surprising). The vendor claims typical Z values of >0.8 for this assay.

While the assay concept is appealing, especially to those with interest in coupled enzymatic methods, questions arise about several points, notably the specificity of screening for inhibitors. As is the case with CB assays, the presence of multiple essential enzymes in the product implies that more than one class of inhibitor may be identified by the screening process; some of these may be protease inhibitors, but Promega has addressed this potential problem by providing a separate control substrate, an aminomethylcoumarin derivative that is also digested by the protease to

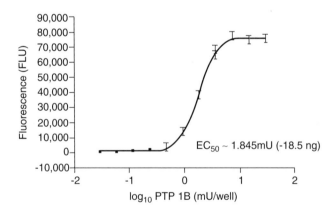

FIGURE 5.9 Assay of protein tyrosine phosphatase 1B using the protease-based system of Fig. 5.8. Clean data are obtained by titrating the enzyme into a solution of the rhodamine-110-labeled substrate (concentration not specified). Used by permission of Promega Corporation.

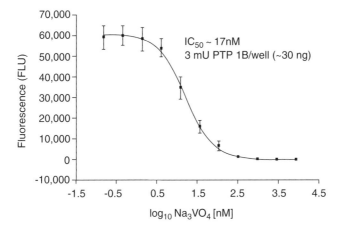

FIGURE 5.10 Determination of IC_{50} for sodium vanadate versus protein tyrosine phosphatase 1B, using protease-based assay of Fig. 5.8 and 5.9. The scatter virtually disappears near the zero-signal asymptote, but this is essentially a damping artifact and the true scatter is likely to be more similar to that observed at low inhibitor concentrations. Used by permission of Promega Corporation.

yield a fluorescent signal. In inhibitor studies, the control substrate is tested side-by-side, thereby revealing any inhibition of the protease. Of course, the same approach can be used in most coupled enzymatic assays; the extra reaction adds to the cost and labor expenditure, but in this case the expense is minimal, since only one set of control-reaction replicates is required for each screening of an entire inhibitor panel. The use of targeted libraries will also ameliorate the probability of generating false hits. However, another issue regarding this assay is the cost, which appears to exceed $1 per well at list price.

5.3.2 Colorimetric Phosphatase Assays

This category is currently dominated by various formulations of the malachite green procedure, but for many years the colorimetric method of choice for phosphatase assays was the synthetic substrate pNPP, a highly labile phosphoester of *p*-nitrophenol that yields the strong yellow color of the latter compound upon hydrolysis. This substrate is of course highly unnatural, as well as very unstable to spontaneous hydrolysis. Moreover, the method was not highly sensitive. Few enterprises engaging in high-throughput screening of phosphatases use pNPP today, although it still shows up in kits for ELISA detection of conjugated alkaline phosphatase. The method does have the advantage that adventitious sources of phosphate do not perturb the signal.

Malachite green (IUPAC designation: 4-[(4-dimethylaminophenyl)-phenyl-methyl]-*N,N*-dimethyl-aniline) does not contain malachite; its name derives from the similarity of color. It has been used for decades to treat infections by parasites, fungi, and bacteria, as well as in commercial dyeing processes. It has also been used as a pH indicator. There is good evidence that malachite green can cause cancer through the formation of covalent DNA adducts, although the active chemical species has not been identified.

Malachite green has been used in assays of inorganic phosphate since the early 1970s (263). While its use in enzymatic assays would seem a logical step from this beginning, the earliest uses for this purpose appearing in PubMed date to the 1980s (264). The use of this method places one in the center of the debate about detecting the inorganic phosphate product rather than the dephosphorylated moiety, since, as mentioned elsewhere, phosphate can come from many sources, including impure assay reagents and hydrolysis of various phosphorylated species present by contaminating phosphatases. Thus, the use of any method of detecting inorganic phosphate as a phosphatase assay requires that one has control and quantitative understanding of other potential sources of phosphate.

The sensitivity, simplicity, and rapidity of the malachite green assay are quite good. A typical formulation yields a limit of detection of approximately 10 pmol (0.2 nM in 50 L) phosphate with a reaction time of under 1 h. This is less sensitive than current CB methods by no more than about one order of magnitude. The method has been used in high-throughput screening procedures in drug discovery enterprises for over 10 years (265–267) and is still successfully marketed. Given what we have seen for antibody-based methods, the per well cost of malachite green is highly competitive, in the range of roughly $0.20, although kits such as Promega's that supply phosphorylated peptide substrates for specific classes of enzymes may be more expensive (∼$1 per well).

Figure 5.11 shows the excellent linearity of this assay. The data do not reveal much of what is going on in the low range, but the scatter appears to be small.

The Molecular Probes Division of Invitrogen has taken a different approach to achieve a colorimetric readout. Their reaction is actually a coupled colorimetric

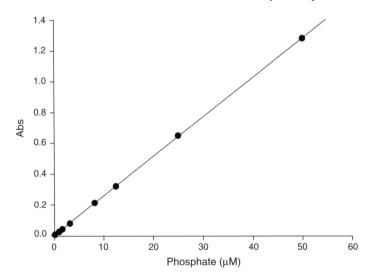

FIGURE 5.11   Results of malachite green phosphate assay. The response to phosphate appears to be at least twofold greater than in Fig. 5.7 from a competing vendor. Used by permission of Anaspec, Inc.

assay, although a fluorometric readout is also possible (268). This interesting reaction involves the enzymatic phosphorolysis of a glycosidic bond, leaving a free purine with a spectrum that differs from that of the glycosylated substrate; the ΔA_{360} is about 11,000 $M^{-1}cm^{-1}$. Thus, the increase in absorbance of the purine substrate is directly related to the initial phosphate concentration. For those who are used to considering the phosphate ion to be an unreactive or inert bystander, the reaction has a surprisingly high equilibrium constant of >130 in the direction of phosphorolysis (268, 269). Figure 5.12 shows the linear range of this assay, which is evidently slightly less sensitive than the malachite green method. The inventors claimed a limit of detection of 2 μM inorganic phosphate, in keeping with this observation; it is not clear whether Invitrogen can do better than this. The technique has an extensive history of citations, despite its apparently modest limit of detection (270, 271).

5.3.3 Chemiluminescent Phosphatase Assays

Most of the effort in designing chemiluminescent substrates for phosphatases has gone into the reporter function (see Chapter 12) and as a result the substrates tend to be either nonspecific or wholly or partially specific to a particular enzyme used in the reporter role. For example, Applied Biosystems offers the CDP-Star® and CSPD® dioxetane-derived chemiluminescent substrates for alkaline phosphatase, based on work done in the late 1980s (272, 273). The same firm provides enhancers of the reactions that are able to increase the luminance signal by fivefold or more. The phenomenon of enhancement requires separate explanation. These substrates have largely

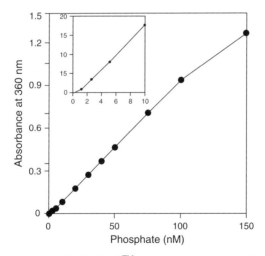

FIGURE 5.12 Results from the EnzChek™ coupled chromogenic phosphate/phosphatase assay. Ribose from a supplied chromogenic reagent is enzymatically transferred to the free phosphate, liberating the chromophore. Although the scale of the graph is in absolute nanomoles rather than micromolar as in the malachite green data above, the method appears to be 5- to 10-fold more sensitive than malachite green. Courtesy of Invitrogen Corporation.

absorbed the market for chemiluminescent assays of alkaline phosphatase, which is substantial, given the prevalence of this enzyme as a reporter. In addition to Applied Biosystems and many others, the substrates are marketed by Roche, KPL, New England Biolabs, and Clontech, and appear for now to have the upper hand over competing chemiluminescent approaches such as the use of luminol (274) and acridinium esters. However, Invitrogen/Molecular Probes has a competing dioxetane derivative (Bold^{TM} APB), developed at Serologics and claimed to be superior in blotting applications. Another approach that may deserve more attention than it has received is conversion of the products of the alkaline phosphatase reaction to dihydroxyacetone phosphate, followed by a light-producing reaction with lucigenin (275).

One of the challenges in developing any technology involving detection of light-emitting molecules in solution is ensuring a large fraction of the excitation energy (whether originating in chemical processes, as in bioluminescence and chemiluminescence, or in exogenous light energy, as in fluorescence and phosphorescence) is reemitted as light, rather than being expended in side reactions that do not produce light. Enhancers increase the light yield by providing an environment in which collisional quenching of the excited molecules is much less likely to occur. Ionic and polar interactions are also undesirable, since they may promote various decay processes. Thus, the ideal enhancer is one that provides a very hydrophobic environment for the excited molecule, allowing it to retain its energy until the stochastic process of light emission occurs. Such molecules as bovine serum albumin and poly(vinylbenzyltrimethylammonium chloride), a polymer of a quaternary ammonium salt, have been described as potential enhancers in this sense (276). A related phenomenon is observed in protein–ligand complexes, in which association of a hydrophobic ligand may enhance the quantum yield of a fluorophore in the protein, although the best-known example, the lysozyme–oligosaccharide interaction, was initially observed by ordinary difference spectroscopy (277). The effect of these enhancers is to reduce the "bleeding" of the excitation energy into forms other than light generation. Both luminescent and fluorescent molecules are subject to these nonproductive side reactions and in both cases their light emission can be enhanced very significantly.

The value of this approach in assays of alkaline phosphatase activity is thus well established, but substrates for other phosphatases have been slow to appear. The dioxetanes with their adamantine moieties have their own idiosyncratic chemistries, and the combination of this with the special requirements of peptide synthesis may give rise to competing constraints, slowing down development in this area. Alternatively, it may be that the adequacy of other methods has prevented commercial enterprises from expending great efforts on additional chemiluminescent phosphatase substrates. A study published in *Methods in Enzymology* suggests chemiluminescent strategies that may be applicable to general protein phosphatases (278).

5.3.4 Radiolabeling Phosphatase Assays

Most vendors no longer intensively market these assays, but New England Biolabs, a venerable institution in assays involving radiolabeling, currently sells kits for protein phosphatase assays using either ^{32}P- or ^{33}P-labeled myelin basic protein (MBP). Phosphorylation on either serine/threonine or tyrosine can be studied specifically. Labeling is accomplished by using an appropriate enzyme, either cAMP-dependent protein kinase (for serine/threonine) or Abl protein tyrosine kinase (for tyrosine). This labeling step must be carried out by the user, including purification of the labeled protein. Phosphorylation may optionally be checked by two-dimensional thin-layer electrophoresis following partial acid hydrolysis of the labeled proteins.

After labeling, the phosphatase reaction is carried out. The radioactive inorganic phosphate released from the substrate is dissolved by adding trichloroacetic acid, is separated from the remaining protein-incorporated label, and is counted. Figure 5.13 depicts the response of this assay system to low-nanogram quantities of T-cell protein tyrosine phosphatase, with a very good linear response in the low range shown in the inset.

Apart from a few traditionalists who believe data only when they are radioactive, it is not clear why anyone would use this method of measuring phosphatase activity. The vendor claims that theirs "is the only kit available that allows the high sensitivity detection of protein tyrosine phosphatases," but this claim appears to be overstated. The stated detection limit of "a few pmol of released phosphate" is readily achieved even by unoptimized CB methods. The list price of the kit is $2.20 per well, but this does not include the fluctuating cost of the labeled nucleotide itself, and of course disposal of radioactivity and various additional labor and equipment costs for cleanup and monitoring must be taken into account. Traditionalists remain, however,

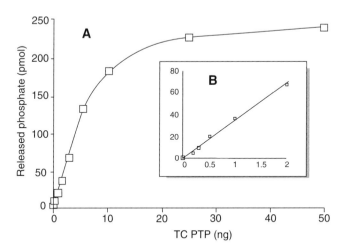

FIGURE 5.13 Results of isotopic-labeling assay of T-cell protein tyrosine phosphatase, using ^{33}P-tyrosine-labeled myelin basic protein as substrate. The substrate was treated with the Abl protein tyrosine kinase in the presence of [γ- ^{33}P]-ATP for label incorporation (~1.5 phosphate/mole of protein) prior to the experiment. Free phosphate was assumed to be the fraction soluble in trichloroacetic acid and counted. Like many methods involving radioactive labeling, the assay is very sensitive, with a limit of detection below 1 ng (i.e., femtomole range of enzyme). Used by permission of New England Biolabs, Inc.

as exemplified by studies performed with radiolabeled phosphatase substrates in the not-so-distant past (279). In some cases, special experiments may require radiolabeling for pulse-chase procedures to determine the specific fate of the phosphate group (280) or for *in vivo* studies of phosphatases (281). It is true, of course, that assays using radiolabeling are easier to troubleshoot, since the label can always be found somewhere (otherwise one must entertain earnest hopes about where it is not).

5.4   CB PHOSPHATASE ASSAYS

The problem we face in contemplating the development of CB phosphatase assays is essentially the opposite of that hindering reverse kinase assays, in which the synthesis of ATP was seen to be thermodynamically unfavorable: here ATP is not involved at all in the initial reaction. Thus, the issue is coupling the production or consumption of ATP to some aspect of the phosphatase reaction. Preferably, this would be accomplished in a way that renders the assay generally useful.

5.4.1   Assays Using Phosphorylated Luciferin

The first system we will examine is the simplest imaginable from a conceptual point of view, one in which the coupled synthesis or hydrolysis of ATP is not needed. Instead luciferin, the luciferase substrate (see Fig. 1.9), is chemically phosphorylated to luciferin-*O*-phosphate, which cannot participate in the light-generating reaction. The phosphatase then hydrolyzes the phosphate-ester bond, liberating luciferin and enabling luciferase to produce light (282). This elegant and simple series, known since the late 1980s, provides an alternative to the expanding range of chemiluminescent methods for measuring AP activity, but unfortunately does not appear to be applicable to other phosphatases that do not hydrolyze this particular substrate. Future developments exploiting clever biochemical manipulations of the luciferase system to generate light from wider varieties of enzymes are to be expected.

5.4.2   Assays Based on the GPL Reaction Series

In contrast to this approach, the author chose to concentrate on the phosphate group itself as a target for CB quantification. Of the known enzymes in nature that carry out oxidative phosphorylation using inorganic phosphate as a substrate, few are readily available from commercial sources. In practice, the same G3PDH–PGK-coupled system used in the cytotoxicity assays described in Chapter 3 has been adopted for detection of inorganic phosphate (Fig. 5.14).

Although PGK is the entity that produces ATP in this reaction series, the central importance of G3PDH should be noted. The very ubiquity and "household enzyme" nature of G3PDH can make us forget what a remarkable enzyme it is. Beginning with a single energy-charged substrate, glyceraldehyde-3-phosphate (G3P), the enzyme is able to produce two high-energy molecules that are employed in later metabolism: NADH, a direct product of the oxidative portion of the reaction, in that NAD^+ is

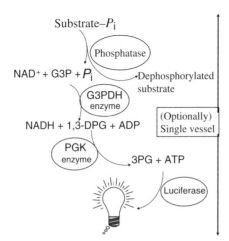

Substrate–P_i

NAD$^+$ + G3P + P_i

Phosphatase

→Dephosphorylated substrate

G3PDH enzyme

NADH + 1,3-DPG + ADP

(Optionally) Single vessel

PGK enzyme

3PG + ATP

Luciferase

FIGURE 5.14 Schematic representation of CB assay for phosphate/phosphatase activity. NAD$^+$, nicotinamide adenine dinucleotide (oxidized form); G3P, glyceraldehyde-3-phosphate; P_i, inorganic phosphate; G3PDH, glyceraldehyde- 3-phosphate dehydrogenase; NADH, nicotinamide adenine dinucleotide (reduced form); 1,3DPG, 1,3-diphosphogylcerate; ADP, adenosine diphosphate; PGK, phosphoglycerokinase; 3PG, 3-phosphoglycerate; ATP, adenosine triphosphate. The scheme can be used as an assay for NAD$^+$, G3P, P_i, or G3PDH, depending on which components are omitted from the reagent mixture. U.S. Patent # 6811990.

employed as the oxidant; and 1,3-diphosphoglycerate, a relatively unstable product that must then be processed by PGK before it breaks down for the glycolytic pathway to proceed. Considering that the enzyme begins with the lowly (from an energetic point of view) free phosphate group in accomplishing this, G3PDH must be considered a marvel of energy transduction. The activity of this enzyme in raising the energetic state of the phosphate group to the point where it is ready to be incorporated in ATP by PGK is the essential element of the coupled bioluminescent pathway depicted.

5.4.2.1 Measurement of Phosphatase Activity by CB Detection of Phosphate The assay scheme depicted in Fig. 5.14 was first used to determine the response to free phosphate (Fig. 5.15). The limit of detection in this unoptimized system (i.e., without the use of ultrapurified reagents) was in the low picomole (10–30 nM) phosphate in 96-well plates, although a lower absolute limit could undoubtedly be obtained both by the use of cleaner reagents and by employing a higher degree of miniaturization. A reaction time of 3–5 min. is typical, and the comments in Chapter 2 about the relative advantages and disadvantages of time-point readouts versus linear fits apply. Nonlinearity observed above the effective dynamic range is due to the fact that one of the reagents has become limiting, nearly always either G3P (which is usually contaminated with phosphate), G3PDH, or PGK. These reagents can contain contaminating phosphate, and the enzyme preparations can also harbor contaminating enzymes that may generate phosphate or ATP from other reaction components, thereby increasing

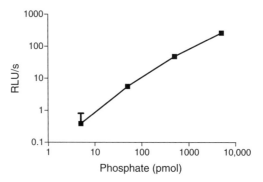

FIGURE 5.15   Response of CB phosphate assay over three orders of magnitude, starting in low picomole range. Total assay time is 3–6 min. The data were reduced by nonlinear regression of the time-dependent readings as in Chapter 2.

the level of the background signal. Therefore, a decision must be made at some point as to what level of background signal is acceptable, since a wider dynamic range may be achievable if a larger amount of the impure reagent is used.

5.4.2.2 Substrates for CB Phosphatase Assays   As with other assay methods involving detection of free phosphate, a wide range of substrates may be employed, including natural substrates. The substrates used in developing the CB assay described were purified α-casein and various phosphorylated peptides without additional modifications. α-Casein is available in a form that has been enzymatically phosphorylated by kinases, but as it proved, chromatographically purified α-Casein has enough natural phosphorylation to serve as an excellent phosphatase substrate (however, a lower grade of α-casein was unsuitable). Dephosphorylated casein is also marketed and can serve as a negative control.

The Fischer tyrosine-phosphatase substrate peptide (283) was used with human leukocyte antigen related phosphatase (LARP), while calcineurin activity was measured with the RII peptide (DLDVPIPGRFDRRV-pS-VAAE) from CalBiochem. There is no evidence of any limitation on the type of phosphorylated substrate that may be employed with this CB assay, as long as potential interferences are respected (see Chapter 3). As is the case with other assays for phosphate, this method enjoys the advantage that the nature of the substrate is largely irrelevant, since it is a free phosphate that is being measured.

5.4.2.3 Results of CB Phosphatase Assays   Figure 5.16 shows the results of a titration of the α-casein substrate against the well-known protein phosphatase from bacteriophage lambda (284). The data were reduced by linear regression of the readouts from the first 2–6 min of the reaction. The reaction is carried out in homogeneous, one-step fashion: all reagents are mixed and the luminescent readout begins immediately. No significant background rate is observed. The results fit a saturation curve, although they do not extend to sufficiently high concentrations of the substrate,

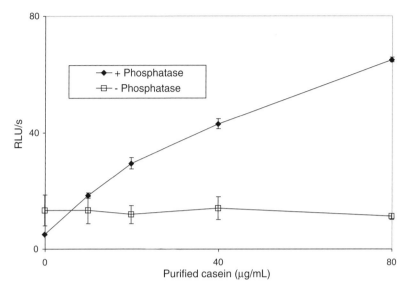

FIGURE 5.16   Activity of the λ-phosphatase measured by CB assay. The phosphatase was assayed with purified but unmodified α-casein as the substrate. The buffer used for the assays of LARP (Fig. 5.17) was modified by adding 2 mM (final) $MnCl_2$. Data from 2–6 min were used for analysis by nonlinear regression.

and calculation of a meaningful K_m for this heterogeneous substrate would have little value in any case. However, the existence of observable saturation implies that inhibitor screening would be possible even with this poorly defined (but very inexpensive) substrate, although again, calculations of the true K_i without a firm knowledge of the K_m could be problematic. Needless to say, other, better-defined and more specific substrates could be used in this assay.

The human leukocyte antigen related phosphatase was tested by using this assay system with the above-mentioned Fischer peptide, which was designed in the laboratory of the renowned Edmond Fischer as a general substrate for protein tyrosine phosphatases. The buffer system and all reagents apart from the enzyme were the same as for the lambda phosphatase, except that most eukaryotic phosphatases do not require manganese, which was therefore omitted. Again, this is a homogeneous, one-step assay method, which yields a result in 5 min or less. The results appear in Fig. 5.17. Again the background signal is negligible. No saturation is observed in this experiment, although an assay with the T-cell protein tyrosine phosphatase (227) using the same reagent cocktails yielded a K_d for this substrate (by nonlinear regression to the Michaelis–Menten equation) of 22 μM, in reasonable agreement with the literature value.

Calcineurin is a phosphatase of an entirely different type, with complex kinetics and cofactor requirements (285). These characteristics present a challenge to the would-be developer of a rapid, homogeneous assay. In the author's hands, it proved impossible

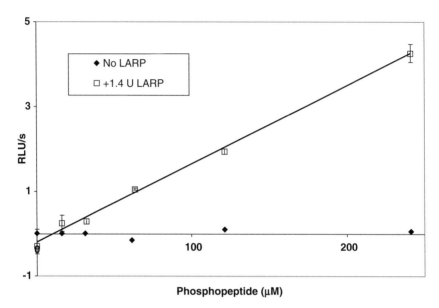

FIGURE 5.17   CB assay of leukocyte antigen-related phosphatase activity. The Fischer peptide substrate (see text) was employed. No saturation is seen in this concentration range. The background rate without enzyme and substrate was subtracted from all data. Data from the first 3 min of the reaction were taken for reduction by nonlinear regression.

to obtain a satisfactory readout using a one-step assay in 3–5 min, probably because of the autoactivation feature of the enzyme. However, a satisfactory level of activity was observed during the interval from 7–12 min, as shown in Fig. 5.18. The rate of luminance increase derived by linear regression is much lower than that seen with other phosphatases; however, fortunately, the background signal is also very low, and there is no doubt about the statistical significance of the results. Figure 5.19 demonstrates inhibition of calcineurin by the autoinhibitory peptide (AIP) (286). This experiment was performed as a two-step assay to allow the calcineurin to autoactivate prior to measurement, and the concentrations of the G3PDH and PGK coupling enzymes were slightly increased. The entire assay run required about 90 min (see figure legend for details). The amount of calcineurin required for this assay is approximately 6% of the quantity typically recommended by vendors for assays employing malachite green.

In summary, it has proved possible to develop one-step homogeneous CB assays for the four phosphatases tested, despite their various characteristics. However, some phosphatases require separate assay development, and in certain cases it may be desirable to compromise by adopting a time-point-based approach. This may hold even more strongly for assays of phosphatases requiring special pH, ionic, or other conditions under which the GPL process does not run efficiently. In these cases, conditions may need to be adjusted following the phosphatase reaction to obtain the CB readout. Nevertheless, the general observation that CB methods can yield highly

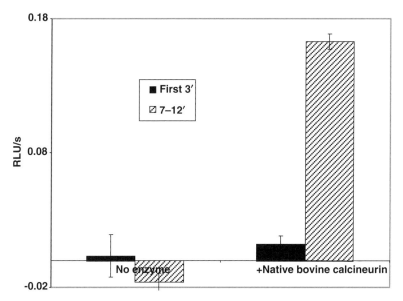

FIGURE 5.18   Autoactivation of calcineurin observed in real-time, one-step reaction with CB assay. Recombinant human calcineurin coexpressed subunits were assayed in a buffer containing calmodulin and $MnCl_2$ as recommended by the supplier (CalBiochem). The substrate was 150 μM RII peptide (DLDVPIPGRFDRRV-pS-VAAE). No activity was seen during the first 3 min, but the enzyme autoactivated rapidly after that point. The absolute luminance readings are low but readily distinguished from background.

rapid, sensitive, and operationally simple assays for a broad range of phosphatases is certainly encouraging.

5.4.3   Challenges in the Development of the GPL-Coupled Bioluminescent System for Phosphate Detection

Reactions in which multiple species participate are often candidates for incorporation into assays of multiple analytes, although it is not always obvious which reactants are suitable or how to exploit the scheme usefully. For example, the reaction series employed in this CB assay is identical to the one used for the cytotoxicity assays discussed in Chatper 3, which are actually G3PDH assays. However, the problems in assay development are so different that the two assay methods have followed entirely distinct paths. This may provide an interesting example of the effects of specific details on the management and outcome of an assay development project. The central difficulty is that to measure free phosphate effectively, one must operate in the virtual absence of extraneous phosphate. All the developers and providers of phosphate detection kits described in this chapter are of course aware of this problem yet in the case of the GPL phosphate assay, the issue presents a greater challenge, because the energy-charged (and therefore somewhat unstable) substrate G3P yields free phosphate upon

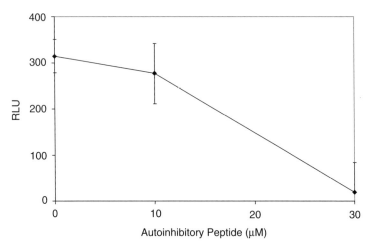

FIGURE 5.19   CB measurement of inhibition of calcineurin by autoinhibitory peptide. The enzyme was first incubated without substrate, with or without AIP for 15 min at room temperature. The substrate was then added and the reaction incubated for 67 min at 30°C. Subsequently, 5-μL aliquots were transferred to 45 μL of calcineurin CB reaction cocktail (112). Single-point luminance was read after 3 min.

hydrolysis. This is an inherent difficulty that can be addressed and minimized, but cannot be directly circumvented in this system. (ADP, which is also a supplied reagent, can be a source of free phosphate as well, but ADP is much more stable to spontaneous hydrolysis than G3P.) G3P is not the only challenge to be addressed. The great majority of the effort involved in the development of this assay has gone into elimination of sources of free phosphate and enzymes that adventitiously produce it.

5.4.3.1   *Handling of Glyceraldehyde-3-Phosphate*   Glyceraldehyde-3-phosphate is susceptible to hydrolysis in aqueous solution at room temperature. The spontaneous hydrolytic rate has been estimated at 2% per hour at neutral pH. However, the rate of this base-catalyzed reaction is considerably reduced at acidic pH. Of course, it is possible to go too low in pH and observe acid-catalyzed hydrolysis, but evidently a pH of 2–3 is favorable for the stability of this molecule. Clearly, this acidic regime is not compatible with the enzymes needed for the reaction; so to ensure the lowest possible level of contaminating phosphate from this source, G3P should be stored appropriately and mixed with other components as close in time to reaction initiation as possible.

G3P should also be assayed for free-phosphate content upon receipt, and, if a reagent lot is being produced, upon use as well. The GPL assay is not suitable for this purpose, since in most formulations G3P is a partially limiting reagent, so that both G3P and contaminating phosphate will increase the signal. (Usually, G3P is partially limiting because of the possibility of phosphate contamination from its hydrolysis,

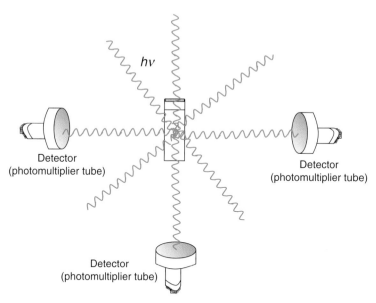

FIGURE 1.1 Schematic representation of autoluminescence. Light (hν) is emitted equally in all directions by the luminous sample, impinging on one or more detectors, which may be photomultiplier tubes, charge-coupled devices, or other types. No lamp is required.

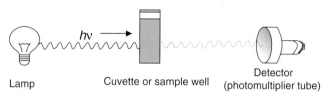

FIGURE 1.2 Schematic representation of spectrophotometry. The lamp illuminates the sample, which absorbs a portion of the light, depending on the concentration of the absorbing species. The remaining light is transmitted through the sample vessel and impinges on the detector. The full intensity of the lamp is known separately from either careful calibration or beam-splitting. The fraction of light remaining indicates the concentration to be determined, assuming the absorbance coefficient is known. Multiple concentrations may be measured simultaneously if all absorbance coefficients are known, data from multiple wavelengths are collected, and independent spectra of the absorbing substances are available.

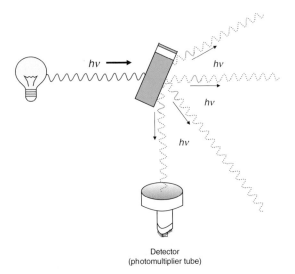

FIGURE 1.3 Schematic representation of fluorometry. Illumination by the lamp in a specific range of wavelengths is absorbed as electronic excitation by sample molecules and reemitted at longer wavelengths. Emission is in all directions, but to avoid receiving light from the lamp, the detector is mounted at an angle to the incident beam. The detector is also tuned by filters or a monochromator to accept only emitted wavelengths.

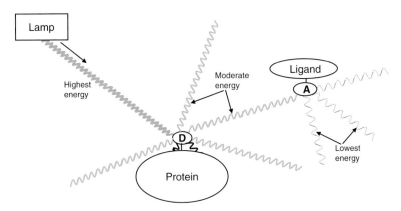

FIGURE 1.4 Schematic representation of FRET. Low-wavelength (high-energy) light from the lamp excites the donor fluorophore (D), which reemits light of moderate wavelength and energy. The acceptor (A) fluorophore is not excited by light at the wavelength of the lamp emission, but can accept fluorescent energy from the donor. After doing so, it in turn reemits light of a still longer wavelength (lowest energy). This process occurs only when the donor and acceptor are proximate, leading to emission at the lowest expected wavelength; when they are distant, only light of the moderate wavelength is observed.

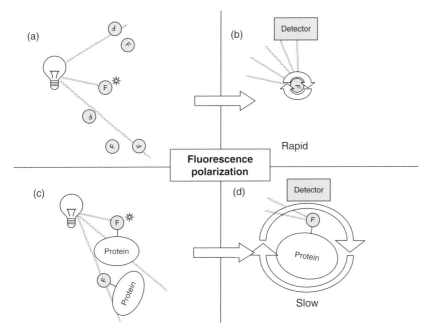

FIGURE 1.5 Schematic representation of fluorescence polarization. The unbound fluorophore of Panel (a) is excited by polarized light from the lamp only if its excitation dipole is properly aligned. Other fluorophores are not excited. (b) The excited fluorophore rotates rapidly because of its small size, emitting light that reaches the detector, which is mounted at a 90° angle from the lamp beam. (c) The protein-bound fluorophore is also excited by the polarized lamp beam only if its dipole is properly aligned. The protein itself and other fluorophores are unaffected. (d) Because the rotation of the complex is slow, the fluorescence decays before the fluorophores dipole is aligned with the detector. Little or no fluorescence is seen. Thus, the fluorescence reading depends on the size of the fluorescing complex.

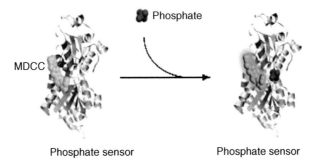

FIGURE 5.5 Space-filling representation of operation of the Invitrogen Phosphate Sensor. The sensor is a phosphate-binding protein from *E. coli* modified with a fluorophore adjacent to the site of phosphate association. The blue fluorescence increases dramatically upon binding. Courtesy of Invitrogen Corporation.

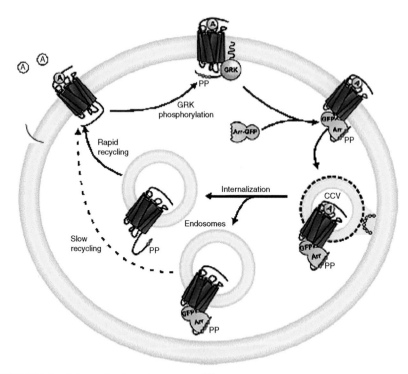

FIGURE 9.1 Schematic view of TransFluor™ GPCR assay technology. Used by permission of MDS Analytical Technologies.

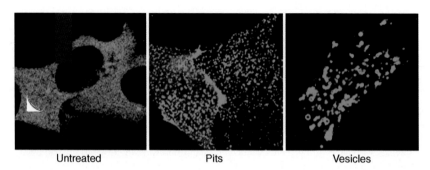

| Untreated | Pits | Vesicles |

FIGURE 9.2 Imaging of cells with stimulated GPCR visualized by TransFluor technology. Used by permission of MDS Analytical Technologies.

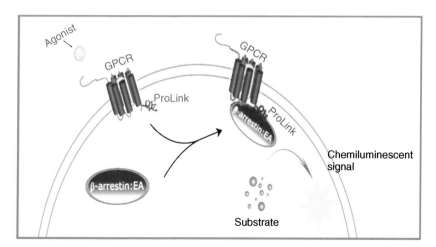

FIGURE 9.3 Assay of GPCR activation by enzyme fragment complementation (EFC). Separate complementary fragments of β-galactosidase are linked to the GPCR and β2-arrestin proteins. Association of the two conjugated proteins brings the fragments into proximity and stimulates the enzymatic activity of the galactosidase, yielding a chemiluminescent signal. Used by permission of DiscoveRx Corporation.

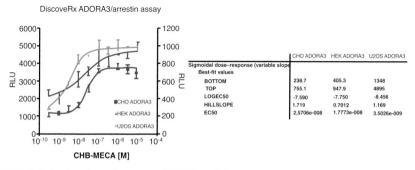

FIGURE 9.4 Sensitive detection of GPCR activity by EFC assay. The receptor in question, ADORA3, had previously been reported not to interact significantly with β2-arrestin (476). Used by permission of DiscoveRx Corporation.

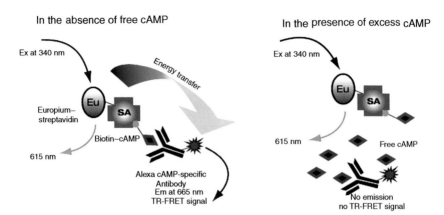

In the absence of free cAMP

Ex at 340 nm

Energy transfer

Europium–
streptavidin

Biotin–cAMP

615 nm

Alexa cAMP-specific
Antibody
Em at 665 nm
TR-FRET signal

In the presence of excess cAMP

Ex at 340 nm

615 nm

Free cAMP

No emission
no TR-FRET signal

FIGURE 9.6 Assay principle of lanthanide chelate excitation or LANCE. cAMP, the molecule under test, is conjugated with the Eu^{3+} donor fluorophore via a biotin linkage. A cAMP-specific antibody conjugated with the acceptor fluorophore associates with the labeled cAMP molecule except in the presence of free cAMP from the process under test (e.g., GPCR activation). Thus, sample cAMP separates the fluorophores and decreases the TR-FRET signal. Used with permission from Perkin Elmer, Inc.

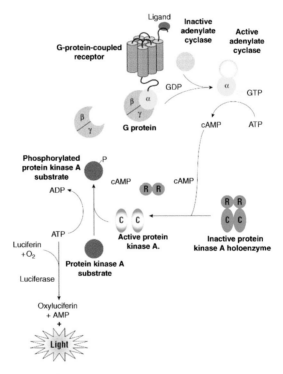

FIGURE 9.8 Schematic representation of CB assay of cAMP concentration using cAMP-Glo™. Productive engagement of the GPCR and the associated rise in cAMP levels cause activation of kit-supplied protein kinase A and consumption of ATP. ATP levels are then measured by Kinase-Glo™ (see Chapter 4), yielding a negative signal relative to cAMP concentration. This extremely complex series of steps is nevertheless robust, showing the potential power of CB reaction series. Used by permission of Promega Corporation.

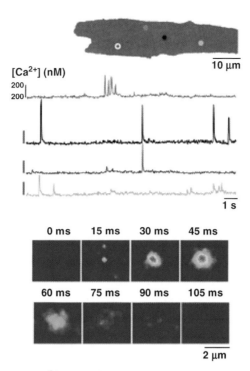

FIGURE 11.2   Bursts of Ca^{2+} release in rat ventricular myocyte labeled with fluo- 4 AM. The outline of a single cell is shown. Courtesy of Invitrogen Corporation.

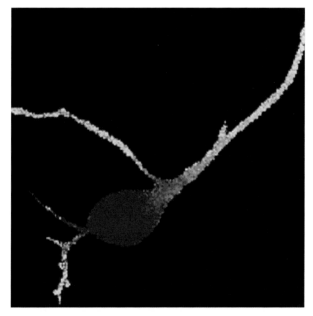

FIGURE 11.4   False-color image of concentration of Ca^{2+} in Purkinje neuron from embryonic mouse cerebellum. Courtesy of Invitrogen Corporation.

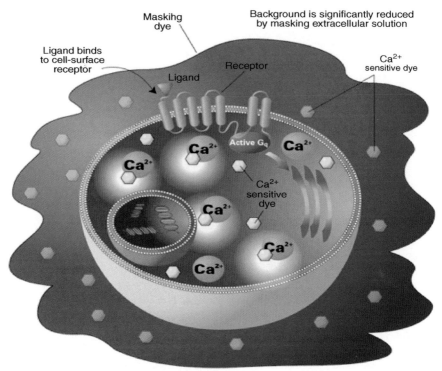

Maskihg dye

Ligand binds to cell-surface receptor

Ligand

Receptor

Background is significantly reduced by masking extracellular solution

Ca^{2+} sensitive dye

Ca^{2+}

Active G_q

Ca^{2+}

Ca^{2+}

Ca^{2+} sensitive dye

Ca^{2+}

Ca^{2+}

Ca^{2+}

A next-generation no-wash, homogeneous, calcium mobilization assay designed for the most challenging and diffecult GPCR targets

Lower background and increased signal intensity
Significant reduction of ligand or target interference

FIGURE 11.5 Schematic view of the events and strategies underlying the homogeneous FLIPR™ Calcium-4 Assay. The key to the no-wash format is the masking dye that suppresses the fluorescence of the dye molecules outside the cell membrane. Used by permission of MDS Analytical Technologies.

which motivates the assay developer to optimize its concentration at an intermediate value.) The malachite green assay (287) is a good choice to measure contamination of G3P by free phosphate. In the author's experience, the phosphate contamination level upon receipt of G3P from the vendor is as high as 13%. This level does not present a difficulty in the cytotoxicity assay and other CB methods employing this reaction chain, but for phosphate assays it is unacceptable.

Fortunately, a convenient and effective way of purifying phosphate away from G3P has been published (288). This system is capable of eliminating some 90% of the phosphate. However, because of the extreme sensitivity of the reaction series, even this level of free phosphate may be too high to allow full exploitation of the assay's potential. A second round of purification may be desirable in some cases, combined with careful handling of the reagent.

5.4.3.2 Spontaneous Hydrolysis of ADP   Although hypothetically the ADP constituent could hydrolyze spontaneously and cause similar problems, this reaction is so slow that the problem is negligible. Indeed, contamination of ADP by ATP after purification by the vendor is much more of a problem, as is the case in the cytotoxicity assay.

5.4.3.3 Buffer Conditions   To reiterate, one major decision point in choosing a phosphatase assay is whether to assess the modified product or the liberated phosphate. One of the important advantages of the latter method is of course applicability to any phosphate reaction. However, some assays for free phosphate require conditions that are incompatible with the phosphatase activity, rendering a continuous assay impossible. The advantages of continuous assays include the possibility of extending a run with a low signal, the ability to draw better conclusions from a large data set, and the potential to recover from a technical error; but these advantages are not always critical. However, in keeping with the goal of "a rapid, one-step, homogeneous assay for every enzyme," it has generally been considered worthwhile to develop assays in which both the test enzyme and all the detection reagents can be mixed and the reaction progress can be monitored in real time. For one thing, the sampling step is saved.

5.4.4   Potential Drawbacks of GPL Phosphatase Assays

The simultaneous reaction series used in this method is relatively complex, increasing the range of possible failure modes. If one is screening for inhibitors, false hits may be generated by compounds that inhibit an enzyme on the light generation pathway other than the test enzyme. This problem is ameliorated, but not eliminated, by the fact that all the enzymes are phosphate-processing enzymes; thus molecules that mimic phosphate and thereby inhibit luciferase, for example, may also inhibit a phosphatase, but not necessarily to a useful degree. Moreover, the problem is worse than the corresponding problem with kinases, since kinases process ATP, as does luciferase; thus the inhibitors exhibit a considerable overlap. There is no reason for phosphatases to have binding sites capable of accommodating ATP. Thus, ATP mimics that inhibit luciferase or PGK may yield false hits that have to be eliminated in secondary screens. Of course, the problem is minimized if the worker is aware of the nature of his or her library and the GPL reaction sequence.

Phosphate contamination of buffers and reagents has already been mentioned as a potential problem. Clearly, ATPases (which are actually phosphatases) and other such enzymes will create a dynamic background signal if present, as will adenylate kinase directly, by synthesizing ATP (and AMP) from two ADP molecules. These problems can generally be narrowed down by swapping reagents, but solving them may involve additional purification and testing steps.

Connected with this is the issue of dynamic range. Given ultrapure reagents, the dynamic range of the assay's response to phosphate is probably limited only by the response of luciferase itself, which is extremely broad (289). However, the caveat is not without meaning, and most CB assays are limited in their range by impurities in one or more reagents, though the range may still be very great. These issues have likely been the main hindrances to the development of a commercial CB phosphatase assay to date, although it is possible that one will be available before this book is printed. One may speculate that the CB field will grow even more rapidly as purification and detection techniques improve.

5.5   CONCLUSIONS

Like other areas discussed in this book, the field of phosphatase assays is far too crowded and complex to allow one to specify the best assay method or even the best strategy for choosing a method. However, one should begin by understanding one's application and eliminating the impossible or undesirable. Some assays may cost too much, despite their elegance. Procedures suitable for alkaline and acid phosphatases may not work, or may not be competitive, in other pH regimes. The available equipment and reagents necessarily present in one's library represent additional constraints. From the point of view of cutting-edge technologies, it is possible to give the nod to several of the fluorometric contenders, including the attractive superquenching approach, as well as novel fluorogenic substrates that also carry the advantages of conceptual and experimental simplicity. CB methods are also on the way, but they have not yet arrived at commercial success.

6

ACETYLCHOLINESTERASE

6.1 INTRODUCTION

Acetylcholinesterase (AChE) is the enzyme that hydrolyzes acetylcholine (ACh), arguably the body's most important neurotransmitter, thereby permitting decay of the ACh-dependent signal and relaxation of the muscle or other stimulated system (for reviews see References 290, 291). This critical enzyme is a target of therapeutic drug development in conditions such as Alzheimer's dementia (292, 293), multiple sclerosis (294), schizophrenia (295), Parkinson's disease (296), and other neurodegenerative conditions. However, AChE is also the target of the so-called "nerve agents," primarily organophosphate compounds such as sarin and VX, as well as a large class of organophosphate pesticides that were designed to inhibit insect but not mammalian AChE. Assays of AChE activity are therefore of interest in medicine, environmental testing, and military and antiterrorism activities.

The reaction catalyzed by AChE is shown schematically in Fig. 6.1. ACh is an ester formed by the reaction of the ubiquitous acetate ion with the hydroxyl moiety of the quaternary amine choline. The catalytic mechanism of AChE has been studied extensively (297, 298). Mixed-mode inhibitors are observed (299), with supporting information based on knowledge of the active site geometry from X-ray crystallography; thus, the inhibition kinetics of the enzyme can be complex. The presence of a covalently bound intermediate in the reaction pathway also lends complexity, since breakdown of this intermediate can be rate limiting in the process, rendering the enzyme's initial turnover qualitatively different from subsequent turnovers.

Coupled Bioluminescent Assays: Methods and Applications, Michael J. Corey
Copyright © 2009 by John Wiley & Sons Inc.

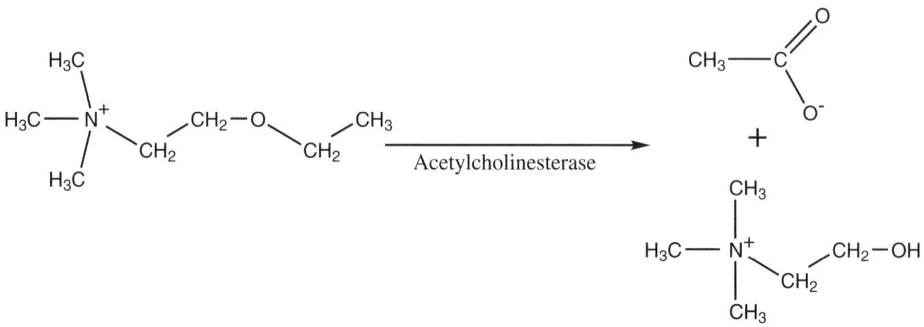

FIGURE 6.1 Schematic depiction of the reaction mechanism of acetylcholinesterase. The enzyme cleaves the important neurotransmitter acetylcholine at the ester bond into acetate and choline.

The medical significance of AChE as a target is a complex picture. In certain conditions, such as certain types of myasthenia gravis, AChE inhibition is the first-line therapy (300), but treatment must be individualized because the autoimmune nature of the condition can be caused by various types of antibodies against a plethora of targets in the cholinergic system (301). In other cases, inhibition of AChE can be regarded as a partial "fix" for conditions actually due to lesions in other biochemical systems. For example, few would hypothesize that excess AChE activity is the fundamental cause of Alzheimer's disease, yet it is true that acetylcholine levels are often depressed in Alzheimer's patients (a situation known as "cholinergic deficit"), which may involve reduced levels of the synthetic enzyme choline acyltransferase. However, AChE activity is also generally depressed, complicating our understanding of the condition (302). The odd conclusion is that the use of AChE inhibitors (303), of which donepezil (among many hundreds of candidates) is currently favored, appears to ameliorate a condition in which AChE activity is already abnormally low. Donepezil, however, may have other modes of action (304). A related situation is evident in Parkinson's disease in which the cholinergic deficit, as determined by *in vivo* PET studies, may be even more severe than in Alzheimer's (305). Brain damage caused by events such as ischemia, whether experimental or pathology related, may also be treated with AChE inhibitors (306).

AChE is also an important target of researchers with very different goals. Both nerve agents and pesticides of the organophosphate class bind covalently to the AChE active site (exhibiting extreme sensitivity to active site geometry (307)), blocking its cognate reaction with the natural ACh or other substrates. The best characterized nerve agents are tabun (ethyl dimethylamidocyanophosphate), sarin (isopropyl methylphosphonofluoridate), soman (pinacolyl methylphosphonofluoridate), and the very deadly VX (*O*-ethyl *S*-(2-diisopropylaminoethyl) methylphosphonothiolate). Ingestion and respiratory or dermal exposure to these compounds even in very small quantities are potentially lethal. Exposure to many of the organophosphate pesticides can cause similar symptoms, and can also be fatal. The toxicity of a large panel of organophosphate pesticides has been tested in rabbits; as of this writing, these data are available from the University of Florida IFAS extension Web site at http://edis.ifas.ufl.edu/PI087#TABLE_1. Toxicity of many of these compounds, both nerve agents and pesticides, can be counteracted by the use of the antidote atropine (308), which blocks muscarinic ACh receptors and thereby suppresses the effects of excessive ACh accumulation. Another class of antidotes includes obidoxime and pralodioxime chloride or HI-6 (309), which associate strongly with the phosphate moieties covalently bonded to AChE, removing them and restoring enzymatic activity. Unfortunately, oxime-resistant nerve gases such as Novichok (newcomer) were developed in Russia in the 1980s. However, the fascinating countermeasure of pretreatment with large amounts of butyrylcholinesterase, which is able to absorb the nerve agents and protect AChE, is now approaching clinical evaluation (310). Clearly, this complex picture of potentially lethal chemical agents, possible antidotes with differential activities, and antidote-resistant toxic compounds warrants the development of a wide array of rapid, sensitive assays, and detection methods.

However, the methods that have been used to assess AChE activity to date are less than ideal in various ways. We will look at established assays and novel methods before turning to the coupled bioluminescent alternatives.

6.2   ESTABLISHED AChE ASSAY METHODS

Ellman published the first modern assay method for AChE in 1961 (311), and the impact of this advance is still felt. The assay scheme uses not the natural substrate ACh, but a synthetic analog, acetylthiocholine (ATCh), a thioester (i.e., the alcoholic oxygen has been replaced with sulfur). This modification permits quantification of the thiol product of the hydrolysis by the use of dithionitrobenzene (DTNB), which is also known, appropriately, as Ellman's reagent. ATCh is an efficient substrate for AChE and appears to recapitulate many of the properties of the natural substrate (312, 313), although some perturbations of AChE activity by the DTNB reagent have been observed (314), leading many workers to employ other thiol sensors in these assays. However, any deviations from normal enzymatic behavior due to the unnatural substrate have not prevented various researchers from completing diverse kinetics studies using ATCh (315, 316). Interestingly, the reaction with ATCh is so rapid that the use of AChE with the ATCh substrate has been patented (104) as a labeling system for immunoassays in direct competition with such standbys as horseradish peroxidase and alkaline phosphatase, which are asserted by the AChE proponents to be slower and less stable. Immunolabeling kits based on this concept are marketed by Cayman Chemical as ACE^{TM}.

Despite the simplicity and convenience of the ATCh assay, enzymologists can rarely resist the impulse to develop methods that employ the natural substrate, and such has been the case with AChE. The use of radiolabeled ACh has been reported (317). A chemiluminescent assay employing ACh was described as early as 1981 (318, 319). This is a coupled assay in which evolved choline reacts with choline oxidase, yielding hydrogen peroxide along with the side product betaine. An enzyme known as horseradish peroxidase then yields a luminescent signal in a reaction involving luminol. Although the disadvantages of using luminol have probably adversely impacted the widespread adoption of these methods, efforts continue in the development of improved chemiluminescent approaches, for example, by using other molecules in place of luminol (320). Ironically, one of the latest approaches goes back to the ATCh substrate, while employing luminol with ferricyanide in a flow analysis system (321). The inherent instability and structural constraints on chemiluminescent substrates continue to hinder progress in these areas.

6.3   RECENT DEVELOPMENTS IN AChE ASSAY METHODS

A substantial portion of the modern armamentarium of assay techniques has been brought to bear on the AChE problem. One interesting approach that is growing in popularity not just for AChE, but for many enzymes without an obvious photometric or fluorometric solution to the assay challenge is to use mass spectrometry (MS) to detect the products of the enzymatic reaction. MS, which began as the ultimate low-throughput, high-resolution tool for molecular analysis, has been the subject of such intense engineering efforts that it can now legitimately be considered a "high-throughput" method, at least if the requisite expertise is present and the effort of establishing the assay has been expended. In 2004, an MS method was announced for screening of AChE inhibitors with an assay time of 4–5 s per sample (322). This implies a read time for 100 samples of some 8 min, much longer than the read times of 96-well luminometer plates, but considerably more rapid than many antibody-based reactions. Unfortunately, the serial nature of MS does not allow throughput performance optimization through further miniaturization, at least with currently available technology. However, like CB technology, MS assays can use natural substrates and do not require labeling steps. Although it is conceivable that interference by molecules or fragments of adventitiously similar molecular weights could occur, in practice this is not likely to be a problem with an experienced operator who is trained to check multiple ions derived from the putative species observed. The major drawbacks of using MS to measure enzymatic action are the expense of the required equipment, the level of personnel training and specialization required, and the knotty issue of throughput.

Sensors that incorporate immobilized AChE may have an important role to play in environmental detection of nerve agents and other AChE inhibitors (see Chapter 15 for more detail on these applications). The biochemical reaction underlying these devices is the same as in the chemiluminescent assay mentioned above: evolved choline enters the reaction catalyzed by choline oxidase, yielding hydrogen peroxide. In the sensor, however, the hydrogen peroxide undergoes a further reaction at the electrode surface, which generates an amperometric signal (323, 324). These devices are able to employ the natural substrate, ACh; if ATCh is used, the choline oxidase component is no longer necessary, because the hydrolysis product thiocholine itself is able to generate a signal upon oxidation at the electrode. It should be noted that while these instruments hold a great deal of promise, actual experience with their performance, reliability, and interference characteristics under environmental conditions is quite limited to date.

Still, amperometric systems with immobilized AChE have been chosen not only for detection of AChE inhibitors, but also for direct quantification of ACh itself (325).

The "lab-on-a-chip" principle, based on work over a decade ago to develop on-chip enzymatic assays (326), has also been applied to AChE (327), motivated in part by the evident need for rapid and highly reliable means of detecting inhibitors with minimal equipment. The reaction series and on-chip separations described in Reference 327 are complicated, leading one to suspect that development of these chip-based assays is a nontrivial exercise in biochemistry, fluid dynamics, and engineering: "Transport and mixing of the reagents occurred by a combination of electroosmosis and electrophoresis using computer-controlled electrokinetic transport." ATCh is used as the substrate. Evolved thiocholine reacts with eosinmaleimide, yielding a thioether that undergoes further on-chip electrophoretic separation and is finally detected by laser-induced fluorescence. The detection time was in the range of 5–8 min, which is competitive with CB assays and superior to most other methods. Detection of VX was performed by off-chip mixing with the enzyme, followed by introduction of the mixture onto the chip. This may be an indication that there is some doubt as to the ability of the chip system to capture environmental inhibitors; however, one must give the researchers the benefit of the doubt in this case. The author is among those who would be reluctant to perform any sort of biochemical assay at all with VX in the air.

6.4   EVALUATION OF CURRENT AChE HIGH-THROUGHPUT SCREENING METHODS

In analyzing high-throughput technologies for screening AChE reactions, we will address first the simple Ellman reaction, assessed by visual spectroscopy. The sensitivity of this method is of course quite limited, requiring dozens or hundreds of units of enzyme and, more importantly, roughly equivalent amounts of candidate inhibitors. At these reagent levels, the reaction is rapid and a 10-min incubation will often be sufficient; this compares favorably, for example, with any antibody-based method. The substrate is the unnatural ATCh, which may lead to complications in interpreting inhibition data, although ATCh has proven to be "less" unnatural than many other designed substrates and has a substantial body of data supporting its use, as described above. Finally, however, the low sensitivity and high reagent requirements of the assay have not induced the vendors to provide it cheaply. The list price of Ellman assay kits for high-throughput screening approaches a princely $3 per well (Bioassay Systems). Purchasing the individual reagents in bulk, for example, ATCh for under $10 g^{-1} and Ellman's reagent for under $15/g^{-1} (both list prices; MP Biomedicals), and optimizing the straightforward assay on one's own may prove to be more economical.

Researchers whose compound libraries are too precious to expend at this rate would like to consider other alternatives. The MS methods mentioned above (322) are roughly competitive with the direct photometric method in terms of throughput, and far superior in sensitivity. The expense of acquiring the training and expertise to use these methods may be more than justified if the organization envisions a large-scale MS operation, in which a range of enzymatic and other processes are amenable to quantification by MS. Since virtually all biochemical reactions of interest involve changes in molecular weights, this approach can lead to development of a core facility with a broad set of capabilities.

The chemiluminescent alternative must be recognized as a serious competitor for both conventional and CB assays. A flow-injection system was described in a recent article (328). Resolution of ACh and choline was determined at a flow rate of 0.1 mL/min and deteriorated by >50% upon doubling of the rate, followed by a slower asymptotic decay. Linearity was claimed from 1 to 3000 nM of each molecule, indicating excellent sensitivity. Throughput was established as 25 samples per minute. This is much lower than what is seen in typical CB assays, but one must consider the possibility that novel designs will allow analysis of multiple reaction streams in parallel at reasonable cost. However, the authors also commented that a more stable immobilized form of AChE was desirable for true, extended high-throughput operations. Of course, conventional microplate-based chemiluminescent assays are also a useful means of measuring AChE activity, though less relevant to a rapid-detection or real-time scenario.

6.5   COUPLED BIOLUMINESCENT ASSAYS OF AChE ACTIVITY

Schemes for coupling the hydrolytic activity of AChE to generation of ATP can be conceived, but have not performed well to date, at least in the author's experience. The only CB AChE assay currently on the market relies instead on the consumption of ATP by processes subsequent to the activity of the enzyme. The methods are described in a pending U.S. patent application (329) and an assay kit is currently marketed as aCella AChE^{TM}. The performance characteristics of aCella-TOX are described as of this writing at the Web site of Cell Technology, Inc. (www.celltechnology.com). Very simply, the reaction series involves production of acetate and choline by AChE, using the natural substrate, followed by ATP consumption catalyzed by any of a choice of coupling enzymes and luciferase-dependent quantification of the remaining ATP. While yielding a negative signal, this approach enjoys the practical advantage in threat detection situations that danger, in the form of an inhibitor of AChE, is indicated by light generation, which may be more easily noticed than the absence of light.

Figure 6.2 shows the results of a titration of AChE with tacrine, a mixed-mode inhibitor of moderate strength that has been used clinically in Alzheimer's dementia. Of note is the fact that the reaction was incubated for only 15 min, followed by a 2-min read. It is also possible to mix the reagents and read immediately, although the data from the first few time points are likely to be inconsistent unless an automated injector is used (see Chapter 2). Operationally the procedure is simple, involving only the mixing of the test reagents, the enzyme aliquot, and the light generation materials. Although the K_i of the inhibitor is not readily apparent from this data set, the author has found the procedure to be sensitive to subpicomole quantities of tacrine, using an unoptimized, precommercial formulation.

Figure 6.3 depicts the inhibition of AChE by the important organophosphate pesticide and environmental contaminant malathion as measured by the same CB assay.

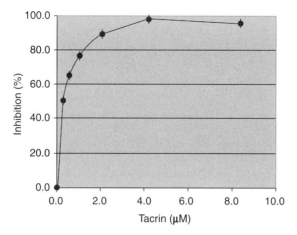

FIGURE 6.2 Titration of acetylcholinesterase with mixed-mode inhibitor tacrine (9-amino-1,2,3,4-tetrahydroacridine). Enzyme and acetylcholine (the natural substrate) were incubated for 5 min, followed by addition of CB reaction cocktail, a further 2-min incubation, and reading of luminance. Used with permission from Cell Technology, Inc., Mountain View, CA.

Again, the entire procedure is complete in 20 min and the sensitivity is good. However, the method yields a negative signal, in that ATP is consumed in the presence of AChE activity, leading to a decrease in evolved luminance. Assays of this type can run into statistical problems of the kind described in Chapter 4. However, in the case of AChE, special considerations apply, since inhibitors of this enzyme can be deadly as well as therapeutic. As discussed in Chapter 15, a positive signal indicating the

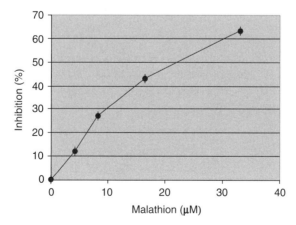

FIGURE 6.3 Titration of AChE with the common organophosphate pesticide malathion. Enzyme, substrate, and inhibitor (malathion) were incubated for 15 min, after which CB reagents were added and the luminance read after a further 2.5 min. Used with permission of Cell Technology, Inc.

presence of an inhibitor may prove to be a desirable outcome in a situation in which nerve gases or dangerous pesticide residues may be present.

6.6   COMPARISON OF COUPLED BIOLUMINESCENT AND OTHER METHODS OF MEASURING AChE ACTIVITY

It is too early to reach firm conclusions about the relative value of the various AChE assay strategies. CB assays, MS, chemiluminescent approaches, fluorometry, and even colorimetric methods may all have roles to play in high-throughput screening for AChE activity and inhibitors. At the moment, it appears that CB methods can deliver at least the highest potential throughput because of their excellent speed and sensitivity and because of the possibility, due to the robust light signal, of microminiaturization to the limit of the equipment. However, strategies based on MS and chemiluminescent substrates will undoubtedly undergo further technical refinements in the near future.

7

MEASUREMENT OF NITRIC OXIDE SYNTHASE ACTIVITY BY COUPLED BIOLUMINESCENCE

7.1 INTRODUCTION

In the late 1980s, the fundamental discovery was made that nitric oxide (NO) mediates the activity of endothelium-derived relaxing factor (330), accompanied by the discovery of the arginine/NO pathway and its biological significance (331). Although NO can arise through other reactions, as described in this chapter, the primary means of biological synthesis of NO is catalysis by the nitric oxide synthase (NOS) enzymes. These enzymes are critical in many areas of medicine and drug development; accordingly, intensive efforts have gone into the development of assay methods for NOS, but the results are still far from ideal, largely because the product of primary biological interest, NO itself, is highly labile.

This chapter begins with an introduction to the biology of this system sufficient to help the reader make sense of the assay methods that follow. We begin with the enzymes themselves.

7.1.1 Nitric Oxide Synthases

The NOS enzymes catalyze the reaction depicted in Fig. 7.1 (332). Continuous NOS activity requires the cofactors FMN, FAD, and tetrahydrobiopterin, in addition to calcium-harboring calmodulin. The enzymatic reaction does not require nitrate or nitrite as a nitrogen source; the nitrogen is instead provided by the common amino acid L-arginine. There are two major classes of NOS, known in most of the literature as "constitutive" NOS (cNOS) and "inducible" NOS (iNOS). cNOS, despite

Coupled Bioluminescent Assays: Methods and Applications, Michael J. Corey
Copyright © 2009 by John Wiley & Sons Inc.

its epithet (which relates to its constant expression, rather than its activity), is sensitive to the presence of free calcium ions, while synthesis of iNOS, which incorporates calmodulin-containing calcium as part of its inherent structure, is stimulated by interferon-γ and lipopolysaccharides and inhibited by glucocorticoids (333). In addition to these well-established phenomena, NOS activity is regulated by an impressive range of proven (334–336) and putative (337, 338) mechanisms. Owing to the importance of NO signaling in many biological systems, the NOS enzymes are regarded as a potential therapeutic target in a number of pathologies, including pulmonary disease (339), diabetes (340), and gastric injury (341), as well as a range of conditions related to septic shock, neurodegeneration, and inflammation (342).

The reaction scheme depicted in Fig. 7.1 does not suggest any facile method of assaying the enzymatic activity of NOS. Conceivably, one could follow the oxidation of NADPH by spectrophotometry, although the limited sensitivity of this method may not be compatible with the achievable enzymatic rates in question. (However, production of NADP$^+$ is in fact exploited in a rather complex coupled assay scheme involving cycling amplification (343)). A radiometric assay utilizes the conversion

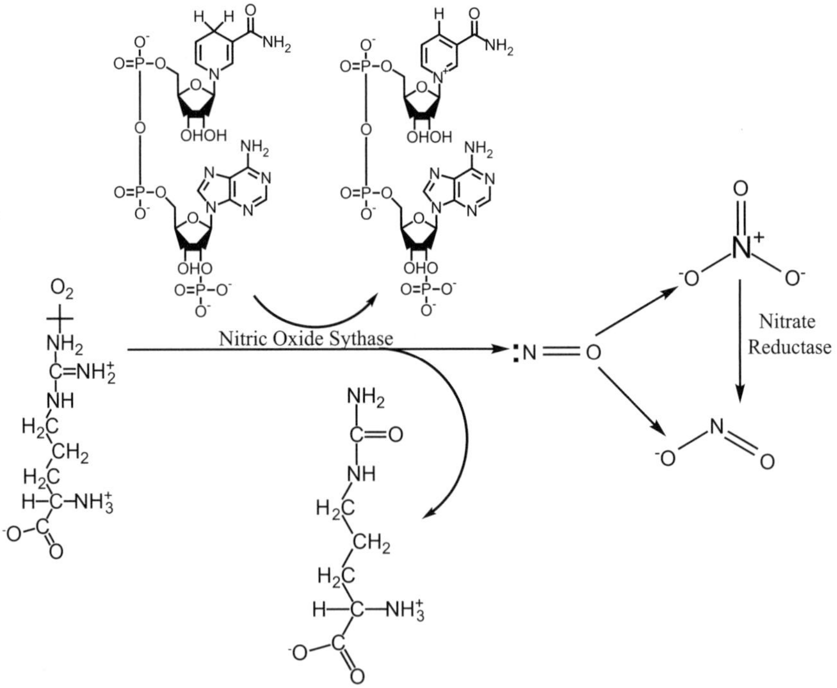

FIGURE 7.1 Reaction scheme of nitric oxide synthase and potential fates of nitric oxide. Arginine is oxidized to citrulline with the participation of the NADPH cofactor (top of scheme) generating the reactive signaling molecule nitric oxide. This is further oxidized to nitrate (NO_3^-) or nitrite (NO_2^-). In quantification schemes, but not *in vivo*, these ions can be readily interconverted by nitrate reductase.

by the enzyme of tritiated arginine to citrulline (344), and kits employing this principle are still sold by various vendors, including Stratagene, Cayman Chemical, and CalBiochem. The claimed sensitivity of this method is very high, although there have been reports of difficulties in achieving this level of sensitivity in practice. In any case, most assays of NOS activity in current use involve detection of nitrite and/or nitrate resulting from oxidation of NO. This approach has the advantage of addressing the only known modulator of the biological effects of the enzyme, albeit indirectly.

7.1.2   The Nitrate Ion in Medicine and NO Biology

Once thought of mainly as a fertilizer constituent, the nitrate ion has proven to be important in vertebrate biology, largely because of its relationship with nitric oxide, which plays a critical role in vascular control. The very nature of NO as a biologically important yet unstable free radical suggests certain aspects of its function: some of the NO molecules generated by processes described below diffuse and exert effects on smooth muscle, but most are destroyed before executing any signaling function. Thus, the transient nature of the species renders direct assays for its presence generally impractical. The generation of NO is therefore almost invariably detected by assessing its breakdown products, nitrate (NO_3^-) and nitrite (NO_2^-).

However, prior to these findings, the medical community was interested in nitrate and nitrite because of their use in diagnosis of bacterial infection. Bacterial reductases reduce nitrate to nitrite, which therefore serves as a marker of the organisms' presence (345, 346). The presence of nitrite in biological fluids is thus a marker of multiple potential conditions and phenomena. The Griess reaction referred to in these reports is specific for nitrite. This has the odd consequence that assays of the more abundant and stable nitrate are often carried out by converting this ion to nitrite prior to running a Griess reaction (see Section 7.2.1). The advent of CB methods may eventually render this rather complicated method unnecessary, at least for nitrate measurements. We shall have more to say about the bacterial reductases in this connection.

7.1.3   NO Biology

It is surprising that even fairly recent literature reports reflect some degree of confusion about the role of nitrate and nitrite in vascular biology, with some articles describing these ions primarily as waste products of NO synthesis carried out by the enzyme nitric oxide synthase (347), while others indicate that the ions actually serve not only as markers of this activity but also as a "sink" of oxidized nitrogen that can generate NO *in vivo* (348). This is hard to explain, because it was clearly recognized in the 1990s that even organic nitrates could be transformed into NO by biological processes (349). Clearly, the precise chemical nature of the organic nitrate, its rate of conversion to and yield of NO, its tendency to generate by-products, and so on, all influence its medical utility. Isosorbide mononitrate (350), isosorbide dinitrate (351), and nitroglycerin (352) are used against angina pectoris and coronary heart disease, with nitroglycerin evidently yielding a lower incidence of headaches as a side effect

(353). The mechanism of action of these molecules is now known to involve conversion to NO. NO exerts its biological effects by stimulating the enzyme guanylyl cyclase to produce cyclic GMP (354), which induces cGMP-dependent protein kinases to phosphorylate the myosin light chain, with consequent relaxation of smooth muscle (355). Many other chemicals are able to stimulate this pathway, including nitroprussides, other nitrate derivatives, acetylcholine, and cGMP analogs. However, *in vivo* control appears to depend heavily on NO generation.

Interest has recently arisen in the concept of using nitrates to treat postmenopausal osteoporosis (353). NO strongly influences the complex processes of bone remodeling (356, 357), and administration of nitrates to rats has been shown to prevent bone loss (358). It is hoped that the use of nitrates will avoid the expense, side effects, and limitations of such approaches as bisphosphonates, estrogen replacement therapy, and selective estrogen receptor modulators such as raloxifene. The cited trial report (353) indicates that nitrates can increase bone mineral density, with intermittent use more effective than continuous use, due to the phenomenon of tachyphylaxis or rapid desensitization. However, data on fracture frequency are not yet available.

NO is also involved in prostaglandin biology, specifically through modulation of the activity of cyclooxygenases, which interconvert these derivatives of arachidonic acid (359). As a result, NO is thought to play a role in cardiopulmonary function and protection from ischemia (360).

This limited discussion touches only slightly on the many biological effects of NO. Dysregulation of NO biology is associated with a number of pathologies, including cancer (361, 362), diabetes (363), arthritis (364), and most notably heart disease (365, 366), along with many other illnesses.

7.1.4 Interrelationship Among Nitrate, Nitrite, and NO

The breadth of both the biological effects of NO and the diseases with which it is associated in both beneficial and detrimental roles strongly indicate the importance of being able to quantify NO; unfortunately, this species is present only transiently and methods of measuring its concentration directly may not be available for some time. One is therefore forced to use assays of its breakdown products, nitrite and nitrate, but this approach has its own difficulties, both technical and interpretive.

7.1.4.1 Decay of NO to Nitrate and Nitrite
Studies employing the stable isotopes ^{18}O and ^{15}N have established that nitrate is the major breakdown product of NO *in vivo*, at least subsequent to inhalation [367–369], although 20–25% of the material is evidently transformed to urea and other substances. However, nitrite is a separate marker that conveys independent information about NO synthesis. The distinction is of scientific and medical value. Nitrate is formed from NO primarily by the action of oxyhemoglobin. Although not considered an enzyme, hemoglobin binds molecular oxygen in such a manner as to change its chemical properties, so that the interaction with NO causes oxidation to the fully oxidized state of nitrate. Spontaneous oxidation in the absence of hemoglobin, however, yields nitrite. A current interpretation

of the differing significance of these ions in biological fluids with respect to their implications for NO biology is as follows: "Circulating nitrite reflects constitutive endothelial NOS activity, whereas excretory nitrate indicates systemic NO production" (347).

7.1.4.2 The Reverse Reactions: Nonenzymatic Production of NO from Nitrate and Nitrite

Under the acidic conditions of the human stomach, NO is efficiently produced by reduction of ingested nitrate (370), potentially yielding carcinogenic nitrosamines, especially in patients with the condition known as Barrett's oesophagus (371, 372). However, beneficial effects of nitrate consumption are also observed, also mediated by its conversion to NO, including a reduction in the oxygen cost of aerobic exercise (373). The significance of nitrate levels in drinking water is a subject of intense current debate, largely due to these contrasting effects, as well as potential differential fates due to population variability. (For example, whole-body production of nitric oxide appears to exhibit a strong inverse correlation with plasma homocysteine levels (374), which are in turn associated with folate deficiency and other life-style factors (375).)

An important remaining question is the significance of spontaneous production of NO from nitrate outside the stomach. One contemporary argument holds that since enzymatic production of NO is complex and energy intensive, regeneration of NO from nitrate and especially nitrite is of adaptive value (348). These workers also point out that it has long been known that injection of NO has effects consistent with its known modes of action at sites that are too far from the injection point to be accounted for by diffusion of the administered NO. Alternative models are considered, but the authors' view is that the nitrite ion serves as a stable reservoir of NO to be reduced and reused in the appropriate tissues. Is nitrate also a source of NO outside the stomach? Nitrite can certainly be reduced to NO, but there is no mammalian enzyme capable of reducing nitrate to nitrite. However, bacteria present in saliva are capable of reducing nitrate, providing another major dietary source for the nonenzymatic synthesis of NO.

7.1.5 Medical Aspects of Nitrate Biology

As mentioned above, intense controversy surrounds the issue of dietary nitrates and their harmful or beneficial effects. Most of the issues pertain to either the generation of carcinogenic nitrosamines or influences on NO biology. Organic foods appear to have less nitrate than food from other sources (376), but in the absence of convincing evidence of the consequences of ingesting nitrate, this information is of little practical use. There is some evidence that vitamin E and selenium may be protective against downstream free-radical products of NO metabolism, notably peroxynitrite formed by the reaction of NO with superoxide (376). What is clear from the nitrate/nitrite/NO story is that their biological effects are profound and pleiotropic, and that an essential component of ongoing studies of these effects is continual enhancement of the methods available for their quantitative determination in various biological matrices and other sources.

7.2   CURRENT NOS ASSAYS

NOS assays in current use may be conveniently divided into three general categories: (1) means of measuring nitrite and nitrate, (2) other assays of NOS activity, and (3) assessment of downstream processes stimulated by NO. We now turn to the first of the liquid-phase categories, which is particularly dominant in the current scene.

7.2.1   Measuring NOS Activity by Assays for Nitrite and Nitrate

Assays and their interpretation should ideally answer two questions: (1) How much is there? and (2) What do the data mean? In the case of nitrite and nitrate measurements, both questions present significant challenges. Interpretation of nitrate concentrations is complicated by many factors, including reaction flux to and from NO and nitrite, the ubiquity of nitrate in both food substances and laboratory scenarios, and the prolonged persistence of dietary nitrate in biological fluids. Nitrite can also be spontaneously created or derived from dietary sources. The first question is also hardly straightforward to address, since many of today's assay methods involve multiple steps and are subject to errors and interferences of various kinds. A systematic analysis of the parameters that are critical in these assays was performed in 1995 (377), leading to the following conclusions: recovery of the NO derivatives was as high as 87%; nitrite was rapidly oxidized to nitrate in plasma; the nitrite measurement alone was meaningless for this and other reasons; and the concentrations in plasma are in the low nanomolar ranges. Clearly, the analyses are challenging. It is possible that CB methods may reduce the complexity and difficulty of some of these assays in the near or medium term. We will first discuss the conventional methods of measuring nitrite and nitrate concentration. However, one of the oldest of practical chemiluminescent reactions is a method of detecting NO. The sample is mixed with a large quantity of ozone, which reacts with NO to produce NO_2, O_2, and a weak light signal in the infrared (378). The method is discussed at the end of this section because, ironically, while it provides a direct assay for NO, its use with biological samples depends on prior reduction of nitrite and nitrate to NO, since the NO present in the sample will likely not reach the detector.

7.2.1.1   The Griess Reaction   Developed in the late nineteenth century by Johann Peter Griess, the Griess reaction (379) is a method of converting the nitrite ion to an azo dye that is readily quantified by visible spectrometry. A reaction between nitrite and sulfanilic acid forms a diazonium ion, which is then coupled to 1-naphthylamine, yielding a soluble red-violet compound with an absorbance maximum of about 540 nm. Since this diazotization reaction is specific for nitrite, analysis of nitrate by this procedure requires its prior reduction by chemical or enzymatic means (if chemical means are used, the reduction must of course be controlled so as not to yield NO, N_2O, molecular nitrogen, etc.).

The Griess reaction appears in today's commercial NOS assay kits in a multitude of varieties. The kit from Cayman Chemical will be taken as typical. The vendor's recommended protocol (following sample preparation, which may be from tissues or tissue culture supernatant) includes five reagent additions, an incubation of 1–3 h for complete reduction of nitrate to nitrite by nitrate reductase (the time depends on the sample type), and an optional 10-min wait for color development before reading. The list price of the standard assay is approximately $1.00 per well (a fluorometric version, apparently with about 10-fold better sensitivity, costs approximately 10% more; see Section 7.2.1). However, this method involves an additional complexity. Unfortunately, the cofactor NADPH, which is required for the NOS reaction, is not compatible with the Griess reaction. Thus, the vendor-supplied materials contain very small (catalytic) quantities of NADPH, along with a system for continuous regeneration of NADPH from $NADP^+$. Cayman therefore offers another assay kit, costing about twice as much, which includes an aliquot of lactate dehydrogenase (LDH) that is added following the NOS reaction to consume any remaining NADPH prior to the Griess procedure. This method involves a shorter incubation for NOS activity, presumably because the higher concentration of NADPH permits a higher rate of catalysis (which, however, raises the question of whether the readout from the simpler method is truly linear with NOS activity when NADPH is limiting). The "LDH" method requires two more steps, not including the recommended heat inactivation of NOS, which is described in the notes but not specified in the protocol.

Other vendors' offerings are similar. For instance, Oxford Biomedical Research offers a kit with a similar limit of detection and a current list price under $2 a well; the "ultrasensitive" version with NADPH recycling costs about 30% more. A CalBiochem kit includes the LDH feature but lists at over $3 a well. Assay Designs, Biovision, eEnzyme, Neogen, R & D Systems, Thermo Scientific, Oxis International, Invitrogen, and Roche all sell similar kits; eEnzyme for the favorable list price of $\sim$$0.62 a well at the 500-well quantity. Promega sells a kit for detection of nitrite alone at (without reagents to convert nitrate) at modest sensitivity for a highly competitive list price of $0.11 a well. Griess reagents can be purchased independently from various sources, including Cayman and Sigma Chemical. The usual disclaimers about rapid price fluctuations apply to this information.

7.2.1.2 Fluorometric Quantification of Nitrite   Fluorometric assays for nitrite are now generally superior to the colorimetric method, assuming one has the equipment and expertise, and interference is not an issue. As we have seen in other chapters, the advent of membrane-permeant reagents has enabled quantification of intracellular NO; as in other cases, these fluorophores are liberated by cytosolic esterases, rendering them resistant to recrossing of the membrane.

A straightforward method for quantification of the major NO breakdown products involves the same enzyme-catalyzed conversion of nitrate to nitrite described above, followed by a dye such as 2,3-diaminonaphthalene (DAN) under basic conditions, leading to a compound (1-(H)-naphthotriazole) that fluoresces with an excitation maximum of 365 nm and emission at 450 nm (380). The CalBiochem kit insert (chosen because this vendor has a wide range of NO-related products, although they may not always be the most economical) gives the assay time as 3–5 h, including the enzymatic conversion, with five reagent additions and two incubations. The limit of detection is ∼10 nM or roughly two orders of magnitude better than that of the colorimetric method. It should be pointed out that the same vendor lists its radiometric assay for NOS activity (using tritiated arginine) as requiring 2 h, with a limit of detection in the "picomolar range." Many other vendors sell DAN and fluorometric kits for quantification of total nitrite/nitrate.

Intracellular quantification of NO by fluorometry is accomplished quite differently. Diaminofluoroscein-2-diacetate (DAF-2DA) crosses the cellular membrane, whereupon it is hydrolyzed by esterases to diaminofluorescein, which is membrane-impermeant; this compound undergoes a rapid reaction with a reaction derivative of NO to yield the highly fluorescent diaminafluorescein-2-triazole (DAF-2T) (381). In addition to liquid-phase NO quantification, this molecule enables direct fluorescence imaging of processes producing NO (382). The limit of detection claimed in the original work is 5 nM, with an increase in quantum efficiency of greater than 100-fold upon reaction with NO. The same group subsequently developed a rhodamine-containing molecule with a similar limit of detection and a superior increase in quantum yield (383), but the fluorescein derivative is currently in widespread use. Since the reactive species is a derivative of NO, it is not clear what fraction of NO produced is actually yielding signal (obviously this will depend on dye loading as well as other parameters, including quantum yield). Moreover, the reaction time is still measured in hours, much of which is needed for loading of the dye. Nevertheless, the advent of this assay method, which answers a question of profound scientific interest about a highly transient intracellular free radical, must be considered yet another triumph of fluorescence technology. It should be noted that DAF-2DA cannot be used for extracellular NO assays, since it is not reactive until the ester bonds are cleaved. DAF-2DA is sold (under varying names and acronyms) by several vendors, including Sigma–Aldrich, Cell Technology, and Cayman Chemical. Strem Chemicals markets an alternative fluorophore for direct detection of intracellular NO involving a chelated Cu^{2+} ion. An important aspect of this new reagent is the claim that it reacts directly with NO, rather than a reactive breakdown product (384). The limit of detection appears to be similar to that of DAF-2DA.

7.2.1.3 Chemiluminescent Assays of Nitrate and Nitrite

Since its early use in NO quantification (385), the light-producing reaction of NO with ozone has yielded a whole subfield of assay development, with numerous laboratories, interested in various biological systems, investigating a range of reductants and other conditions to enable accurate quantification of nitrite and nitrate [386–389]. This method was originally used to determine that NO was not the sole player in endothelium-derived relaxing factor signaling (i.e., there was a protein or other factor involved in addition) (390). Although this approach appears to have adequate sensitivity, the drawbacks are, unfortunately, the need to re-reduce the nitrite and nitrate to the unstable NO for the measurement and the cost of the instrument; for example, the widely used Sievers 270B costs around $20,000 and is limited to NO analysis.

7.2.1.4 Coupled Bioluminescent Assays of NO and Its Derivatives

No CB assay for NO or its derivatives has yet been reported in the literature. However, such an assay has been enabled by the author in collaboration with S. Dhawan. This method, which is a CB nitrate assay, is depicted in Fig. 7.2.

The reader will pardon an anecdote that is perhaps illustrative of the current state of research in CB technology outside of its core areas. The author arrived in California to begin the development of the CB nitrate assay, only to find that the key reagent, sodium nitrate itself, had not arrived from the vendor. It appeared that a plant nursery was to be found nearby. We obtained a suitable product and conducted our first trials, which yielded a crude but statistically significant response to nitrate content (Corey and Dhawan, unpublished data). Thus, what may have been the first ever demonstration of coupled bioluminescent detection of nitrate was performed with a sample of commercial tomato fertilizer. Like most other CB assays, this assay was extremely rapid (3–5 min), although measurement of NOS activity would of course require conversion of NO to nitrate, which would lengthen the process.

Since this method detects only nitrate (nitrite is a side product of the reaction), its use in NOS assays and NO determination would depend on the assumption that nitrate is the only important end product of NO chemistry or, alternatively, that a known proportion of NO undergoes transformation to nitrate. Although much of the product literature for reagents directed to detection of nitrite indicates that detection of total nitrite and nitrate is a superior method of assaying NO, this view is of course affected by the nature of what is being marketed. If a nitrate assay proves to be the most convenient, it should be possible to ensure a high level of conversion of NO to

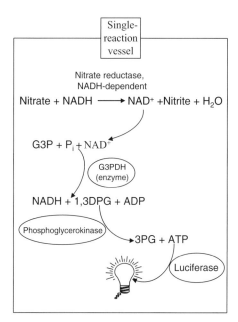

FIGURE 7.2 CB scheme for detection of NAD^+. The reaction components are virtually identical to those of the G3PDH-based aCella-TOX^{TM} cytotoxicity assay described in Chapter 3, except that NAD^+ is made the limiting reagent instead of G3PDH.

nitrate by supplying oxyhemoglobin in the reaction vessel. Alternatively, nitrite may be converted to nitrate by chemical or enzymatic means, or a cycling system involving conversion of nitrite to nitrate by independent chemical or enzymatic means, followed by the signal-yielding CB reaction, may be developed for highly sensitive detection of catalytic quantities of nitrate. However, such a system would probably require the use of an oxidant other than NAD^+ for the recycling of nitrite, since NAD^+ is being quantified.

7.2.2 "Direct" Assays of NOS Activity

One's judgment of whether an assay is "direct" may depend on one's scientific role. To a mechanistic enzymologist, only a real-time display of the catalytic machinery of the enzyme in action may be sufficiently satisfactory to earn the word (this has been achieved for some enzymes [391–394]). Here, we refer to assay techniques that provide a readout based on the production or consumption of a direct participant in the canonical NOS reaction. Technically, the radiometric assays employing tritiated arginine are "direct" measurements of NOS activity, since the product is quantified, and have already been addressed in Section 7.1.1. The fluorescent intracellular method discussed in Section 7.2.1 is likewise a highly effective means of assessing NO production, although the earlier fluorophore is now known to interact with a derivative of NO. Methods based on chromatography and HPLC, for example, this technique of assessing the state of the tetrahydrobiopterin cofactor (395), do not appear to have sufficiently high throughput for routine use in drug discovery and are not treated here. The reader is directed as well to other interesting and novel technologies for a detailed study of NO-related processes, such as the development of electrodes sensitive to minute quantities of NO (396, 397). Capillary electrophoresis methods are sensitive to ~100-amole levels of direct and indirect products of NOS activity (398).

7.2.3 Coupled Enzymatic Assays of NOS Activity

Like the field of CB assays itself, the number and breadth of potential coupled NO assays, both bioluminescent and otherwise, are practically unlimited. The effects of NO are highly pleiotropic and many downstream enzymes and other effects could be selected. Patch clamp studies in smooth muscle cells are of course a kind of coupled biological assay that can be useful in mechanistic investigations (399, 400). GPCRs (see Chapter 9) are involved in NO signal transduction and may yield a downstream readout of biological interest (401). NOS activity has been coupled to tyrosine hydroxylase (402), the NADPH/diaphorase system (403), and reporter genes (via guanylyl cyclase activation) (404). The availability of arginine analogs that are general NOS inhibitors, such as NG-monomethyl-L-arginine (405, 406) and NG-nitro-L-arginine methyl ester, as well as inhibitors specific for cNOS and iNOS (407, 408), with any of these coupled systems enhances the possibilities for elucidating NO signal transduction mechanisms and deconvoluting the effects of the two major forms of NOS. The list could be extended indefinitely, but the assay developer dedicated to testing novel drug candidates is interested primarily in rapid, simple methods that answer questions as closely related to the biochemical system of interest as possible.

7.2.3.1 *Assays of NOS Activity Involving Guanylyl Cyclase*   The last possibility mentioned in the previous paragraph introduces the topic of coupled assays involving guanylyl cyclase. Again, the biological effects of cGMP and consequent activation of protein kinase G (PKG) are varied (although in the above-mentioned case, a mutant cyclase creates cAMP, rather than cGMP), and the assay developer has a choice of many potential readouts. However, the coupling of these processes provides an opportunity to develop a CB assay that is much like the cAMP assay described in the context of G-protein-coupled receptors (Chapter 9). Activation of PKG would be assessed by one of the CB methods described in that chapter, either by depletion of ATP or, conceivably, by its synthesis. Competitive non-CB methods are found in time-resolved fluorescence kits from Cisbio and other vendors; again, cGMP assays are available from many sources, including Biovision, CalBiochem, Cell Technology, Cayman Chemical, DiscoveRx, Molecular Devices/MDS Analytical Technologies, Oxford Biomedical, and PerkinElmer. The Biovision assay, for example, provides a tracer cGMP coupled to horse radish peroxidase. The conjugate is immobilized to the plate by an anti-cGMP antibody, but is displaced from an antibody by cGMP formed in the assay, yielding a negative signal dependent on cGMP on addition of a substrate for the peroxidase. A clever possibility is coupling of guanylyl cyclase activity to calcium flux, creating a real-time NOS assay (409); this might also lead to a CB assay employing aequorin (see Chapter 11).

7.3 CONCLUSIONS

Although many assays for NO and NOS activity are available, none of the current methods is ideally suited to the needs of the high-throughput screening establishment. The methods have adequate sensitivity, but are generally slow and extremely expensive, or are too cumbersome to develop in the drug discovery environment. Given the possibility of simple, sensitive, and very rapid CB assays, it appears that NO quantification and NOS activity measurement may be a fruitful area for advances of this kind.

8

THE COUPLED BIOLUMINESCENT PYROPHOSPHOROLYSIS ASSAY

8.1 INTRODUCTION

CB applications are not restricted to screening for enzyme inhibitors and measuring cytotoxicity. With its READIT®-coupled bioluminescent assay (410), Promega has entered the turbulent arena of assessment of DNA complementarity. READIT is a highly sensitive means of performing a range of important tests: single-nucleotide polymorphism (SNP) analysis and general genotyping, translocation analysis, allele identification and correlation analysis, and even detection of genetically modified organisms in various contexts. As a true CB assay, with the associated advantages, READIT is expected to continue to present strong competition for the assay dollar in this field.

For the benefit of those who have not yet encountered SNPs and the modern world of genetic variability, we begin with a brief review of the role and significance of DNA complementarity assays in modern molecular biology and clinical practice. The description and competitive analysis of the READIT assay follow.

8.2 GENETIC VARIATION IN MODERN MEDICINE

Thousands of disease states and many adverse drug reactions have been traced in recent years to genetic factors. Some diseases are caused by "inborn errors of metabolism," often attributable to discrete mutations in essential genes. Others are exacerbated by the genetic makeup of some individual patients, while still other genotypes appear to confer immunity to disorders that may be fatal to the vast majority; resistance to HIV infection conferred by a mutation in the CCR-5 chemokine receptor (411) is a dramatic example of this widespread phenomenon. Moreover, reactions to therapeutic (and recreational) drugs are strongly modulated by mutations in the cytochrome P450 enzymes of the liver (e.g., see References (412, 413)). Thus, the current extremely high level of interest in human genetic variation is based on the desire to understand and treat genetically linked diseases, as well as the need to administer drugs in a safe and effective way to genetically diverse patients. Forensics, population studies, and wildlife biology and ecology will also benefit from improved techniques for assessing genetic variation.

Although the common perception of mutations to those outside the field is of small, discrete alterations in DNA, genetic diversity and pathogenic lesions can assume many forms, including insertions and deletions, trinucleotide repeat variants, translocations and other rearrangements of chromosomal elements, and elimination or duplication of entire chromosomes. Different types of variation require different methods of assessment. However, if one distinguishes the pure research required to identify the changes in DNA associated with a condition for the first time from the subsequent testing for those changes, it becomes clear that the former necessitates the use of inventive or even completely novel means of determining the contents of a genome; the latter, relying on the former, enjoys the luxury of choosing among high-throughput methods.

The variety of mutational types is mentioned above, but for the purposes of this chapter, we will use the single-nucleotide polymorphism or SNP (pronounced "snip") as our model. A SNP is the smallest possible heritable change in DNA (apart from epigenetic events, which have variable stability across generations (414)). It is simply a mutation in a single base pair of DNA. SNPs may or may not lead to functional changes in genes, which in turn may or may not lead to pathologies and/or differential drug reactions. Many SNPs are entirely harmless, yet cystic fibrosis, sickle cell anemia, and the fatal Tay–Sachs disease are caused by SNPs. Because a single assay technique can be used to study many SNPs ("score" SNPs), there has recently been considerable corporate interest in development of these techniques. A broad classification of approaches to assessing known genetics variations would include "resequencing," or the iterated sequencing of corresponding regions of the genome from many individuals; restriction fragment length polymorphism (RFLP) analysis; and various kinds of methods based on DNA complementarity.

Resequencing is commonly done, perhaps because it is conceptually simple, but it is an extremely slow and wasteful means of studying SNPs, except in unusual cases where many potential SNPs lie in a short region of DNA. Even in these cases, complementarity methods may yield better results at lower cost. RFLP analysis was of tremendous service in the 1980s and 1990s as a means of demonstrating the existence of variations and their association with various phenotypes. Moreover, much of the early forensic work was performed by RFLP analysis. However, this method is too costly and tedious and insufficiently precise to compete with modern means of SNP analysis. Finally, assays based on DNA complementarity are enjoying tremendous scientific and commercial success. We now turn our attention briefly to these methods.

8.3   DNA COMPLEMENTARITY

The fundamental principle here is that DNA strands that are complementary, that is, which are paired with each other according to the Watson–Crick rules, have physical properties that are distinct from those of unpaired strands. Pairing causes the strands to associate in ways that are compatible with enzymatic action or physical detection by other means, such as differential hybridization. Development of an assay for complementarity therefore consists of designing a suitable probe, which may be either DNA or a chemically altered species that behaves similarly to DNA in terms of Watson–Crick pairing, and then engineering a means of detecting probe complementarity that exploits the nature of the probe. In these days of elegant nucleotide chemistry and exquisite control of conditions, both hybridization and primer extension methods are useful and competitive ways of determining complementarity. Other methods described below, including the CB method, use still another concept, that of differential DNA degradation. (One source of confusion regarding degradative reactions is the existence of two distinct mechanisms: one is the reverse of DNA synthesis while the other is not.) Brief descriptions of these alternative strategies are provided.

8.3.1   DNA Hybridization

In the context of SNPs, "hybridization" usually means determination of the melting temperature (T_m) of a complex of a DNA strand derived from the sample of interest and a short synthetic strand of DNA or a DNA derivative designed for detection of a particular SNP. The common procedure is to amplify the DNA first by PCR, followed by determination of the T_m using the more copious amplified sample. The T_m is directly related to complementarity, as well as the length and composition of the probe. Simple formulae used for decades to predict T_m (such as $2 \times [AT] + 4 \times [GC] - 5$) have been replaced in practice by more complex equations that claim greater accuracy (e.g., see http://www.cnr.berkeley.edu/~zimmer/oligoTMcalc.html). For these studies, however, the absolute T_m is not of great importance, since the test DNA is always run side-by-side with standards representing each possible outcome. If its behavior matches any of the standards, its DNA is assumed to match it as well; if it matches none, the hybridization procedure is a failure and exhaustive resequencing is indicated. Dissociation of the test DNA at the T_m can be measured by any of many methods, including FRET, FP, radiolabeling techniques, or surface plasmon resonance (415).

A method which stands in its own category is Acyclo Prime™, marketed by PerkinElmer. Here, the incorporation of a single nucleotide opposite the SNP site is detected by FP. The four (or fewer) nucleotides supplied in the reaction are conjugated to distinct fluorescent dyes. When the modified nucleotides are incorporated, the molecular weight (and therefore the rotational correlation time) of the complex containing the dye moiety greatly increases, yielding an FP signal as described in Chapter 1. This elegant method allows detection and identification of the SNP in a single reaction, but it is not yet clear how its sensitivity relates to that of methods that permit further amplification of the signal subsequent to the PCR step.

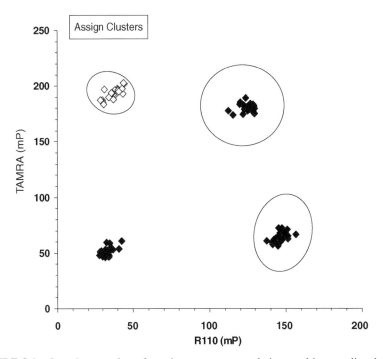

FIGURE 8.1 Superb separation of two homozygous populations and intermediate hetero-zygous population using ordinary PCR amplification, followed by allele-specific dye incor-poration and separation by FP in combination with flow cytometry. Labeled primers must be purchased for each SNP of interest to use this method. Used with permission from PerkinElmer, Inc.

Figure 8.1 shows the impressive signal separation possible for distinguishing heterozygous and differing homozygous samples.

Additional brainpower has been invested in the chemistry of the probes themselves. The minor groove binder invented at Epoch Biosciences (416) and more recent devel-opments, including probes whose fluorescence is directly modulated by association with complementary DNA (417), indicate the vigor of research efforts in this area. Peptide nucleic acids (PNA) are also used (418).

8.3.2 Primer Extension in SNP Analysis

Complementarity in DNA also gives rise to another phenomenon that can be exploited for SNP detection. DNA polymerases, the enzymes that copy DNA, must be primed, that is, must encounter a region of double-stranded DNA adjacent to a region of single-stranded DNA, before they are able to extend the priming strand into the single-stranded region. If the final ($3'$) base pair of the double-stranded region does not follow the Watson–Crick base-pairing rules, primer extension will proceed at a very slow rate or not at all. Thus, independent primers can be designed with exact complementarity to each (or any) of the potential alleles. Each will prime DNA

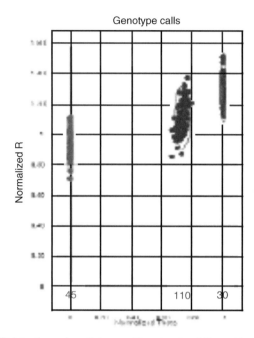

FIGURE 8.2   SNP detection using allele-specific extension followed by hybridization-based capture on beads. A mismatch at the 3' end of the primer impedes the polymerization reaction, so that only matching primers are extended and detected. In the final step, the beads in a specialized flow cytometer-like instrument (Illumina® BeadStation). Specific primers (not labeled) and beads for hybridization are among the consumables. Created based on a genoplot from Illumina, Inc. with permission.

synthesis when hybridized to its cognate allele, but not with the others. Here, then is a case in which a minimal change in DNA (the SNP) can be amplified to yield an arbitrarily large binary signal (extension versus no extension), assuming appropriate conditions are used. Not surprisingly, a wide range of fluorescent and other methods has arisen to take advantage of this opportunity (419–421), including developments in the ever-versatile field of mass spectrometry (422).

As an example, the GoldenGate™ assay marketed by Illumina uses allelespecific extension reactions coupled with a universal PCR protocol to multiplex up to 1536 simultaneous analyses at independent loci. Figure 8.2 illustrates the homozygote/heterozygote separation possible with this method. The combination of the preparation, hybridization, and reading steps using the BeadStation™ reader appears to require a total of 3 days. The list of vendors offering similar products and services is broad, and the constantly changing market landscape would render fruitless any attempt to create an exhaustive list.

8.3.3   Primer Degradation in SNP Analysis

Just as DNA polymerization and primer extension can be made to depend upon complementarity, so exonuclease activity (primer degradation) is strongly affected by primer-template association. Here, it is important to distinguish between the two polarities of exonuclease activity, and our example will be the famous DNA polymerase I (Pol I), for the discovery of which Kornberg justly won a Nobel Prize (423). As a dual-function exonuclease, Pol I can digest DNA in either direction; the two ends of a DNA strand are named 5′ (pronounced as five prime) and 3′ (three prime) to specify the location on the ribose moiety of the free end of the phosphate backbone. The directions are therefore named 5′–3′ and 3′–5′. Here, we are concerned with the latter, the degradation of DNA proceeding in the opposite direction from synthesis. The crucial point, however, is that this is not the only chemical reaction whereby DNA can be degraded in this direction. The other important process of this type is known as "pyrophosphorolysis." The distinction is clarified in Fig. 8.3. As can be seen in the figure, pyrophosphorolysis (Panel B) is the true reverse reaction of DNA polymerization (Panel A). Assisted by the catalytic power of the polymerase, the pyrophosphate

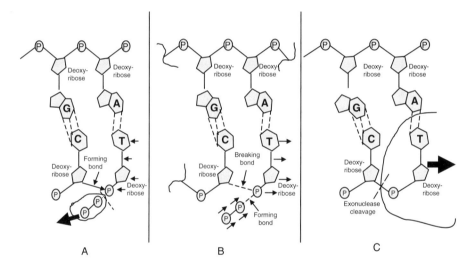

FIGURE 8.3 Schematic representations of three enzyme-catalyzed reactions at the 3′ (growing) end of DNA. A: DNA polymerization. Thymidine (T) triphosphate, the nucleotide undergoing incorporation, joins the end of the growing DNA strand by bonding covalently to the deoxyribose moiety of cytidine (C) through the α phosphate group, and also by forming weaker hydrogen bonds with adenine (A). The β and γ phosphates of T simultaneously depart as the single ion pyrophosphate; this hydrolysis and subsequent breakdown of pyrophosphate provide the chemical energy that drives polymerization, which is catalyzed by DNA Polymerase. B: Pyrophosphorolysis. This reaction, also catalyzed by DNA polymerase, is the exact reverse of polymerization. Pyrophosphate in solution attacks the phosphate ester bond of the DNA backbone, causing breakage of the bond to deoxyribose and liberation of the nucleotide triphosphate (in this case, thymidine triphosphate) in an energy-rich form suitable for coupling into a bioluminescent reaction series. C: 3′-Endonuclease digestion of DNA. This reaction is different from those in the other two panels and is carried out by a separate enzymatic activity. The phosphoester bond of the DNA backbone is cleaved, followed by departure of the liberated nucleotide monophosphate. Pyrophosphate is not involved, and the free nucleotide is not a useful substrate for CB assays.

ion attacks the phosphate backbone, breaking the phosphoester bond and liberating the newly formed nucleotide triphosphate for reaction with the kinase. The $3'-5'$ exonuclease activity, by contrast (Panel C), reverses the polymerization process by effectively breaking the DNA backbone, but neither nucleotide triphosphates nor the pyrophosphate ion is involved in the exonuclease reaction. After exonuclease cleavage, the nucleotide monophosphate (properly a nucleoside monophosphate) diffuses away and the reaction is complete.

The critical corollary is that the exonuclease reaction cannot be coupled in any simple way to a bioluminescent reaction, because no product of the process is useful as either an energy source or a signaling entity of luciferases. Pyrophosphorolysis, however, is a different matter. The deoxynucleotide triphosphates that are the input molecules to the polymerization reaction are of course the *products* of the reverse reaction. It is precisely the ability of the polymerase to reincorporate the energy-rich pyrophosphate moiety into the nucleotide triphosphate via pyrophosphorolysis that makes the READIT reaction series possible. By providing a large excess of pyrophosphate and an adequate concentration of the polymerase, the reaction can be driven in the direction of DNA degradation and simultaneous formation of deoxynucleotide triphosphates. The nucleotide thus formed is still not useful as an energy source for luciferase, but as we shall see, a kinase is provided which transfers the terminal phosphate of each type of nucleotide triphosphate to ADP, which is also supplied, yielding ATP for the light-generating reaction. As the reverse reaction of polymerization, pyrophosphorolysis is carried out by any DNA polymerase, even those whose exonuclease activities have been crippled by recombinant means (424–426). Even though they are useless from the point of view of pyrophosphorolysis, however, exonuclease activities play a role in other genetic analysis strategies, which may incorporate mass spectrometry (418, 427) or hybridization-dependent digestion of internally quenched fluorescent probes (428, 429).

Both pyrophosphorolysis and exonuclease activities can be strongly affected by mismatches at the $3'$ end of the DNA strand, and can therefore be exploited to detect complementarity. The basic principle is that a single mismatch can give rise to a strong signal, because the entire primer is either intact or degraded, depending on the initial event. We now take up specific consideration of READIT®, a CB pyrophosphorolysis assay.

8.4   THE READIT® PYROPHOSPHOROLYSIS ASSAY

From a technical point of view, the READIT technique has very little in common with the competing methods. For this reason, the decision has been made to present the technical aspects of the assay first in this instance, to enable the reader to appreciate the contrasts with other methods, followed by the competitive analysis.

8.4.1   READIT®: Description and Technical Aspects

The READIT assay takes advantage of the pyrophosphorolysis reaction and coupled phosphate transfer, followed by ATP-dependent luciferase luminescence, to yield an assay for DNA complementarity. (The reader should note the distinction between ATP (adenosine triphosphate), the substrate for firefly luciferase, and dATP (deoxyadenosine triphosphate), a substrate for DNA synthesis and a potential product of pyrophosphorolysis but not a potential luciferase substrate.) The enzyme catalyzing the phosphate transfer must be capable of engaging both nucleotide phosphates and deoxynucleotide phosphates as substrates. Although the identity of the actual enzyme used in the READIT® reaction is proprietary information concealed under the trade name READIT Kinase®, various natural or engineered enzymes could be used for this process, including those mentioned in Reference (430).

Finally, the ATP serves as a substrate for luciferase, generating light in proportion to the flux through the polymerase and kinase reactions. The kinase reaction can proceed in either direction, and in fact is more free to do so than pyrophosphorolysis because the free-energy change of transferring a phosphate group from a deoxynucleotide to a nucleotide is small, yielding an equilibrium constant close to 1. However, if the initial reaction mixture is rigorously free of ATP, the process will be essentially unidirectional in the direction of ATP formation for a considerable period. Kinetically, this requires the READIT Kinase® to be present in excess, so that ATP synthesis is proportional (preferably equal) to the quantity of dNTPs generated by pyrophosphorolysis. Of course, the luciferase components must also be present in excess to yield a linear response to the DNA analyte.

Since the pyrophosphorolysis reaction depends on the match between the two nucleotide residues at the 3' end of the strand being digested, the reaction will stop if a mismatch is encountered. Because of the extreme sensitivity of CB methods, it is not difficult to envision a number of ways of exploiting this phenomenon. DNA insertions, deletions, and chromosomal translocations can be detected and quantified by using probes that span the expected sequence variations. The method is also well suited to detection of SNPs, using essentially the same strategy, with a separate reaction for each anticipated allele. Sequence-specific primers can be utilized to seek very small quantities of viral or other DNA in real time, yielding a signal only if the target DNA molecule is present, although this idea is not currently being promoted by Promega to the same extent, perhaps because there are other good methods of detecting very small amounts of contaminating DNA of known sequence.

8.4.2 Performing the READIT® Assay

Tested protocols for READIT® are available from Promega at http://www.promega.com/readit/. Here, we are concerned with operational parameters that enter into a competitive evaluation.

We will begin by assuming that our researcher is interested in genotyping of the *lofundin* gene, which has two alleles differing at a single nucleotide position: a common form with no discernable biological effect, and a rare mutant form that also has no effect on the phenotype of the individual but causes copious grant money to pour forth out of government agencies. She wishes to distinguish among three possibilities in patient samples: homozygous wild-type *lofundin*, homozygous mutant, and heterozygous. She is currently evaluating the various technical options for her study.

Although there are amplification-free options, such as the luminescent up-converting phosphor technology (431) and surface plasmon resonance (432, 433), she has considerable experience with the PCR and chooses to pursue a strategy incorporating an amplification step followed by (or coupled with) detection of the product. Readers unfamiliar with PCR are encouraged to consult any of several recent titles introducing the field (434, 435).

In order to use READIT®, she needs to know the sequence of the allelic substitution, of course, but she also needs to determine nearby sequences for designing DNA primers (probes). There is a vast literature on primer design and other aspects of PCR; for our purposes, it is sufficient to mention that the vendor supplies software that minimizes the various failure modes, such as excessive or artifact prone secondary structure and adventitious sites of association. Moreover, the probes should be designed to anneal at >55°C, a practice that is widespread but not universal. In addition, because the READIT® process includes an exonuclease step to reduce background signals, the synthesized strand must be protected from digestion by this enzyme. This is accomplished by incorporation of an unnatural phosphorothioate linkage between nucleotide residues at the 3′ end of the probe, which is hydrolyzed only very slowly by exonucleases. Primers usually represent a very small portion of the expense of these experiments, and incorporation of phosphorothioate linkages would increase this cost by only about 50% at current prices.

8.4.3 Competitive Position of READIT®

Because this CB assay differs widely from competitive methods, one would expect technical advantages to appear on both sides. Other methods definitely have the

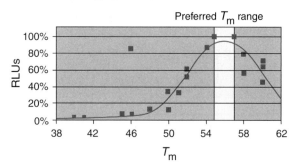

FIGURE 8.4 Effects of offpeak melting temperatures on READIT™ signal. The plot shows the maximum signal attainable with a given probe of defined T_m. Although the optimal T_m band is defined, probes that melt at temperatures well outside of this range (~52–60°C) still perform well in this assay. Used with permission of Promega, Inc.

advantage of being better established in this medically critical area. However, robustness is an important parameter in such assays as well, and the Promega technical literature makes the point that READIT® performs well even when the achieved melting temperature is not precisely optimal (see Fig. 8.4). Supplied probe-design software appears to be straightforward and is held to yield probes with high probabilities of succeeding in routine use. The READIT® CB method must be considered a lively competitor in the growing field of SNP detection and analysis.

9

COUPLED LUMINESCENT ASSAYS OF G-PROTEIN-COUPLED RECEPTORS

9.1 INTRODUCTION TO G-PROTEIN-COUPLED RECEPTORS

G-protein-coupled receptors (GPCRs) are multimolecular signal-transducing structures that convey information from the extracellular environment to the cytosol, leading to modulation of cellular activities. Their activities are usually initiated by the association of a cognate ligand molecule with the receptor, although certain specialized GPCRs respond to events such as the impingement of light on specific sensor molecules. GPCRs are currently the most important enzymatic target of high-throughput screening procedures. Their central role in biological signaling is consistent with the elegance and efficiency with which GPCRs couple extracellular information with the controlled production of second messengers such as cyclic adenosine monophosphate (cAMP), calcium (Ca^{2+}), and inositol triphosphate (IP_3), thereby inducing activation of a broad range of intracellular activities. Their biological importance has also placed them at the center of today's drug discovery enterprise. A large proportion, estimated at 40% (436), of all prescription drugs currently marketed, including most of the best sellers, modulate GPCRs. However, the signaling networks associated with GPCR activities are so complex and involve so many levels of cross talk that in approaching drug development with these targets in mind, one must often discard simple models of "blocking" a particular biological mode of action with a specific reagent and conceive of the intervention as "tuning" of the pathways. The antipsychotic Zyprexa^{TM} (olanzapine) from Eli Lilly, which primarily targets the D2 dopamine receptor, causes a relatively modest level of undesirable extrapyramidal effects; this "atypical" nature appears to derive from its relatively weak association

---

with the D2 receptor, allowing fairly normal dopamine signaling via intermittent occupancy. This is a single example of the ongoing effort to hit a narrow affinity target, a recurrent theme in GPCR pharmacology, and success can usually be measured only with biological readouts. Nevertheless, while *in vitro*-derived insight into isolated systems may not be sufficient to elucidate GPCR biology, it is still necessary. Development of such enhanced agents and those with improved specificity depends of course on the availability to medicinal chemists of structural information as well as powerful minimization and docking software, but additionally on high-throughput tools that enable accumulation of large structure-activity data sets. Further examples of successful GPCR-modulating drugs are blockers of the 5-HT4 receptor, including Janssen's Prucalopride™ and the lower affinity Zelnorm™ (tegaserod) from Novartis, which are used in treating irritable bowel syndrome and severe constipation. Among the many other conditions in which GPCR modulation may prove beneficial are cardiac pathologies, mental diseases, asthma (437), renal disease (438), addiction, and cancer (439, 440). This brief list, however, conveys only a limited notion of the myriad biological contexts in which GPCRs are found, including transduction of vision, hearing, and other sensations; many forms of homeostasis; and various aspects of metabolism (441). The tremendous extent and versatility of GPCR functions is still being actively elucidated in many laboratories.

9.1.1 Structure and Function of GPCRs

9.1.1.1 G Proteins G proteins are proteins that associate with guanidine and its derivatives, especially guanidine diphosphate (GDP) and guanidine triphosphate (GTP). Although the term "G protein" refers to this slightly more inclusive class, which encompasses the "small" GTPases, the set of heterotrimeric or "large" G proteins composed of G_α, G_β, and G_γ subunits is more frequently considered in the context of drug discovery. GPCRs are multimolecular complexes of G proteins and receptors in which the state of the receptor (unassociated or associated with antagonist versus associated with agonist) governs the activity of the G protein and thereby modulates the signaling process. (For an excellent review of G-protein signaling pathways, see Reference 442. Also peruse the Signal Transduction Knowledge environment at stke.sciencemag.org for much additional current information on GPCRs and other signaling entities.)

The exchange process between the two ligands, GDP and GTP, confers the G-proteins' signaling capabilities, since the exchange leads to alterations in the structure of the G protein itself. Association of the G_α moiety with GDP is generally considered to represent the quiescent state, while association with GTP leads to downstream events. In the canonical mechanism, the G_α-GTP complex dissociates from the $G_{\beta\gamma}$ dimer, but recently acquired evidence has broadened our understanding of the molecular mechanisms of G-protein action to include subunit rearrangements and interactions with other effector molecules. In some cases, the $G_{\beta\gamma}$ dimer also has separate activities. Further work in this area will have important implications regarding the generality of particular GPCR assays. In any case, a general characteristic of GPCR signaling is *amplification*, in the sense that the independent ligand-associated receptor molecule, once freed from a single activated G_α subunit, is able to associate with and activate additional G_α subunits in concert with $G_{\beta\gamma}$. Thus, a receptor–ligand-association event can lead to numerous G_α–GTP complexes, each of which can give rise to many second-messenger molecules, further expanding the signal.

Several types of G_α subunits are known. Each modulates its own specific kind of downstream signals. The distinct natures of the signaling cascades presents both opportunities and challenges to the assay developer, since each type of signaling may require an entirely independent approach to measurement. The above-mentioned activities of $G_{\beta\gamma}$ may also be of interest.

Among the most important functions of G_α is formation of cAMP from ATP (with liberation of pyrophosphate) by the G_α subtype known as $G_{\alpha s}$ (or simply G_s). This protein stimulates membrane-associated adenylate cyclase, which catalyzes the indicated reaction. (The involvement of ATP in this process also implies the possibility of direct measurement by CB methods.) cAMP is a second messenger of major importance. Its known modes of biological action involve activation of protein kinase A (PKA; see Chapter 4), which promotes the signaling process by phosphorylation of a wide range of potential downstream targets. Many of the major themes of our current understanding of biological signal transduction originally emerged from work on $G_{\alpha s}$; for instance, synthesis of cAMP and consequent activation of PKA lead to phosphorylation of phosphorylase kinase and initiation of the glycogen-hydrolysis cascade, leading to production of available glucose, in a reaction series elucidated by the eminent Fischer and Krebs (443). This work established the importance of protein phosphorylation as a biochemical switch. Similarly, the concept of second messengers came from work on $G_{\alpha s}$ (444). As a molecule strongly implicated in signaling but only peripherally involved in direct metabolic flux, cAMP may be considered a prototypical second messenger, and as such is an ideal target for the attention of assay developers. Despite the entropic cost of forming the cyclic molecule cAMP, the conversion of ATP to cAMP is energetically facile due to the evolved enthalpy of hydrolyzing the phosphate ester. cAMP may be regarded as a "wasteful" molecule from an energetic point of view, since the free energy of the internal ester bond is generally not recaptured by metabolic processes, but this very property makes it an ideal small-molecule second messenger, since adventitious (non-enzymatic) formation of cAMP is improbable and it plays virtually no role as a metabolic intermediate. cAMP-dependent signaling is ubiquitous in biology and assay researchers are greatly interested it its precise quantification; the possibility of employing CB methods for this purpose has arisen (see http://www.promega.com/campglo/).

$G_{\alpha i}$ subunits generally inhibit adenylate cyclase, but, as with the other G-protein subunit types, this simple description conceals an extraordinary wealth of diversity within this family. For example, specific isoforms of $G_{\alpha i}$ participate in exquisitely specific modulation of developmental effects (see Reference 441 for one example). $G_{\alpha i}$ is a more recently characterized form of inhibitory G protein. It is sensitive to pertussis toxin, like $G_{\alpha i}$, and shares other features of $G_{\alpha i}$, but exhibits different expression patterns and higher activity under some conditions (445). This functional variability among closely related proteins raises a potential difficulty for assay developers. If the enzymatic activities and biochemical consequences of these G-protein complexes are similar, distinguished only by the available molecular players present *in vivo*, how can *in vitro* assays be expected to resolve these effects and yield lead compounds that are specific modulators of the desired target reactions? Fortunately, G proteins are not all the same, and screening strategies that are capable of separating activities against the various members of the group are possible. However, no model system is identical to the biological entity being modeled. Drug development is a process of iteration and refinement, rather than all-revealing single experiments.

$G_{\alpha q}$ (also known as G_q or $G_{\alpha q/11}$) subunits play roles in still other signaling networks. The membrane-associated phospholipase C (PLC) is stimulated by the G protein to hydrolyze phosphatidylinositol-4,5-bis-phosphate (PIP2) to two molecules, inositol triphosphate and diacylglycerol (DAG), both of which have powerful and complex second-messenger effects. Another consequence of $G_{\alpha q}$ activation is release of intracellular stores of calcium, another critical second messenger of GPCR-transduced signals. The dependence of intracellular calcium flux on G-protein activities provides yet another avenue for development of CB GPCR assay techniques, since the aquatic luciferase aequorin produces light in response to calcium ions. This phenomenon is a very active area of CB assay development, as is discussed in Section 9.2.2.2.

Finally, an inherent aspect of the signaling role of G proteins is limitation of the lifetime of the stimulatory (or inhibitory) signal. This is accomplished "automatically" through an intrinsic feature of these proteins: the ability to hydrolyze GTP. This provides a natural termination to the process, one which allows the G_α protein to reassociate with the other subunits, and it also presents opportunities for CB assays, at the same time introducing complications in the form of time dependence of the pool of total GTP.

9.1.1.2 *Receptors that Act in Concert with G proteins*

Receptors that act in concert with G proteins have common structural features (446, 447), including seven membrane-spanning domains, beginning with the extracellular amino terminus and ending with the intracellular carboxy terminus. Each of the termini can be as large as a typical protein (although in some cases they consist of only 6–7 residues), and the entire structure of these single polypeptides, with all the transmembrane segments and the connecting loops, is capable of being immense and intricate. The ligand-association site is near the extracellular portion of the folded protein and incorporates residues from several of the transmembrane segments. The mechanical function of the receptor is conceptually simple but intricate at the molecular level: on associating with its cognate ligand, the receptor undergoes a conformational change that brings about the exchange of GDP for GTP in the contacting G-protein, leading to its activation. However, the various GPCR classes exhibit separate binding mechanisms, which can include multiple steps and even conformational changes in the ligands, especially protein ligands (448). Activating ligands span a remarkable range, including lipids, amines, nucleotides, eicosanoids (i.e., derivatives of arachidonic acid, such as prostaglandins, thromboxanes, and leukotrienes), and other small molecules, including synthetic pharmacophores; all sizes of peptide structures from individual amino acids to proteins; and even light, which stimulates the receptors involved in visual signaling.

In keeping with this diversity, the receptor moieties of GPCR complexes also represent an enormous range of forms, including elements of hundreds of critical biological systems. An informative classification system that has gained acceptance is to consider the class of sensory GPCRs separately; this class includes visual, auditory, tactile, and olfactory sensors and constitutes approximately 60% of known GPCRs. This class is of limited current medical interest, primarily because restoration of function of defective GPCRs is generally beyond our current abilities. The other 40% are conveniently divided by accumulated information about homology, ligand type, and 3D structure according to the so-called GRAFS system, incorporating the Glutamate, Rhodopsin, Adhesion, Frizzled/taste2, and Secretin subclasses (449). Of these the Rhodopsin class predominates and includes, for example, the adrenergic receptors of the type originally discovered. The following brief set of examples is intended to be representative, rather than exhaustive. The glycogen-hydrolysis cascade has been mentioned in connection with $G_{\alpha s}$, but other molecular targets of PKA phosphorylation include Raf1, an upstream element in growth control and oncogenesis, as well as calcium channel proteins. Neurotransmitters such as dopamine and hormones such as luteinizing hormone and vasopressin and their cognate receptors have been shown to act via $G_{\alpha s}$. The obvious implication is that since ligand engagement of the receptors has varying effects despite the fact that production and release of cAMP and calcium are common features of these GPCRs, the specific cellular phenotype plays a central role in determining just what the second messengers do. However, the effects are also modulated by the differences among the types of G proteins. Interleukin-8, a chemokine involved in angiogenesis, has been shown to act via the $G_{i/o}$ proteins, which inhibit adenylate cyclase. Acetylcholine, a crucial neurotransmitter (see Chapter 6), with its receptor utilizes both these G proteins and G_q. Lysophosphatidic acid, a by-product of phospholipid metabolism that has been recruited by evolution for signaling purposes, affects both the inhibitory G proteins and the $G_{12/13}$ family.

Consideration of the whole structure of the multimolecular GPCR complex within the membrane, possibly associating productively with adenylate cyclase and other downstream effectors and transducers, reveals the challenge that GPCRs represent to assay development. The ligand-induced conformational change in the receptor itself is difficult to measure and characterize, even if one could isolate active receptors. The obvious approach to this problem is to make use of the machinery nature has supplied in the form of the coupling enzymes that yield molecules amenable to measurement. However, the stability of the whole business in the absence of membranes is hardly assured. Therefore, as we shall see below, assays of GPCR activities tend to take the form of cell-based assays. An alternative that is employed less frequently is to isolate or reconstitute membrane fragments containing the elements of interest in active form. Again, the purpose of the enterprise is generally to observe the behavior of the receptor itself; the rest of the complex molecular and/or cellular apparatus is there simply to enable this observation.

9.1.2 GPCRs in Medicine

In addition to the ubiquity and central importance of GPCRs, their attractiveness as targets for therapeutic interventions is also enhanced by the fact that agonists, antagonists, and to some degree allosteric modulators (450) are chemically accessible and have a high probability of being effective, since there are generally no requirements for crossing membranes or surviving the enzymes of the cytosol. The short list of examples of conditions currently treated by GPCR modulators may be expanded as follows. Angiotensin II receptor blockers are used against diabetes-induced renal damage, hypertension, and heart disease (451). Antagonists of α-adrenergic receptors are indicated for benign prostatic hyperplasia (452), while blockers of β-adrenergic receptors are useful in treating congestive heart failure (453). Agonists of dopamine receptors are employed against Parkinson's disease, whereas antagonists are used against schizophrenia and other psychoses (454).

Although the agents available and the progress that has been made have favorably transformed the treatment of many important medical conditions, the work remaining to be accomplished is vast. The reader will note the relatively small number of receptor types appearing in these lists of success, this despite the evident existence of 800 GPCRs, of which about 300 are considered "druggable," and only approximately 30 of these are currently targeted by marketed pharmaceuticals (455). The majority of targeted receptors are those with monoamine cognate ligands, such as acetylcholine, dopamine, the catecholamines (e.g., epinephrine), histamine, and serotonin. However, advances in lipid chemistry and GPCR science have also led to development of various eicosanoid derivatives for therapeutic use, including ramatroban and seratrodast for such conditions as rhinitis and asthma (456, 457), misoprostol for use in combination with the famous progesterone antagonist RU486 as an abortifacient and alone in treatment of ulcers (458), various synthetic prostacyclins for hypertension (459), and others. This illustrates how chemistry, GPCR biochemistry, and medical insight can enjoy synergistic results in terms of new therapeutic categories of drugs. Nevertheless, most GPCR modulators in the current pharmacopeia have emerged from initial observations of activities in *in vitro* systems, without the highly cooperative and rational combined efforts of multiple scientific viewpoints that should be possible today. This leaves a tremendous field of opportunities for pharmaceutical development and potential benefits to patients in terms of improved collaboration, more efficient and effective assay techniques, and the promise of the remaining underexploited receptor classes.

However, although the reasons for the current intense commercial interest in GPCR modulators are obvious, even if we had in hand suitable modulators of high specificity and variegated potency for all these GPCRs, the challenge of genetic variability would immediately present itself (460). Inborn variants in GPCRs cause or contribute to heart disease (461, 462), hypertension and asthma (463, 464), obesity (465), and mental disorders (466, 467). A famous variant in CCR5, a chemokine receptor and a GPCR, increases AIDS susceptibility (468, 469). A much longer list of diseases associated with rare, monogenetic GPCR variants is found in the excellent review (460). It is likely that only a portion of the aberrant functions of variant GPCRs can be modeled in preformed or generic *in vitro* systems, and these studies may also require different sorts of assay methods, in the sense that quantitative accuracy may assume greater importance than high throughput in these cases. This further multiplies the need for diverse approaches. Additional complications in studying GPCRs *in vitro* are introduced by the complex reality of biologically critical covalent receptor modifications (470) and emerging discoveries in the area of GPCR accessory proteins (471). Such intricacies tend to direct the assay developer even more firmly toward cell-based approaches.

9.2 GPCR ASSAY METHODS

As is the case with other targets, the distinction between luminescent and "other" assays of GPCR function is somewhat artificial. The biochemistry may be virtually the same, and the choice of a luminescent or fluorescent tracer for cAMP may seem like a minor distinction, although the caveats of Chapter 1 regarding the challenges of any fluorescent method still apply. On the other hand, the advent of novel reagents such as aequorin has raised the inherent debate almost to the philosophical level. All the issues of assay time and complexity, reagent quality and stability, fidelity of the model, hit quality, cost, required training, equipment, and so on come into play, forcing the assay developer to make judgments based on many complex parameters and eternally incomplete information. However, the mass of information also reflects the availability of a wide range of options for measuring the activities of GPCRs.

Because of the many biochemical products of GPCRs, this chapter is organized differently from other chapters, in which CB assays and the alternatives are discussed in separate sections. Here, it makes more sense to organize the information by the general biochemical strategy: that is, whether cAMP, calcium, GDP/GTP, arrestins, or other features of GPCR biochemistry yield the primary molecular signals. Each of these sections will therefore include a discussion of the non-CB possibilities, followed by bioluminescent assays if they are available. It is hoped that this approach will be of utility to those workers who already know which molecular targets are of special interest.

9.2.1 Arrestins as Markers of GPCR Activity

Though conceptually complex, this assay class is addressed first because of the intense current interest it has engendered.

It is perhaps ironic that this very promising approach to assessing GPCR activities depends on the desensitization process, that is, the reactions that result in termination of the ligand-induced signal. This would appear to be a subsidiary event, potentially removed in time from the activation events that are the major part of the "story," but desensitization has proven to be such a robust and general phenomenon that it currently represents a strong competitor among the set of assays directed to the entire range of GPCRs.

Desensitization commences with the association of cytoplasmic arrestin with the receptor in its active, liganded state. Arrestin ends the signaling activity and induces endocytosis of the entire complex, without generally leading to degradation or damage to the receptor; thus, the whole metabolically expensive molecule can be recycled back to its transmembrane workstation for another round of signaling. Of course, this is only a single level of control of GPCR signal transduction, which also encompasses hydrolysis of the GTP moiety by the intrinsic GTPase activity of G proteins, as well as downstream feedback processes. However, desensitization is of extreme importance in GPCR biology and medicine, in part because it is associated with the problem of tolerance to opiates and many related phenomena. From the point of view of the assay developer, it may be regarded as the only readily measured dynamic aspect of GPCR biochemistry that appears to be universal. The most widespread approach to the assessment of GPCR desensitization at present is Transfluor™ (472), based on technology originally developed at Duke University (473). The method employs cell lines engineered to express both GFP-tagged arrestin and high levels of the GPCR of interest. Fig. 9.1 depicts the scheme underlying the assay, which depends critically on the migration of β-arrestin in promoting desensitization in response to GPCR stimulation. Fig. 9.2 is an image of a tissue slice undergoing GPCR stimulation and desensitization, made with the TransFluor technique.

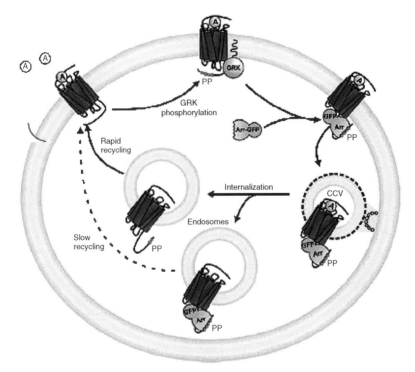

FIGURE 9.1   Schematic view of TransFluor™ GPCR assay technology. Used by permission of MDS Analytical Technologies. (See the color version of this figure in the color plates section.)

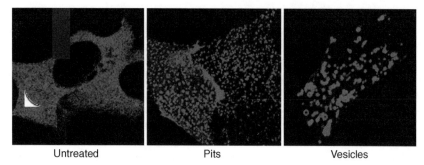

| Untreated | Pits | Vesicles |

FIGURE 9.2   Imaging of cells with stimulated GPCR visualized by TransFluor technology. Used by permission of MDS Analytical Technologies. (See the color version of this figure in the color plates section.)

Competing with this fluorescent approach is a luminescent technique developed by PerkinElmer. Like Transfluor, this method relies on the proximity of the GPCR and the labeled arrestin, but in this case the signal is generated by the BRET™ (or BRET2™) system, in which photons generated by *Renilla* luciferase, which is generated as a recombinant fusion protein with the receptor of interest, are transferred to a nearby mutant GFP molecule (which is fused to arrestin) for subsequent reradiation at a wavelength characteristic of the emission spectrum of GFP (474). Thus, the system luminesces at the wavelength of the luciferase or of GFP, depending on whether the two proteins are near each other. Since the expressed luciferase is held by the cell, this provides means of measuring the association between β-arrestin (which is conjugated to GFP) and the cell surface. This intriguing combination of luminescent and fluorescent technologies deserves to be exploited more fully than it has been to date, in that it provides many of the advantages of both FRET on the one hand (such as sensitive spatial resolution of the interactions of labeled molecules) and bioluminescence on the other (e.g., no lamp or precise collimation is required). Still, the requirement for near-simultaneous measurements of both the GFP emission at 510 nm and the unmodified luciferase emission at 395 nm represents an equipment limitation, although PerkinElmer markets the EnVision™ multilabel reader for this purpose. Moreover, given the stated constraints, it is perhaps unrealistic to expect nature to provide luminescent and fluorescent molecules precisely engineered for this ingenious application, and in practice the luciferase substrate used in the assay is the proprietary modified coelenterazine DeepBlueC™, which is readily oxidized by the *Renilla* enzyme to yield light of which the spectrum overlaps that of GFP. The BRET method avoids many of the problems of FRET, including autofluorescence and other interference sources, as well as all the difficulties resulting from the lamp, such as photobleaching, scattering of light by particulates, and a noisy detection chamber. However, the BRET2 signal decays very quickly, placing a premium on rapid mixing and measurement when this technique is used. At present, BRET and BRET2 are currently receiving little active marketing attention by PerkinElmer, but it is to be hoped that further technological developments will enhance their competitive position. It is tempting to treat this elegant approach as a CB method, but according to our rigorous definition, which involves a separate chemical step coupled to the luminescent output, this reaction does not qualify. Still, its potential utility spans many assay types beyond kinase studies, including proteases, two-hybrid-type assays, affinity measurements, and other research endeavors in which the proximity of proteins must be measured. More will be seen from this fascinating luminescent alternative to FRET.

Finally, given the history of enzyme complementation in studies of protein interactions, it is hardly surprising that this method is also represented in the competition for GPCR–arrestin dollars. DiscoveRx has developed PathHunter^{TM}, a system employing complementary fragments of β-galactosidase with an electrochemiluminescent substrate (Fig. 9.3). The bulk of the β-galactosidase moiety is expressed in a fusion with arrestin, but a short complementary peptide is expressed following the GPCR of interest. DiscoveRx claims exceptionally high sensitivity for this method (Fig. 9.4), as well as the ability to work with "challenging" receptors. The technique appears similar to one reported by Applied Biosystems personnel (475).

9.2.2 Calcium Quantification in GPCR Assays

Like arrestin-based assays, calcium is another area of GPCR assay development in which fluorescent and luminescent technologies collide. The variety of available assays is wide and the technical considerations correspondingly intricate.

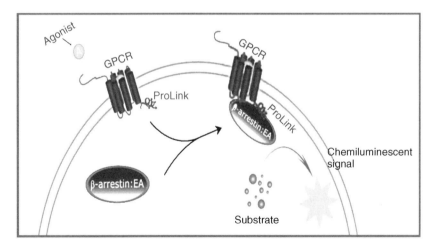

FIGURE 9.3   Assay of GPCR activation by enzyme fragment complementation (EFC). Separate complementary fragments of β-galactosidase are linked to the GPCR and β2-arrestin proteins. Association of the two conjugated proteins brings the fragments into proximity and stimulates the enzymatic activity of the galactosidase, yielding a chemiluminescent signal. Used by permission of DiscoveRx Corporation. (See the color version of this figure in the color plates section.)

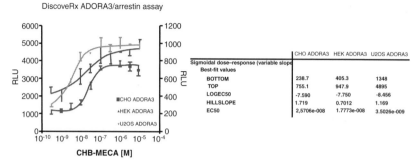

FIGURE 9.4   Sensitive detection of GPCR activity by EFC assay. The receptor in question, ADORA3, had previously been reported not to interact significantly with β2-arrestin [476]. Used by permission of DiscoveRx Corporation. (See the color version of this figure in the color plates section.)

9.2.2.1 Fluorescent Calcium Quantification in GPCR Assays

Due to the extraordinary variety of assay methods available even within this restricted category, we begin with a conceptually simple approach: preloading of cells expressing the GPCR of interest with a fluorescent dye that responds to calcium. The dyes are usually derivatives of 1,2-bis(2-aminophenoxy)ethane-N,N,N′,N′-tetraacetic acid (BAPTA), a powerful calcium-chelating agent, that incorporate additional "shrubbery" which yields a fluorescent response to the captured ion. Appropriate salts of the dyes may be used to improve membrane penetration, but a more advanced approach is to supply the dye in a chemical form (usually an acetoxymethyl ester) that is converted by intracellular enzymes (esterases) into a membrane-impermeant form upon contact with the cellular cytosol. Originally it was necessary to wash the cells following dye loading to remove extracellular dye molecules, but with superior dye technologies (especially an improved ratio of bound to unbound fluorescence yield) it is now possible to add the dye reagent, incubate, and perform the experiment. Moreover, "ratiometric" readings smooth out variations in dye loading. Ratiometric readings employ two independent wavelengths for the readout, but the varying wavelengths can be on either the excitation or the emission side, depending on the Stokes-shift characteristics of the dye and other practical matters, such as interferents. The principle is that loaded dye may be quantified at one wavelength, providing a normalization standard for the "shifted" signal upon association with calcium—obviously this can be generalized to other dye systems. The wavelength chosen for normalization can be one to which the dye responds (or at which it fluoresces) in both free and bound states, but it does not have to be, assuming the shifted wavelength provides reliable quantification of bound dye molecules. With many of the chemical and biochemical problems solved, variations in the quality parameter among these dyes largely come down to the size of the differential fluorescence signal that can be achieved upon calcium binding. It should be noted that in some protocols, the growth medium may be removed prior to loading of the dye. Usually, the plates are centrifuged before this step (if they are not, the step depends on settling and may be problematic), but the process is still labor intensive and can lead to uncertainties about the number of cells remaining. If growth media are not removed, artifacts such as a fluorescence "dip" due to dilution effects may be observed, although these phenomena as well as background fluorescence can be attenuated by such technologies as masking dyes (477).

A commercial example illustrates the concept of calcium detection by preloaded small-molecule dyes. ABD Bioquest markets the Screen Quest^{TM} Fluo-8 No Wash Calcium Assay for this purpose. The vendor claims that this is "the brightest fluorescent calcium dye currently available" with affinity very similar to that of other dyes. The cells to be screened are incubated with the dye, which readily crosses membranes; however, as with most of the currently available dyes, intracellular esterases then cleave blocking groups, yielding acidic moieties whose negative charges prevent the dye from migrating back out of the cell. The manipulations of interest are then performed, presumably leading to intracellular release of Ca^{2+} and a 100–250-fold increase in the fluorescence signal of the dye (excitation at 490 nm, emission at 514 nm). Although there are always concerns about artifacts when cells are loaded with foreign materials prior to experiments, in this case the metabolic participation of the dye appears to be minimal and no evidence of adventitious effects has appeared to date. The cost as of this writing is an attractive $0.12 per well in quantity for the kit requiring removal of the growth medium, or $0.19 per well for the kit with fetal bovine serum compatibility.

With the advent of peptides capable of binding calcium-sensitive fluorophores with high affinity, the next generation of fluorescent assays of calcium flux may involve a hybrid between preloading and genetic expression of part of the molecular sensor (478). In fact, another advanced alternative to preloading of chemical dyes is expression of the entire sensor molecule within the target cell (479). Using a baculovirus delivery technology (480), Invitrogen scientists induced expression of a recombinant calcium-sensing protein (Premo Cameleon^{TM}) in HEK 293T cells. The histamine response of the H1 GPCR was then successfully measured using a ratiometric readout. The inventors note that while this method still involves manipulation of the target cells, expression of the calcium-sensing protein is much more durable and stable than dye preloading, enabling more effective planning and use of workstations. Moreover, the cells have had two days to recover from the transfection process prior to the experiment. The mechanical function of the sensor protein is fascinating in itself. Association with calcium causes a conformational shift in the linker peptide related to the chelation process, thereby bringing the fluorescent donor and acceptor into proximity and causing a FRET signal (see Chapter 1).

9.2.2.2 Calcium Quantification in GPCR Assays by Aequorin

We turn now to the major alternative to fluorescence methods in calcium detection, the remarkable aequorin system (discussed at greater length in Chapter 11). Aequorin is a protein from the jellyfish *Aequoria victoria*, which yields flash luminance upon association with Ca^{2+}. Its mechanism of action involves oxidation by dissolved O_2 of its bound substrate coelenterazine, yielding a light-producing, ring-opening decarboxylation reaction and leaving an amide, coelenteramide (481). A significant difference between this reaction and those of the beetle and bacterial luciferases is that in the case of aequorin, the two energy-supplying molecules (O_2 and coelenterazine) are not generated by the biochemical process under study. Instead, the cofactor Ca^{2+}, which is not absolutely obligatory for luminance but enhances the luminescent reaction by about a million-fold (482), is the switching molecule of interest to the assay developer. This obviates one set of technical concerns, those related to the production and maintenance of the energy source, at the cost of introducing others related to the aequorin system.

The aequorin light reaction (483) and its calcium dependence (484) have been known since the 1960s, but the marriage of GPCR biochemistry and aequorin-based calcium detection was consummated only as recently as 1993 (485, 486), although clever uses of aequorin in quantifying calcium predate this by nearly a decade (487). The quantitative range of calcium detection is favorable for physiological use ($\sim$50 nM to 50 μM (488, 489)). It is fair to say that the advent of widespread use of aequorin in GPCR assays is a recent development, and protein engineering of aequorin to introduce desirable properties is ongoing (490, 491). The reader is advised to check the very latest in aequorin-related technology before deciding on investments in specific products.

Belgium-based Euroscreen SA was the commercial pioneer in developing aequorin-based GPCR assays. Originally, a major problem in performing high-throughput screening of GPCR modulators with calcium detection, especially screening of agonists, was the fact that luminometers, even those with integrated injectors, did not have the capability of injecting different compounds into each well. Since the flash luminance occurs rapidly following mixing of the agonist with the GPCR-expressing cell, this represented a significant technical barrier. However, in 2000, Euroscreen announced the development of a suitable system. Instead of the test compounds, cells were injected into wells preloaded with the chemicals to be screened. An essential element of the system was a series of cell lines expressing GPCRs of interest and incorporating "promiscuous" G proteins such as $G_{\alpha 16}$ to enable switching of GPCRs that are naturally associated with adenylate cyclase stimulation and cAMP formation to induction of PLC activity and intracellular calcium release (491). The actual identity of the cells was relatively unimportant, and in practice the versatile CHO-K1 cells (CHO, Chinese hamster ovary) were generally employed for their facile transfection and strong expression characteristics. The assay costs were remarkably low (estimated at the time at $\sim$\$0.07 per 96-well-plate well, including all materials).

However, the widespread growth in adoption of the aequorin technology awaited a more suitable mating of biochemistry and detection instrument. In 2003, Euroscreen SA announced the commercial availability of a package of their aequorin-based technology and the CyBi^{TM} Lumax SD flash-luminescence reader for high-throughput screening of GPCRs. Euroscreen was acquired by PerkinElmer in early 2007.

The cell lines that are the basis of the technology express both the GPCRs of interest and the apoaequorin, that is, aequorin without its obligatory substrate coelenterazine. Prior to use in an assay, the cells must be washed and loaded with coelenterazine, a procedure for which a 4–18 h incubation at room temperature is recommended. The cell suspension is allowed to recover for an hour at room temperature and diluted appropriately. For an agonist assay, the cell suspension is injected and luminance is read immediately for 20–40 s; for an antagonist assay, a 15-min incubation to allow the modulator to occupy the GPCR ligand site precedes injection of the agonist and reading. Fig. 9.5 (provided by the vendor) depicts typical dose–response curves for agonist and antagonist activity.

(a) HI: Agonist response on the VICTOR, loading 4 h
CHO-H_1 agonist assay, 96-well suspension (50,000 cells per well)

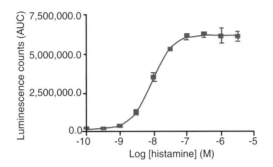

Agonists	pEC$_{50}$ (M)	Signal: background
Histamine	8.07	83.6

(b) HI: Antagonist response on the VICTOR, loading 4 h
CHO-H_1 antagonist assay, 96-well suspension (40,000 cells per well)

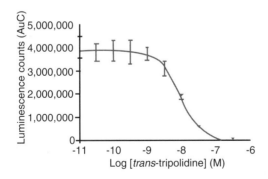

Antagonists	plc$_{50}$(M)
Trans-tripolidine	7.99

FIGURE 9.5 Results of aequorin-based assay of calcium release in GPCR activation. The errors appear to increase at higher luminance levels, although the residuals appear to be very small. See text for methods information. (a) Agonist response. (b) Antagonist response. Used with permission from PerkinElmer, Inc.

9.2.2.3 Comparison of Fluorescent and Luminescent Methods of Calcium Detection

It is unnecessary to reiterate the points made about luminescent and fluorescent measurements in Chapter 1, but issues specific to these assay methods arise. Measurement of flash luminance requires specialized equipment and training. Both methods require preloading of compounds not naturally present in these cell types, although coelenterazine is at least a natural compound, albeit a rather labile one. Autofluorescence is not an issue with coelenterazine. The throughput advantage might initially appear to belong to the luminescent method, but the read time of 20 s or more per well dictates either that reading a plate requires a considerable amount of instrument time, or that an effective pipelining strategy be employed (multiple injectors, imaging, reading of multiple wells simultaneously, incubation and reading occurring at once, or some combination of these).

In 2003, Schering-Plough scientists performed a side-by-side comparison of the two methods with LOPAC^{TM}, a library of 1280 pharmacologically active compounds available from Sigma–Aldrich. The Calcium 3 dye was used with FLIPR^{TM} plates for the fluorescent wing (Calcium 3 has now been superseded by more responsive dyes) (492). The Z value for the fluorescent assay peaked at a high of 0.8 at 5000 cell per well and declined at higher concentrations, dipping below the rather arbitrary quality "cutoff" of 0.5 at 25,000 cells per well, presumably because of autofluorescence. The aequorin assay performed acceptably (Z $\simeq$ 0.5) at 5000–15,000 cells per well and improved to 0.6 at higher concentrations. Cell availability was therefore indicated as a potential issue. Black plates were found to be far superior to white for the luminescent assay because of cross talk. The aequorin assay exhibited a much higher signal-to-noise ratio (166 versus 2.3), although this effect might have been significantly smaller if more advanced fluorescent dyes, which were not yet available, had been used. Despite this apparent quality difference, however, the two methods yielded similar IC$_{50}$ values for the MCHR2 antagonist tested (416 nM for aequorin versus 603 nM for the fluorescent method). (These data are not available for reproduction herein, but see slide 11 at this URL: http://www.euroscreen.be/library/documents/products_review/Bryant_SP.pdf.) However, a reduction of $\sim$30% in the IC$_{50}$ may be considered to be a significant indicator of a superior assay, since the receptor itself appears to be binding the antagonist more tightly, potentially implying conditions that are closer to the native situation in the luminescent assay. The error bars and residuals also appear more favorable for the aequorin assay, although this cannot readily be quantified by visual inspection. The luminescent assay was also much more robust with respect to addition of potential interferents, both solvent (DMSO, which is an important solvent, since it is very frequently used for dissolving compound libraries for screening procedures) and, not surprisingly, fluorescein and other adventitious fluorophores. "Convenience" issues were similar, although the luminescent method required four times as many cells for optimal performance. Loading of the cells with coelenterazine was considered easier. The statistical performance of the luminescent assay was generally superior, with smaller CVs and larger dynamic ranges in many cases.

The performance and sensitivity issues were perhaps the most significant. The correlation between antagonist activities as determined by the two assays was good.

No strong antagonist identified by one assay was missed by the other assay. However, the results with agonists strongly diverged from this finding. The two strongest agonists found by the fluorescent method were essentially inactive by the aequorin assay. This is especially surprising since the luminescent method appeared to show superior performance with the single agonist chosen for extended analysis. However, further questions about the performance of the luminescent assay with agonists were raised by additional experiments, in which identified antagonists were tested for agonist activity. Of the three compounds identified as dual agonists/antagonists by the fluorescent technique, none was confirmed by aequorin as an agonist. The reason for this remains unexplained. It is probably not due to some kind of signal saturation, since the luminescent assay is able to perform well with a subset of agonists. The workers summed up their experience with the conclusions that the aequorin method yielded somewhat better statistics, was somewhat easier to perform (especially cell preparation), and gave a stronger signal relative to noise. However, although it was not mentioned in the conclusions, some concern must remain that the performance of the aequorin assay with agonists is questionable, at least in this evaluation.

9.2.3   Quantitative cAMP Assays of GPCR Activity

Cyclic adenosine monophosphate is generated when a $G_{\alpha s}$-type G-protein is stimulated to activate adenylate cyclase, which cleaves ATP and forms the cyclic phosphoester. Thus, cAMP measurements are a useful readout for essentially all GPCRs that incorporate $G_{\alpha s}$-type G proteins. The range of methods currently available to quantify cAMP for assessment of GPCR activities includes standard fluorescence measurements, fluorescence polarization, competitive time-resolved fluorescence, competitive enzyme complementation with a luminescent readout, radiometric methods, competitive ELISA, and coupled bioluminescence. This extreme diversity of methods provides an enormous range of options, while adding to the complexity of choosing a system to use.

The general principle is competition by cAMP formed in the course of the experiment with labeled cAMP for association with a cAMP ligand, usually an antibody. The variations arise in the form of the label and the nature of the detection method. However, essentially all of these methods depend on such processes as diffusion (usually of large molecules) and ligand displacement, both of which can be slow. Thus, the term "high throughput" must be considered in a relative sense in weighing these methods against others. It is generally unrealistic to expect a readout within seconds, or even a few minutes. All methods involving competition for labeled cAMP are presented first (including competitive radioimmunoassays). Coupled bioluminescent assays are addressed at the end of the section.

9.2.3.1   Competitive cAMP Radioimmunoassays   The principle of the radioimmunoassay is straightforward: a radioactively labeled antibody is presented to the target, nonbinding material is washed away, and radioactive disintegrations are counted.

In the competitive cAMP assay, radiolabeled cAMP (the label is usually ^{125}I) is supplied with a cAMP-specific antibody. Unlabeled cAMP generated in the course of the reaction displaces the label. When the anti-cAMP antibody is captured (the modern method of capture is microspheres suitable for use in a scintillation proximity assay or SPA), the amount of label remaining, and therefore the signal, varies inversely with the quantity of cAMP formed. If the amount is small, one can expect a fairly linear (though negative) response to generated cAMP; however, a large cAMP flux will readily drive the signal into sublinear territory due to binding-site saturation unless a large excess of the expensive antibody and perhaps equally expensive labeled cAMP are provided. Provision of large amounts, however, raises the possibility of back reactions (although ^{125}I-cAMP may not be a substrate for many of these). The sensitivity claimed by PerkinElmer, for example, is under 1 pmol per well with a CV under 10%. Thus, sensitivity is not an issue, and, with the advent of SPA, a homogeneous assay can be achieved; the length of the incubations (a minimum of 2.5 h) and the problems associated with the use of radioactivity are the major disadvantages of these methods.

9.2.3.2 Competitive cAMP ELISA

The competitive ELISA for cAMP quantification is typified by the cAMP HTS Immunoassay Kit from Chemicon. This is rated as faster than the radiolabeled method mentioned above, including two 30-min incubations, but in dealing with ELISA one must not fail to consider the wash procedures. As mentioned elsewhere, the difficulty of effective ELISA washing and the amount of labor required in providing clean wash solutions, running the procedures correctly, and maintaining the equipment add significantly to the cost in both financial terms and employee morale. Five washes are recommended for this assay. Since the principle is competitive displacement of an enzyme-linked cAMP molecule, the assay yields a negative signal. Various other problems, such as edge effects, evaporation, and cross-contamination of samples, are associated with ELISAs, unless one takes the expensive step of using plate covers. However, the chemiluminescent readout from the synthetic alkaline phosphatase substrate yields a fairly good sensitivity of about 10 pmol per well.

9.2.3.3 Fluorometric cAMP Assays

Straightfoward fluorescence assays for cAMP are rarely used these days, although they are easily conceived. Instead, the fluorescence signal is refined in such a way as to eliminate the notorious wash step, enabling the generation of a fluorescent signal in a homogeneous format. Common readouts are fluorescence polarization (FP) as well as time-resolved fluorescence (TRF), especially time-resolved fluorescence resonance energy transfer (TR-FRET). (These concepts are explained in more detail in Chapter 1.) Briefly, FP is an assessment of the size of a fluorescent complex by determination of its rotational correlation coefficient, which is accomplished by measuring how long it takes to achieve depolarization of a fluorescent excitation pulse. Labeled cyclic AMP is a good target for an FP approach, because it is much smaller than a ligand such as an antibody—thus, competition by unlabeled cAMP causing displacement of the antibody will dramatically decrease the size of the effective fluorophore (from the size of the labeled-cAMP–antibody complex to that of the labeled cAMP alone). Such a method was employed in the [FP]^2 CampFire^{TM} assay marketed until recently by PerkinElmer. This assay responded to antagonists active over three log orders of IC$_{50}$ and tolerated DMSO up to 1%. Z factors were high. However, the usual lengthy incubation (0.5–24 h) is recommended to allow reagent dissociation and reequilibration. PerkinElmer no longer markets this assay, evidently believing that their Surefire^{TM} kinase assays are superior.

TRF has been found to be a useful alternative to FP for measuring proximity, although unlike FP, TRF requires two fluorophores or fluorescently labeled compounds with appropriate spectral overlap. The advantage is that the TRF principle is selective for the specific fluorescent interaction of interest, reducing interference from other fluorophores in the matrix.

Fig. 9.6 depicts the LANCE^{TM} scheme of competitive displacement of a labeled anti-cAMP antibody from a biotinylated cAMP molecule, reducing the TR-FRET signal (493). The technology is well established in many fields of assay development, including kinase assays, antibody detection, and an expanding range of others. There is a very wide Stokes separation between the Eu^{3+} donor (excited at 340 nm) and the tracer, which is conjugated to an Alexa^{TM} dye (emitting at 665 nm) licensed from Molecular Probes (now part of Invitrogen); however, in the absence of effective FRET, the donor emits at 615 nm, which is nevertheless well separated from the FRET signal. Incubations are 90–120 min, but readings can be taken up to 20 h later. Fig. 9.7 depicts dose–response curves using cAMP alone, the β2-adrenergic agonist epinephrine (yielding a negative signal as expected), and the combination of epinephrine with the strong antagonist propanolol.

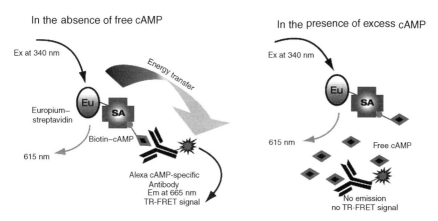

FIGURE 9.6   Assay principle of lanthanide chelate excitation or LANCE. cAMP, the molecule under test, is conjugated with the Eu^{3+} donor fluorophore via a biotin linkage. A cAMP-specific antibody conjugated with the acceptor fluorophore associates with the labeled cAMP molecule except in the presence of free cAMP from the process under test (e.g., GPCR activation). Thus, sample cAMP separates the fluorophores and decreases the TR-FRET signal. Used with permission from Perkin Elmer, Inc. (See the color version of this figure in the color plates section.)

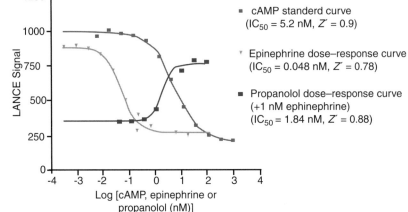

FIGURE 9.7   Data showing response of LANCE cAMP quantification to cAMP titration, agonist (epinephrine), and antagonist (propanolol). Used with permission from PerkinElmer, Inc.

PerkinElmer is far from being alone in the TR-FRET market for cAMP–based GPCR assays. In the HTRF^{TM} technology from Cisbio, an Eu^{3+}-labeled anti-cAMP antibody is used with tracer cAMP conjugated to "d2," an advanced fluorescent acceptor whose improved characteristics (especially smaller size) are asserted to improve EC_{50} determinations. In the "one-step" assay, all reactive species except the europium conjugate are mixed for a 30-min activity incubation, followed by addition of the donor fluorophore ("cryptate conjugate") in a lysis buffer and a further 30-min incubation, presumably to allow displacement and reequilibration. The method requires an instrument capable of TRF. The kit is marketed in three forms, "femto 2," "dynamic 2," and "HiRange," to address various cAMP concentrations to be encountered; the specifications cover the range of 0.3–121 nM. In an assay against a vasopressin receptor, the calculated EC_{50} figures corresponded reasonably well with the literature (494). One advantage Cisbio claims for this kit versus kits employing the biotin–streptavidin system (the target is unspecified) is avoidance of potential complications from biotin present in biological samples or growth media.

9.2.3.4 Chemiluminescent Competition-Based cAMP Detection

MSD markets a cAMP assay that employs their proprietary carbon electrode microplate technology (MULTI-ARRAY or MULTI-SPOT^{TM}). Generated cAMP displaces tagged cAMP molecules from anti-cAMP antibodies coating the plate, whereupon an electrical signal causes the remaining label to emit an electrochemiluminescent (ECL) signal. The technique features "multiplexing": receptor is immobilized in four spatially distinct locations within each well, corresponding to four electrodes within the plate. Separate signals are obtained for each electrode. Measurable inhibition constants are subnanomolar, typically with very small errors. The assay geometry allows the excellent control procedure of immobilizing a similar receptor that is not expected to yield a signal in the same well. The assay involves a 1-h incubation, a wash, addition of labeled ligand, another 1-h incubation, a further wash, and rapid reading of the plate (70 s), for a total assay time of 2.5 h. The interesting multiplexing feature allows multiple readouts from individual wells, potentially contributing to throughput, although the assay is nonhomogeneous and rather slow, and any accuracy advantage accruing from multiple measurements within the same well may be small in these days of highly replicable reagent dispensing.

9.2.3.5 Comparisons of Competition-Based cAMP-Detection Methods

It would be too much to hope for a global side-by-side comparison of all the methods under the same conditions, let alone a broad study encompassing calcium-based and other methods as well. However, useful work has been done, although, not surprisingly, the results are in conflict for some parameters. The results of side-by-side comparisons of the DiscoveRx EFC technology with LANCE and HTRF are currently available at the following URL: http://www.discoverx.com/literature/merck_camp_presentation_from_mw06.pdf. Note that the comparison was performed by Merck scientists, making it hard to be sure about the degree of independence of the non-peer-reviewed results. LANCE yielded unrealistically high EC_{50} values for the both agonist and antagonist, although the activities of both compounds were at least detected. The HTRF procedure yielded a stronger EC_{50}, albeit with a poorer signal-to-background ratio, for the agonist, but failed with the antagonist. EFC gave good results with both molecules. 3456-well plates were used in these assays, indicating, among other promising points, the excellent signal strength of all three assays.

It should be noted that in 2006, PerkinElmer performed a separate comparison of its own LANCE, the Cisbio TR-FRET method, and EFC, concluding that EFC is less sensitive by an order of magnitude and that only LANCE yielded good signals after a 24-h incubation, allowing effective "off-line" plate reading. Thus, as usual, quality depends on one's choice of parameters, as well as, perhaps, one's identity.

Among other firms that offer cAMP immunoassays are Applied Biosystems, Assay Designs, Biovision, Calbiochem (colorimetric, but sensitive), Cayman Chemical, Cisbio, Molecular Devices (MDS Analytical Technologies), and Sigma–Aldrich.

9.2.3.6 Coupled Bioluminescent Methods

Since cAMP is a product of enzymatic activity on ATP, it should in principle be possible to reverse this activity to yield ATP in a cAMP-dependent fashion. The author has attempted to make this work as a practical cAMP assay. Unfortunately, the reaction is highly endergonic (although entropically favorable), and little success was achieved in developing an assay for practical use. The law of mass action suggests that if it is possible, it will require high concentrations of adenylate cyclase and inorganic pyrophosphate in an ATP-free solution, along with a fairly high concentration of luciferase and its substrates.

The cAMP-quantification methods described in the previous sections exploit the structure and/or biochemistry of cAMP, with little regard for its second-messenger status. Such is not true of the cAMP-glo^{TM} assay method from Promega (Fig. 9.8).

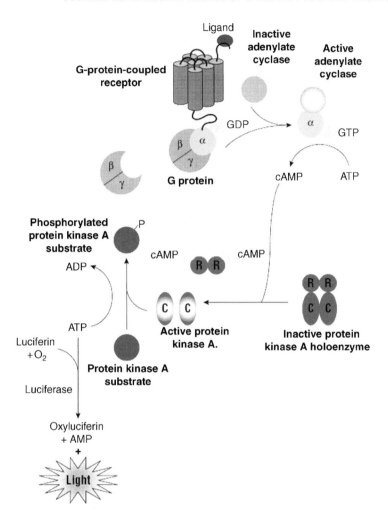

FIGURE 9.8 Schematic representation of CB assay of cAMP concentration using cAMP-Glo™. Productive engagement of the GPCR and the associated rise in cAMP levels cause activation of kit-supplied protein kinase A and consumption of ATP. ATP levels are then measured by Kinase-Glo™(see Chapter 4), yielding a negative signal relative to cAMP concentration. This extremely complex series of steps is nevertheless robust, showing the potential power of CB reaction series. Used by permission of Promega Corporation. (See the color version of this figure in the color plates section.)

This technique exploits the exquisite sensitivity of protein kinase A to activation by cAMP. Addition of minute quantities of cAMP to a vessel or well containing the other PKA substrates (protein target and ATP) as well as luciferase and its substrates yields a negative light signal. Apart from the choice of cAMP as the limiting reagent, this is essentially the same assay as that described in Chapter 4, in that kinase activity is measured by following the consumption of ATP by means of luciferase. As in that assay, a negative signal is generated, and the considerations mentioned there apply here as well. Note that Promega markets the kinase kit separately as Kinase-Glo^{TM}, although it involves the same technology.

The cAMP-Glo protocol begins with a 15-min lysis step. Subsequently, a reagent solution containing the PKA is added for a 20-min incubation to allow detectable consumption of ATP. Finally, the luciferase components (the Kinase-Glo reagents, which also appear to contain a stop reagent for the kinase) are added in a third step and the reactions are incubated for 10 min and read in a plate luminometer. The total assay time of 45 min is shorter than the time required for other cAMP assays. Promega asserts that the half-life of the reaction (presumably using proprietary genetically engineered luciferase) is greater than 4 h, allowing processing in batch mode and performance of the assay without injectors.

Given our general knowledge of CB assays, it should be possible to manipulate the assay configuration to yield slightly different results if desired. For example, detection of even smaller concentrations of cAMP may be possible by reducing the amount of PKA (to reduce background activity of PKA and/or the activity of other ATP-consuming processes) used in the initial enzymatic step and lengthening this step. If a continuous assay is desired, it is likely that an ordinary preparation of luciferase reagents (without a kinase inhibitor) could be substituted for the Kinase-Glo reagent, although by doing this one sacrifices any advantages enjoyed by use of the enhanced luciferase enzyme.

The limit of detection for cAMP in the standard assay appears to be in the low nanomolar range. This is comparable to that exhibited by the cAMP competitive assays, though not superior. Fig. 9.9 shows the results with the forskolin agonist, apparently yielding a very low coefficient of variation (the error bars are undetectable except

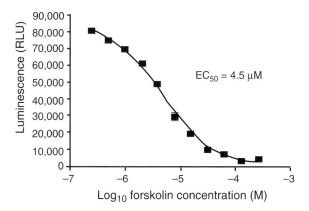

FIGURE 9.9   Response of the cAMP-Glo assay of GPCR activation to the agonist forskolin. Increasing concentrations of the agonist stimulates greater levels of GPCR activation and raises the cAMP concentration, leading to activation of protein kinase A and consumption of ATP, finally reducing the luminance signal. The data quality is above average for such a complex assay. Used by permission of Promega Corporation.

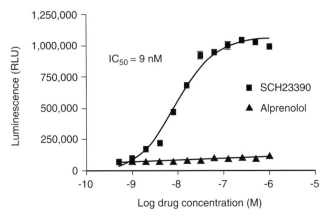

FIGURE 9.10 IC_{50} determination with cAMP-Glo. HEK293 cells expressing the D1 excitatory dopamine receptor were treated with the antagonist SCH23390 in the presence of 100 nM SKF38393 (agonist). Control data were taken with the inactive alprenolol in place of the antagonist, allowing free activation of protein kinase A and nearly complete exhaustion of ATP. The data indicate a strong IC_{50} value for SCH93390, although without using multiple concentrations of agonist it is difficult to say whether this is an absolute or merely phenomenological value; also, the curve shape suggests the presence of unexplained cooperativity in the system. Used by permission of Promega Corporation.

at one point). The measured EC_{50} is similar to that measured by other means. The residuals are slightly worse in the reaction with the antagonist SCH23390 (Fig. 9.10), although the coefficient of variation for the individual points is still extremely small. Clearly, the method is suitable for assaying both agonists and antagonists, although no strong advantages over competitive methods are obvious, apart from the slightly shorter reaction time, the ability to use an instrument without a lamp and without careful collimation of the light path, and the absence of interference from fluorescent compounds.

9.2.4 GPCR Assays Involving Inositol Triphosphate Detection

Inositol triphosphate is yet another second messenger implicated in GPCR signaling. The primary source of IP_3 is hydrolysis of phosphatidylinositol-4,5-bis-phosphate to IP_3 and another second messenger, diacylglycerol, by the enzyme phosphatidylinositol phospholipase C-β, which is activated by G_q- and G_o-type GPCR signal cascades. IP_3 further modulates intracellular release of Ca^{2+}, another second messenger, and exerts further profound effects on intracellular signaling, including modulation of activation of MAP kinases (495) and estradiol effects (496). Production of IP_3 itself is a useful readout for assessing GPCR activities, but the greatest utility of these assays may be in attempts to gain an integrated picture of the downstream effects of activation of a given GPCR, including the full range of potential second messengers.

The traditional assay for IP$_3$ formation as a consequence of GPCR activity involves feeding the cells the tritium-labeled myo-[^{3}H]inositol, along with inhibition of inositol phosphatase to enable accumulation, in a highly labor-intensive step requiring extensive cell handling and washing (497). Radiolabeling-displacement assays and others involving chromatography were eventually replaced in the drug discovery enterprise by modern HTS methods.

Like other small molecules we have examined, IP$_3$ is readily quantified by competitive fluorescence technology. The HitHunter^{TM} platform from DiscoveRx involves displacement of fluorophore-labeled IP$_3$ from IP$_3$-binding protein by free IP$_3$ generated by GPCR activation (Fig. 9.11). As we have seen before, the liberation of the fluorophore as a small molecule from the much larger molecular complex with its binding protein causes a dramatic decrease in FP. A standard curve for IP$_3$ detection shown in Fig. 9.12 indicates low-nanomolar sensitivity.

Cisbio has taken a different approach in the IP$_3$ arena by developing a TRF assay for inositol-1-phosphate (IP$_1$). Vendor data demonstrate the rapid accumulation of this metabolic breakdown product of IP$_3$ following production of IP$_3$ by GPCR activity. The IP$_1$ TRF assay is precisely analogous to Cisbio's cAMP assay (described in Section 9.2.3.3). The claimed limit of detection for IP$_1$ is 15 nM. It is, of course, possible for the concentration of a breakdown product to be considerably higher than that of the source if system flux is very high, but this should not be assumed. However, Cisbio literature makes the convincing argument that while the existence of IP$_3$ is transient, that of IP$_1$ is not, and the latter is a superior assay target for that reason.

9.2.5 GPCR Assays Involving GTP

The objective of GPCR-activity assays involving GTP is not direct quantification of GTP, but rather measurement of GTP association with G proteins, since that is the

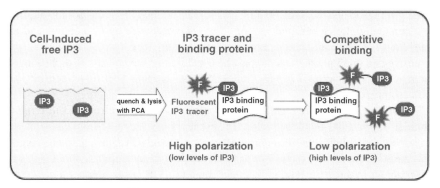

FIGURE 9.11 HitHunter^{TM} assay of IP$_3$ as surrogate for GPCR activation. The supplied conjugate of IP$_3$ and the fluorescent tracer are displaced by IP$_3$ from the sample, greatly reducing the molecular weight of the fluorophore, increasing its rotation rate, and diminishing the FP signal. Thus, the assay yields a negative response to IP$_3$ concentration. Used by permission of DiscoveRx Corporation.

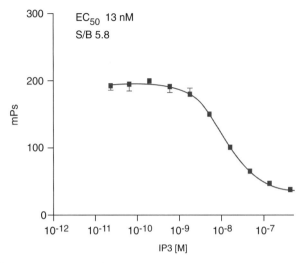

FIGURE 9.12   Standard curve of IP$_3$ using HitHunter FP assay. Used by permission of DiscoveRx Corporation.

event essential to activation of the proteins. Methods involve radiolabeling abound. A widespread approach is the use of ^{35}S-labeled GTP, which associates with G proteins, and therefore with cells expressing GPCRs, upon activation of the GPCRs. Association is quantified by filtering and counting or, in the more modern variant, by SPA (498).

The first major nonradioactive GTP-binding assay for GPCR study was a variation of the DELFIA^{TM} TRF platform from PerkinElmer (499). The key molecular element of the assay is an Eu^{3+}-labeled GTP molecule, which associates with the G protein upon effective stimulation. Time-resolved autofluorescence is measured following filtration and washing (Fig. 9.13). As the product literature points out, there is a nonnegligible background rate of GDP/GTP exchange by G proteins, and this rate should be measured and subtracted from the stimulated rate. The assay is labor intensive and nonhomogenous but has the considerable advantages of eschewing radioactivity and incorporating only a single "exotic" reagent. DELFIA is sensitive and well established.

As of this writing, the DELFIA assay appears to have a clear lead in terms of nonradioactive GTP-binding assays. The issue of whether a CB assay is possible arises, but the technical hurdles are significant. Although GTP can clearly be coupled into the ATP/luciferase system by terminal transferases, quantification of GTP is not the problem in this case. The most direct route to a CB assay, and not a very direct one, might be to obtain the GPCR of interest in a membrane preparation in sufficient quantity to allow sequestration of GTP upon activation of the G protein. The pool of free GTP could then be measured by a combination of the transferases and the luciferase system, yielding a negative luminescent signal. Although considerable labor would be required in preparing active GPCR-containing membranes, this assay does

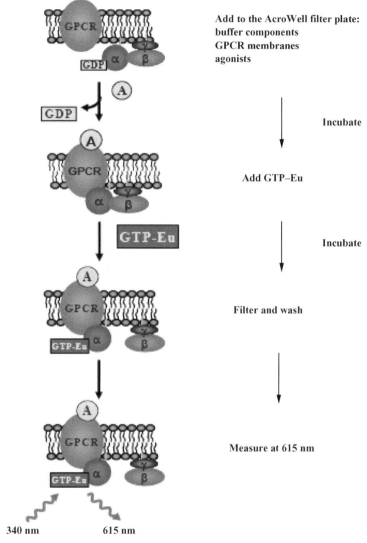

Add to the AcroWell filter plate:
buffer components
GPCR membranes
agonists

Incubate

Add GTP–Eu

Incubate

Filter and wash

Measure at 615 nm

340 nm   615 nm

FIGURE 9.13 Assay of GTP binding by G protein using time-resolved fluorescence method DELFIA. Activation of the G protein leads to the exchange of bound GDP for the GTP–Eu conjugate. Remaining free conjugate is then washed away and the binding is measured by TRF. Used with permission from PerkinElmer, Inc.

not seem quantitatively out of reach, considering the fact that the DELFIA method also relies on measuring individual GTP molecules sequestered by G proteins. The CB method is likely to be much faster than DELFIA, and, if we may generalize from other systems in which fluorescent and luminescent methods can be compared, the sensitivity may be at least equivalent. However, DELFIA does yield a positive signal.

9.2.6   GPCR Assays Involving Reporter Genes

One of the major thrusts in GPCR assays is measurement of their downstream effects, usually in whole cells. This approach has the great advantage of yielding information about the relationship of receptor engagement with events of widely understood biological significance, rather than second-messenger flux or G-protein activities, which may have varying implications from cell to cell. Any cancer biologist knows the consequences of Jun N-terminal Kinase (JNK) activation (as far as they are known). This is similar in a way to the argument made in Chapter 6 regarding the superiority of methods of measuring actual inhibition of acetylcholinesterase, rather than the structural approach of querying the presence of specific dangerous molecules. The biology is the true end point of any assay.

The range of potential downstream targets of GPCRs is too great to fall within the scope of this chapter, but we will permit ourselves a brief look at a few areas, beginning with mitogen-activated protein kinase (MAPK) activation. These kinases are profoundly involved in growth-signaling pathways. At least four groups of MAPKs are activated by the gonadotropin-releasing hormone receptor (GnRHR), a GPCR whose downstream signals are mediated primarily by activation of protein kinase C via G_q-type G-proteins: extracellular signal-related kinase (ERK), JNK, p38MAPK, and big MAPK (BMK) (500).

It is unnecessary to duplicate information given in Chapter 12 about reporter assays, which are means of measuring the transcriptional activation of specific gene promoters. The principles are no different in the field of GPCR study, although the sheer quantity of potential target genes is so great that one suspects a good-sized biotech industry could emerge merely to supply appropriate reporter constructs. Examples of academic researchers who have used these reporter assays in analysis of GPCR activation and its consequences would fill pages of citations; we give here only a few examples (501–503). The Rees et al. offering is interesting because the paper relates specifically to development of a MAPK reporter-based screening system for agonists and antagonists of the CXCR1 chemokine receptor, a GPCR.

In the high-throughput screening context, where the emphasis is less on elucidating complex downstream networks and more on reliable and versatile readout strategies, elements of GPCR signaling that are more proximate to the initial events may nonetheless give rise to reporter gene assays. Transduction of second-messenger signals is often mediated by proteins that serve as adaptors between the small molecule and the promoter elements known as response elements. The cAMP response element-binding (CREB) protein is an example of such an adaptor. CREB is a current target of "knockout"-type therapeutic approaches such as antisense and decoy binding targets, for example, in the induction of Bcl-2-mediated apoptosis in cancer (504). CREB is a general mediator of the critical cAMP pathways, yet it is also close enough in molecular biological terms to the regulatory machinery to be of clear biological relevance. The response to this implicit demand is to engineer a fusion of the response element and a luciferase reporter gene, which can be used with the whole range of products marketed for these assays. On the other hand, not surprisingly, established technologies such as LANCE and others have been adapted to detect phosphorylated

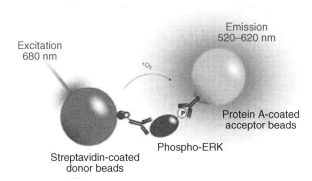

FIGURE 9.14 AlphaScreen®/SureFire™ resonance energy transfer assay of activation of extracellular signal-related kinases (ERKs) as surrogate for GPCR activity. Two antibodies or other ligands are necessary, one immobilized on the acceptor bead and the other conjugated to the donor bead, often by a biotin–streptavidin linkage. The proximity of the donor and acceptor beads, brought about by the antibody-dependent linkage through the phosphorylated ERK (analyte), leads to rapid migration of singlet oxygen across the gap and light emission. Used with permission from PerkinElmer, Inc.

CREB itself. Phosphorylation of CREB is associated with its marking for degradation by proteasomal processes, representing yet another evolved method of modulating the effects of GPCR signaling (505). Many kinases are activated by GPCRs, and observing phosphorylation as a strategy for investigating the effects of GPCRs is a "downstream" strategy that is growing in popularity and receiving increasing attention from industry.

Deserving of separate mention is the recent announcement of the combination of the SureFire™ technology developed at TGR Biosciences with the PerkinElmer AlphaScreen® platform (see Fig. 9.14 for a schematic view of the method as used in detecting ERK phosphorylation (506)). AlphaScreen itself is a bead-based resonance energy transfer technology original developed by Dade Behring and marketed as LOCI™ (luminescent oxygen channeling immunoassay) (507). LOCI is still used clinically, while PerkinElmer has rights to research applications (ALPHA stands for "amplified luminescent proximity homogeneous assay"). AlphaScreen involves separate Donor and Acceptor beads, both of which are smaller (250 nm) than, for example, the beads used in the scintillation proximity assay. PerkinElmer claims the advantages of better suspension characteristics, lower clogging, higher surface-area-to-volume ratio, and so on, for this feature. The principle of signal generation depends on generation of singlet oxygen by the donor bead, which is conjugated with the photosensitizer phthalocyanine. This moiety reacts with molecular oxygen upon excitation at 680 nm to form singlet oxygen, which has a half-life of a few microseconds and is estimated to diffuse about 200 nm during its lifetime. If the singlet oxygen species contacts a proximate Acceptor bead before decaying, energy transfer to thioxene molecules conjugated to the bead causes emission of light in the range of 520–620 nm; otherwise no signal results. The proximity depends, of course, not only on random diffusion, but also, critically, on the presence of interacting species immobilized on the beads.

Donor beads are often provided with conjugated streptavidin, allowing the user to employ his/her own biotinylated ligand to address that side of the interaction. The Acceptor beads are often provided with immobilized antibodies on the surface; the antibody can be against a ligand of interest, or, more commonly, against a tag that can be attached to the desired ligand through chemistry or genetics/protein expression.

From the point of view of optics, AlphaScreen offers several advantageous features. The readout is in time-resolved mode, which effectively eliminates many sources of fluorescent background. The absolute range of the excitation pulse is well removed from most sources of autofluorescence, and so on. The fact that the emission observed is actually higher in energy than the excitation beam is unusual (impossible, for ordinary fluorescence) but is not difficult to handle for most instrumental setups that are capable of time-resolved fluorescence measurements. High sensitivity and very low background are claimed, and the possibility of extreme miniaturization allows cost to reach a few cents per well.

Earlier AlphaScreen kits were directed to the detection of small molecules such as IP_3. The IP_3 kit was complex, involving (1) a biotinylated IP_3 analog to associate with the streptavidin-conjugated Donor beads; (2) an IP_3-binding protein tagged with glutathione-S-transferase (GST) (these two reagents are supplied in the AlphaScreen IP_3 Supplement Kit); and (3) an anti-GST antibody immobilized on the Acceptor beads for use with any GST-fusion binding protein. Successful performance of the assay required a long series of events to occur without error. However, the complexity, at least, of the new-generation kits has been reduced to about seven vials, including buffers and user-supplied reagents, with a straightforward nine-step protocol, although the assay still requires several hours for all the required diffusions and reactions to occur. The reagents are light sensitive and the reaction must be carried out in a dark environment.

Fig. 9.15 shows the results of an AlphaScreen assay rapamycin inhibition of phosphoinositol-3-phosphate kinase (PI3K) pathway kinases. The data appear to be of average quality for the methods we have looked at.

Note several interesting features of this technology, especially in comparison with such methods as FRET. FRET is a useful means of measuring molecular separation for only a small range of distances (depending on the chosen fluorophores), since the orbital overlap integral of the donor and acceptor species falls off as the sixth power of the distance. The AlphaScreen technology depends on molecular diffusion instead (or, rather, the competition between diffusion and the lifetime of singlet oxygen), implying an inverse second-power relationship between separation and signal. Thus, one might expect a wider range of useful separations, but this issue is complicated by the fact that the size of the beads itself is on the order of the measurable separation.

The complex phenomenon of activation of serum response factor (SRF) by various GPCRs will serve as our final example. SRF plays a role in differentiation of smooth muscle. Six different types of G proteins have been found to activate SRF, depending on receptor and ligand type (508). A serum response element reporter fusion construct was used as the primary readout to help in deconvoluting these effects. Genetic engineering and the widespread use of luminescent assays enable virtually

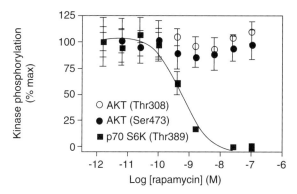

Confluent mouse NIH-3T3 cells were serum starved on and stimulated with
PDGF for 10 min in the presence of various concentrations of rapamycin
inhibtor. Phosphorylation of AKT (Ser473), AKT (Thr308), and p70 SoK (Thr369)
kinases were measured using AlphaScreen *SureFire* phosphokinase and
AlphaScreen Protein A detection reagents on an EnVision Alpha plate reader.

FIGURE 9.15 Assay of effects of rapamycin (inhibitor) on kinases of the phosphatidy-
linositol-3 kinase pathway. Only phosphorylation of the specific kinase target (threonine 389
of p70 S6K) is affected. Used with permission from PerkinElmer, Inc.

any combination of biochemical perturbation and assessment of the transcriptional
effects by measuring luciferase activity.

9.2.7 Novel GPCR Assays and Other Strategies

With such an extraordinarily broad range of GPCR assays available, a few outliers
are only to be expected. The ALIS^{TM} approach from NeoGenesis encompasses both
affinity selection and mass spectroscopy. A library of 5,000,000 compounds is divided
into 2000 pools of 2500 compounds each. GPCRs are expressed in high quantity and
obtained *in vitro* in active form, able to bind agonists and antagonists. Compounds
from the pools that associate with the GPCRs are isolated from the pools by affinity
selection and identified by high-throughput mass spectroscopy. These leads may then
be tested in other ways. Of course, there are many other approaches that exploit
physical associations between small molecules and GPCRs. This is in a sense the
opposite of the "downstream" philosophy, and it is to be expected that many of these
ligands will not prove to be useful as biological response modifiers. As of this writing,
NeoGenesis does not appear to be an active presence on the Internet and may be out
of business. A related approach is of course targeting of the libraries, using structural
and docking information from the ever-improving field of bioinformatics. Peakdale
Molecular is an example of a firm with an expanding panel of compound libraries
derived from *in silico* pharmacophore modeling and "virtual screening."

Other firms enter the GPCR market in "service" roles. Athersys offers the
RAGE^{TM} technology, a means of bringing about chromosomal overexpression of
genes of interest. Application of this method to GPCRs leads to cell lines expressing

controlled levels of the receptors without the complications of extrachromosomal DNA. Chemicon has developed high-expressing cell lines for "promiscuous" G proteins, which allow a calcium mobilization readout for all GPCR types. Developments in the field are many and rapid, and it is inevitable that new ones will appear on a regular basis.

9.3 SUMMARY

The role of CB assays in GPCR research is currently undergoing redefinition. It is rather surprising that the aequorin system came into the picture so late, given the biological understanding that has existed for several decades, but, as is often the case, the technical challenges have been significant, and since other methods are rapid and sensitive as well, the need has not been as strong as in other areas. The coupled PKA assay currently yields a negative signal, although a positive signal should be achievable by use of the Welch methods. In its present form, however, the method represents an elegant alternative that enjoys many of the common advantages of luciferase-based coupled assays. Finally, the issue of whether cAMP itself can be driven thermodynamically uphill to yield ATP and constitute an effective assay method remains unresolved, largely due to the thermodynamic problem.

Several levels of choices exist for the assay developer focusing on GPCRs. Availability of equipment, in-house expertise, and personal preference may influence the choice of physical readout, but the time, cost, and informative value of the methods presented differ greatly, and for an enterprise thinking of undertaking GPCR screening for the first time, or even for an established group, consideration of newer methods such as aequoporin or recombinant calcium sensors may prove valuable. The choice of biochemical readout is inseparable from the strategic goals of the organization, such as whether a general GPCR capability is desired. In this case, measurement of second messengers that are restricted to certain GPCR types, such as cAMP or IP$_3$, may not be the complete answer. However, the availability of hybrid GPCRs in which the G proteins are exchangeable for G proteins of other types may soon resolve this issue for many or all of the second messengers, enabling, for example, GPCRs that normally yield a cAMP readout to induce calcium flux or IP$_3$ synthesis. Other issues are whether the arrestin/endocytosis process can yield a superior level of biological understanding in a cost-effective manner and whether a complete picture of second-messenger activities and tracing of the fates of both G protein components are desired. All of these will enter into a decision, or lead to the conclusion that only multiple, semiorthogonal approaches will yield a sufficient level of understanding to support the intended drug discovery effort. The choice of GPCRs and G protein types to be studied is of course part of this and depends on many factors, which may include in-house expertise in chemistry and biochemistry, medical need, corporate niche and patent position, and so on. The GPCR field is so vast, the number of GPCRs awaiting productive attention with a view to pharmaceutical development is so great, and the potential value of effective interventions in GPCR pathology is so high, that a commitment to invest in this area may provide unforeseen benefits for decades.

10

COUPLED BIOLUMINESCENT PROTEASE ASSAYS

10.1 INTRODUCTION

Like the kinases and phosphatases, proteases are a large and highly heterogeneous class of enzymes that has been the target of copious investments of effort and treasure with the goal of developing advanced assay technologies. The literature on proteases is voluminous in the extreme. However, the summary of proteases and their properties herein is somewhat abbreviated relative to the contents of other chapters, simply because the importance and market penetration of CB protease assays have been limited to date.

10.1.1 Proteases

Proteases and the similar (and overlapping) category of peptidases are enzymes that cleave the peptide bond, the covalent bond that links adjacent amino acid residues in those versatile biological machines known as proteins. In short, proteases *digest* proteins, leaving either shorter peptide chains or individual amino acids behind. The peptide bond is stronger than an ordinary single carbon–nitrogen bond because of resonance stabilization associated with the free electron pair on the nitrogen, which allows the O–C–N structure to rehybridize, yielding a metastable species with a double bond between the carbon and the nitrogen (the oxygen is negatively charged and the nitrogen positively charged in this alternative structure). Thus, proteases must cleave strong bonds, and the ability to cleave the peptide bond has been viewed informally as the final test of any human-designed structure

Coupled Bioluminescent Assays: Methods and Applications, Michael J. Corey
Copyright © 2009 by John Wiley & Sons Inc.

asserted to have enzyme-like characteristics, a test that has been exceedingly difficult to pass (104).

10.1.1.1 Importance of Proteases   Proteases play literally hundreds of different roles in disease processes, of which it is possible to list only a few here. The notorious β-amyloid peptide, a causative agent of Alzheimer's disease, is produced by the aberrant action and/or inhibition of proteases (509, 510). Human immunodeficiency virus (HIV) requires protease activity for productive infection, and this activity is an important therapeutic target (511, 512). A failure of protease inhibition is a critical causative factor of emphysema in smokers, leading to unchecked digestion of essential lung tissue (513). Proteases play important roles in cancer metastasis (514). Diseases of various types and severities are caused, exacerbated, controlled, or prevented by the action or inhibition of proteases. Moreover, as the most important available tool for specific cleavage of proteins, these enzymes are also indispensable in the life sciences and are the targets of intensive protein-engineering efforts.

10.1.1.2 Types of Proteases   The traditional classification of proteases is by mechanism of action, although the mechanistic categories correlate with functional groups only in a general sense. The great Hartley proposed divisions in 1960 (515) so insightfully that further developments have merely confirmed his scheme (516). Thus, the four major classes are as follows:

- *Serine proteases* invariably display the canonical "catalytic triad" of serine, histidine, and aspartate residues in a fixed spatial orientation, each of which is essential for efficient catalysis (517–519). These elements act in a "charge-relay" system in which the serine is deprotonated to yield a strong oxyanionic nucleophile in the active site by mechanisms that took many years to elucidate and underlie much of our current understanding of catalysis by hydrolases (520, 521). Later work established the bases of substrate specificity in these enzymes (522, 523). The functional variety of serine proteases is vast, ranging from digestive enzymes secreted by the pancreas, such as trypsin and chymotrypsin, to the critical activation factors of the coagulation and complement cascades (see Reference 524 for an excellent general view of the regulatory activities of proteases). Serine proteases are generally active against a multiplicity of target proteins because their substrate specificity is usually limited to recognition of a single residue in the context of a folded protein; thus, these enzymes would cause general damage to many proteins if their activity were uncontrolled. They are therefore almost invariably synthesized as proenzymes, or *zymogens*, that is, enzymes that are inactive until an inhibitory sequence is cleaved away by the action of another protease molecule (525) (which can be the same protease in some cases). However, even serine proteases that have evolved exquisite specificities, such as Factor I of the complement cascade, which cleaves only a single protein target (the activated C3 protein of the same cascade), are initially synthesized as zymogens (although Factor I exists in active form in circulation) (526).

- *Cysteine proteases* resemble serine proteases in that their mechanism of action depends on deprotonation of an amino acid residue, which then acts as a nucleophile. However, the deprotonated sulfhydryl (thiol) group is not as strong a nucleophile as the oxyanion of deprotonated serine, and various electronic effects within the active site are postulated to account for the activity of cysteine proteases in general and their distinctive characteristics (527, 528). The cysteine protease group includes degradative enzymes such as papain, but also the extremely important cathepsins (529) and caspases (proteases that cleave peptide bonds involving aspartic acid; formerly known as "interleukin-converting enzymes" or ICE proteins), both with signaling roles. The caspases in particular are implicated in the process of programmed cell death (apoptosis), a major element in the control of cancer growth, and are an important target of both assay and drug development.

- *Metalloproteases* (or *metalloproteinases*) require metal ions, generally Zn^{2+}, for catalytic activity. Zinc-requiring enzymes include thermolysin, the critical drug target angiotensin-converting enzyme, endopeptidases (530), neurotoxins, and a wealth of carboxypeptidases, among others (531). As we will see below, the metal requirement places special constraints on CB and other assay methodologies that are sensitive to chelating agents.

- *Aspartic proteases* have aspartate residues within their active sites that are required for activity, but they are distinct from serine proteases in that the remainder of the catalytic triad essential to the latter is absent. The family includes the important enzyme renin, which catalyzes the critical first step in the conversion of angiotensinogen to angiotensin II and thereby plays a crucial role in the control of blood pressure (532, 533). The important digestive enzyme pepsin is also a member of this class and is implicated in the formation of duodenal ulcers.

10.2 PROTEASE ASSAYS

Means of measuring protease activities were developed long before their mechanisms of action were understood. Among the traditional approaches that cannot be recommended for high-throughput operations, or indeed for any but the most specialized studies, are measurements of acid evolved by the liberation of carboxylate groups (employing pH indicators or, worse, pH titrimetry, one of the most tedious assay procedures the author has ever been required to carry out); quantification of free amino groups by means of a ninhydrin (or other) reaction; HPLC, capillary, or other separation of peptide products (these are of course useful for initial characterization of the products of a novel protease); and the less barbaric but still cumbersome approach of using mass spectrometry to analyze these same products. One must, however, add the caveat that these methods may well spring back to life, even in high-throughput applications, if they can be automated on chips or by other means.

In general, however, today's assay developer interested in inhibitor screening of a characterized protease has a range of increasingly attractive strategies to choose from.

The main variations are detection of cleavage of a peptide, which is often related to or derived from a natural substrate, and measurements of chromogenic or fluorogenic products of the hydrolysis reaction. The latter methods employ unnatural substrates in which the chromophore or fluorophore is coupled to a peptide, amino acid, or other small molecule, usually by an amide or ester bond (the peptide bond is a form of amide bond). Proteases may be assayed for activity against bonds that are not amides; the risk is that since most alternatives are weaker bonds that are more easily cleaved, effects that are able to block the peptidase activity but do not completely inhibit the weaker esterase activity, for example, may not be detected. Moreover, the rate constant of the esterase reaction is typically greater, which may lead to misleading conclusions about kinetics or the strength of inhibition. Nevertheless, the use of more labile substrates may be an acceptable approach for lead identification or high-volume kinetics studies.

10.2.1   Chromogenic Protease Assays

It is possible to conjugate a chromophore to any peptide, so proteases that act on free peptides (this includes essentially all proteases apart from those with highly unusual substrate preferences, such as Complement Factor I) can be assayed with such a conjugate. The chromophore most commonly used is *p*-nitroaniline, which is intensely yellow.

One example will suffice. Caspase-3 is one of the apoptotic cysteine proteases of prevailing widespread interest (534, 535). Clear evidence that chromogenic substrates are alive and well in the marketplace is the plethora of molecules being offered for assays of caspase-3, despite the availability of much more sensitive fluorogenic substrates. Chromogenic assays for caspase-3 activity are offered by at least the following firms: AnaSpec, Assay Designs, Biovision, CalBiochem, Invitrogen/Molecular Probes, MBL International, Millipore, Promega, R&D Systems, and Sigma–Aldrich. Frequently, a fluorogenic substrate is provided in the kit along with the chromogenic substrate, and as a result an effort to reduce cost by using a chromogenic substrate may not meet with success, unless the substrate is purchased as an independent chemical, from Invitrogen/Molecular Probes, for example. The lowest list price found for a caspase-3 colorimetric assay kit was just over $2 per well; the kit included DEVD-*p*NA, the *p*-nitroaniline derivative of the DEVD peptide, as the only substrate. A 2 h, 37°C incubation with the substrate is recommended, followed by the readout at 405 nm. The method is more than adequate for detecting induction of caspase activity in cell culture, where materials are not a major limitation.

10.2.2   Fluorometric Protease Assays

Since the sensitivity advantage of the fluorescent readout is about two orders of magnitude compared to chromogenic methods and the cost may not be higher, the only obvious reasons to avoid the approach are lack of expertise, lack of equipment, or a known source of fluorescent interference.

Technically, one must distinguish, as elsewhere, between fluorogenic assays, in which fluorescence is increased, typically via liberation of a fluorophore by enzymatic action, and the more general category of *fluorometric* assays, in which a fluorescence change of any kind is measured. However, for proteases the two are virtually synonymous, since so many of the strategies employed easily yield a positive signal, often with very little background.

10.2.2.1 Protease Assays with Fluorogenic Substrates

We continue to use the popular caspase-3 as our example enzyme. The most prominently marketed fluorogenic substrates for caspase-3 are Ac-DEVD-AMC, Z-DEVD-AMC, and Z-DEVD-R110 (536). The "Z" stands for a carbobenzoxyl group attached to the amino terminus of a peptide to block it from participating in side reactions, while "R110" is the fluorophore rhodamine 110. Rhodamine 110 has been around for some time, but it remains a good choice for peptide conjugation because its fluorescence is almost totally eliminated by amide bond formation. Invitrogen/Molecular Probes markets the EnzChek® Caspase-3 Fluorometric Assay Kit incorporating the Z-DEVD-R110 substrate at a list price under $3 per well; this, however, provides an excellent example of what a little effort expended in assay development can do, since the same substrate is available from the same vendor in bulk form at a list price of 20 mg for $1319. One should be able to develop a method to yield results in a high-throughput screening enterprise using as little as 1–5 µg of this substrate, implying that 5000–10,000 wells can be assayed for this price for a small sacrifice in convenience (possibly even a smaller quantity would work in a 384-well or 1536-well format (537)). The cost savings over the kit purchase appear to be in the range of 10–30-fold. However, it should be noted that the Invitrogen kit also includes the caspase-3 inhibitor Ac-DEVD-CHO, which can be purchased separately as well.

Figure 10.1 shows the results of a caspase-3 assay used to follow induction of apoptosis by camptothecin. Although error bars are not presented in the figure, it demonstrates that a strong and consistent fluorescent signal can be obtained rapidly under these conditions.

What can be done for caspase-3 is possible for most other proteases as well, and the market offers a wealth of general to highly specific fluorogenic protease substrates, usually with the same fluorophores (or the related 7-amino-4-trifluoromethyl coumarin, AFC, which has a different pH profile from AMC) and blocking groups we have already seen. Derivatives of the peptide AAD are fairly specific for granzyme B, at least with respect to the members of the caspase family (although many other proteases recognizing aspartate residues will of course hydrolyze this as well). Among hundreds of other such products are the derivatized QAR peptide for special proteases such as spinesin and prostasin, modified VPR for kallikreins and coagulation factors, ERTKR for furin, and simpler sequences such as FR for papain and LR for certain cathepsins.

Invitrogen/Molecular Probes markets as general protein substrates a patented line of proteins, including casein and BSA, heavily conjugated with BODIPY dyes. Technically, these are "fluorogenic" substrates, but, since the actual mechanism of the increase in the fluorescence signal involves relief of intramolecular quenching rather

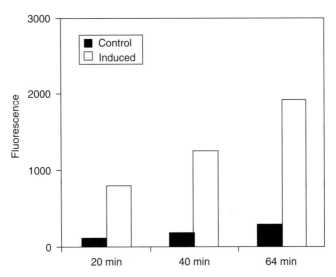

FIGURE 10.1 EnzChek™ assay of protease activity. The results depicted are from an assay of caspase-3, a marker of apoptosis. Apoptotic cells, with higher protease activity, had been treated with the apoptosis inducer camptothecin (10 μM) for 4 h. The fluorogenic peptide substrate Z-DEVD-R110 was used, where Z is the "carbobenzoxy" group and R110 is rhodamine-110. Courtesy of Invitrogen Corporation.

than covalent liberation of a fluorophore, we defer discussion of these substrates to Section 10.2.2.2, where we will find other peptides as well.

10.2.2.2 Protease Assays Depending on Fluorescence Resonance Energy Transfer Fluorescence resonance energy transfer (FRET) is the phenomenon of irradiation of one fluorophore by another, resulting in absorption by the receiving fluorophore of the released photon and either quenching (i.e., the energy is dissipated in nonfluorescent processes) or reemission of the photon, nearly always at a longer wavelength. FRET is discussed in more detail in Chapter 1.

Because the enzymatic action of proteases typically yields two peptides or proteins, either of which could be conjugated to a fluorophore, one would expect FRET to be a natural fit for protease assays, and, indeed, the assay community seems to concur with this conclusion. The primary and rather obvious strategy employs a peptide with a FRET donor and quencher pair spanning the protease cleavage site (538). The distances involved are well within the range of strong FRET quenching. Commonly used fluorophore pairs include EDANS (donor)/dabsyl (quencher) and Alexa 488 (donor)/QSY7 (quencher) (539), but hundreds of pairings are possible and development of new FRET pairs is a very active field. The FRET phenomenon depends inversely on the sixth power of the radius of separation, giving rise to an extreme sensitivity to distance and, usually, a very strictly limited physical range of action; since different fluorophores have different effective ranges, however, a robust effort on the chemistry side may yield entirely novel reagents for new applications. The FRET approach is proposed now not only for canonical high-throughput screening, but also for high-volume profiling of protease specificities (540).

A further clever and useful development within this family of concepts is the development of FRET-based peptide substrates with fluorophores that are actually proteins, such as the cyan or yellow fluorescent proteins, enabling the entire substrate to be constructed by genetic engineering techniques and synthesized by the cellular machinery (541). This strategy is generally limited to preliminary efforts within academia at present, but could replace the entire concept of the synthetic FRET substrate, depending on how difficult the manipulations prove to be and whether adventitious biological effects are observed.

However, improvements are still possible within the current paradigm. Recently, effort has gone into elimination of even the slight artifacts represented by autofluorescence in the visible range and other optical effects of these dyes by developing quenching systems in the near-infrared range (540) for critical applications such as assays for HIV-1 protease activity. The fluorescent donor developed at LI-COR Biosciences is IRDye^{TM} 800 CW, which is paired with the novel near-IR quencher QC1, whose absorption spectrum is held to overlap extremely well with the donor emission, yielding a very clean quenching effect. One suspects that such developments will lead to improved assay quality across the field.

An alternative to the use of a labeled peptide is to retain the labeling concept, but use a natural protein as the substrate instead (542). A CB assay of this type was presented in Chapter 5, using the natural phosphorylation of α-casein as the hydrolytic target. Casein can also be a fluorogenic protease substrate if it is conjugated to a dye such as BODIPY (543). A straightforward approach is simply to conjugate fluorescein isothiocyanate (FITC) to casein, yielding a derivatized protein that was successfully used to screen inhibitors of the β-secretase enzyme, which may be critical in Alzheimer's disease (544).

Finally, there is evidence that FRET assays of protein substrates can provide limited and potentially flawed information, a situation which is ameliorated by adding two-color cross-correlation data to the analysis (545).

10.2.3   Exotic Fluorometric Protease Assays

It comes as no surprise that the exciting fluorometric technologies we have seen applied to the measurement of other enzyme activities have also appeared as contenders for the protease dollar. The time-resolved strategy is closely related to FRET, as described above, but frequently uses the europium ion as the donor (546). LANCE is also back, with a time-resolved fluorometric caspase-3 assay (547). These methods of course require instruments capable of time-resolved fluorescence measurements, and the comments made elsewhere about antibody-based methods (see Chapter 4) still apply to LANCE.

The superquenching strategy we saw with kinases and phosphatases has also been applied to protease assays (548), as has the similar polyelectrolyte amplification scheme (549).

10.3   COUPLED BIOLUMINESCENT PROTEASE ASSAYS

The engaging concept of a CB protease assay has been known since the 1980s, yet has been pursued commercially only within the past 10 years. Promega, a leader in bioluminescence, has recently proposed *two* such assays that employ entirely independent approaches. A U.S. patent issued only a year before this writing describes two "submethods" of one of these approaches, which is in general the use of appropriately modified luciferins. The other set of techniques employs mutant luciferases that are protease substrates themselves. We begin with the first of these approaches.

10.3.1   Protease Assays Using Luciferin Derivatives

Amine derivatives of luciferin, the obligatory luciferase cofactor, were made and studied in the 1960s (550), and a range of chemically derivatized luciferins were synthesized and used for CB assays as early as the late 1980s (551), including phenylalanyl and argininyl derivatives for carboxypeptidase assays. Further progress was made in the 1990s when *N*-acetyl-L-phenylalanyl aminoluciferase was used in a highly sensitive chymotrypsin assay (552, 553). Finally came the arrival on the market of several commercial CB protease assays from Promega (554), which include kits for caspases 3 and 7 (one kit), 2, and 6; several proteasome kits for trypsin-like, chymotrypsin-like, and caspase-like activity; and dipeptidyl peptidase IV or DPPIV (555), a protease of interest in dental research. According to the vendor's data, the sensitivity improvement of these assays over standard fluorometric methods appears to be at least two orders of magnitude, with excellent linearity in the low range. Unfortunately, the data are not available for publication at present, but they can be examined at the following URL, which also links to other material about the CB protease assay: http://www.promega.com/figures/popup.asp?fn=4799ma&partno=G 8350%2C+G8351&product=DPPIV%2DGlo%26%23x2122%3B+Protease+Assay. The given protocol requires two reagent additions and 1–4 h incubations. The cost of the kit if ordered in bulk can be reduced to an attractive range (relative to the fluorescent kits) of ~0.40 per well.

10.3.2   Coupled Bioluminescent Protease Assays Employing Recombinant Luminescent Proteins

As appealing as is the above story, it does not appear to represent the final word in CB protease assays. The luciferin cofactor is not the only essential element of the light-producing reaction that is subject to modification. Luciferases are proteins, and as such can be covalently modified, genetically engineered, or both. The same is true of aequorin and other luminescent proteins. One manipulation that suggests itself in this context is the introduction of a protease cleavage site into these proteins, preferably one whose cleavage modifies the activity of the luminescent protein. To date there have been only a few developments of this kind, but given the market and the promising data, additional progress is almost certain to occur.

An early event in the evolution of luminescent proteins as protease substrates occurred with the filing of an American patent application in 2000 that describes the engineering of the *Renilla* enzyme to include a caspase-3 cleavage site (the canonical DEVD) that destroys the light-generating ability of the luciferase when hydrolyzed (556). The patent is assigned to Chemicon International and the resulting product is marketed by Millipore as Cleavalite^{TM}, although the information available online is limited and no reference to Cleavalite was found in PubMed as of this writing.

In the same time frame, the jellyfish photoprotein aequorin was successfully engineered to include a cleavage site for HIV-1 protease, an enzyme of great medical interest (557). The same group has introduced other marvels of aequorin engineering, including some of the immunoassay reagents mentioned in Chapter 11, as well as a "BRET" (bioluminescent resonance energy transfer) system employing aequorin (558) (see Reference (559) for an interesting review of some of the possibilities).

Promega scientists know a lot about luciferase (as well as possessing some of the best patent protection), and it is hardly to be expected that they would be left out of this recombinant race. In late 2007, the successful engineering of a beetle luciferase to include an *activating* protease cleavage site was announced at a meeting (560). Again the data are not yet available for publication, but the presentation is available online at http://www.promega.com/drugdiscovery/Posters/Wigdal_GloSensor_Protease_Assay. pdf. Of note are the range of digestion conditions successfully tested (pH 5.2–9.1), the specificity of several hundred- to over a thousand-fold achieved with respect to other proteases and substrate sequences, and especially the promising variety of proteases successfully tested, including several caspases, PSA, enterokinase, HIV-1 protease, and SARS-derived enzymes. The engineering of a luciferase that requires activation by a protease is an impressive piece of biochemistry (briefly, it was accomplished by connecting the native N- and C-termini with a linker containing the protease recognition site and creating new termini elsewhere in the molecule) and serves as a demonstration of achieved progress in protein engineering. However, it appears from the Promega data that, now that the problem has been solved, further engineering to produce new mutant luciferases as substrates for other proteases may not be terribly difficult, since the cassette approach used allows substitution of virtually any sequence in the linker.

10.4 SUMMARY

Protease assays must be viewed as an area of great potential growth for CB methods. Fluorogenic compounds have taken the early lead in terms of acceptance and data generated, but CB assays of this type enjoy advantages that will be difficult to overcome with fluorometric methods in the long run. Both synthetic and especially genetically engineered substrates can possess much more exquisite substrate specificity. The sensitivity advantage appears to be in the range of two orders of magnitude, consistent with that seen in other areas where comparison is possible. Finally, the cost appears to be highly competitive and even challenging to the fluorometric strategy, although what the eventual prices will be for the coming assays based on genetically engineered proteins as substrates remains to be seen, since preparation of these reagents can be expensive.

11

COUPLED LUMINESCENT ASSAYS INVOLVING AEQUORIN

11.1 INTRODUCTION TO AEQUORIN

The calcium-sensitive luminescent aequorin proteins were discovered in 1961–1962 by Shimomura, one of the giants of bioluminescence research. The discovery itself has been described in References 561, 562. The source of the most important aequorin is the jellyfish *Aequorea victoria*, although the term *Aequorea aequorea* Forskal (for Petrus Forskal, who named the species in the 18th century) is preferred by Shimomura himself (3). The jellyfish was abundant in the inland waters of the American Pacific Northwest until about 1990 but declined rapidly after that date for unknown reasons. Fortunately, aequorin had already been cloned (563) and was soon expressed in active form (564). However, handling and purification problems delayed the process of obtaining a high-resolution structure by X-ray crystallography until 2000 (565).

11.1.1 Aequorin as the First Photoprotein

Aequorin was the first *photoprotein* to be discovered. A photoprotein, as distinct from a luciferase, is a protein that emits light in a direct stoichiometric relationship with its own concentration; in other words, a protein undergoes long-lasting internal changes in producing light, before ending the reaction in a form that does not readily generate further light. A luciferase molecule can produce many thousands of photons before undergoing an unregulated degradative process, usually oxidative, that terminates the reaction. Luciferase is an enzyme; photoproteins are not enzymes. The obvious consequence is that for an efficient detection reaction, photoproteins, unlike luciferases,

Coupled Bioluminescent Assays: Methods and Applications, Michael J. Corey
Copyright © 2009 by John Wiley & Sons Inc.

must be supplied in quantities roughly equivalent to the quantities of analyte (calcium ion in the case of aequorin).

11.1.2   The Aequorin Reaction

The mechanism whereby aequorin generates light, which was substantially elucidated only after the structure was available, makes it clear why the protein does not exhibit enzyme-like catalytic properties. Production of light involves complete dissociation of essential elements of the aequorin machinery (see Fig. 11.1). CO_2 is released, along with the reaction product coelenteramide (produced from the obligatory cofactor coelenterazine in a ring-opening reaction similar to that described for luciferin in Chapter 1), leaving inactive apoaequrin. The reverse process, the reassociation of the reaction components to the active complex, is entropically unfavorable, like nearly all such multimolecular associations, and is therefore very slow in comparison to the timescale of light generation in response to calcium. There also appear to be one or more reduction steps involved in regeneration, since modification of free cysteine residues in the protein affects the rate, but not the final result, of the process (566). Shimomura points out, however, that aequorin can be forced to act as an enzyme

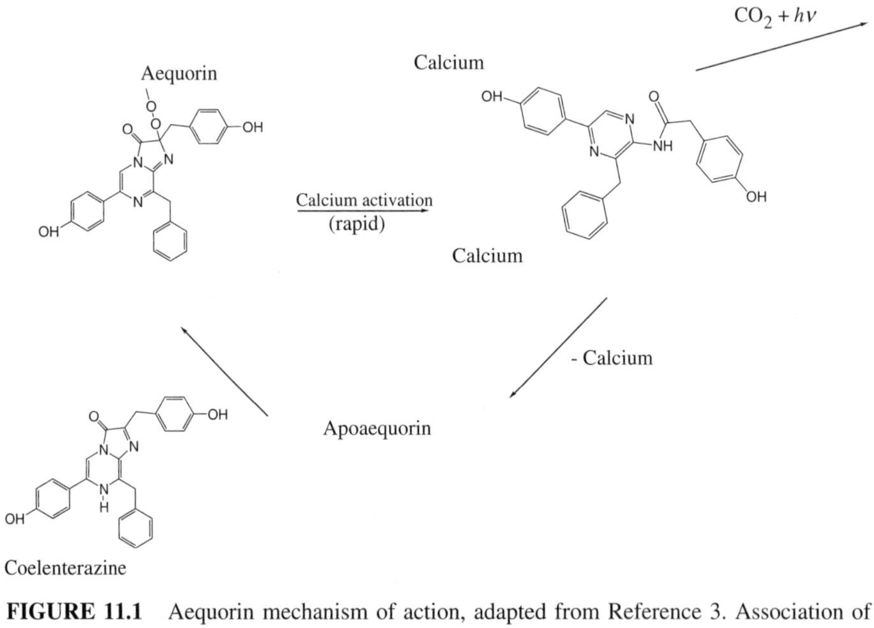

FIGURE 11.1   Aequorin mechanism of action, adapted from Reference 3. Association of two calcium ions causes a dramatic conformational change in the protein, triggering a light-producing decarboxylation and release of coelenteramide. Dissociation of the calcium ions is then followed by slow reassociation of coelenterazine and regeneration of the functional photoprotein. A constant supply of coelenterazine or a means of recycling coelenteramide to coelenterazine must be provided for continuous luminance.

by supplying the components needed for regeneration in the presence of activating Ca^{2+}, but the turnover rate of the ersatz "luciferase" is only $1-2\,h^{-1}$ (3). Under these conditions, the reaction terminates when coelenterazine is exhausted.

The aequorin reaction was recognized as distinct from the luciferase reactions quite early, not only because the protein does not resemble luciferases, but also because the luminance exhibited entirely different properties. Instead of being dependent on the presence of an energy source, aequorin responded to calcium, which undergoes no chemical change during light generation or any other phase of aequorin activity. In other words, calcium was acting as an effector, rather than a substrate. The calcium-sensing aspect of aequorin biochemistry is due to the presence in the aequorin structure of three EF-hand calcium-binding domains that alter the protein conformation upon association with calcium, permitting the light-generating reaction and dissociation of the products. It should be noted that other cations, including some of the lanthanides, induce light production by aequorin and are potential interferents, but fortunately these ions are rarely met with in biological systems (although some fluorescent schemes use lanthanides).

11.2   DETECTION OF CALCIUM IN PRACTICE

The diversity of calcium signaling in humans is vast, to the point that it has served as a model system for the development of mathematical models of complex signaling pathways (567). Apart from G-protein-coupled receptor (GPCR)-related studies, calcium detection and quantification contribute to fields such as the following: neuronal function (568, 569); calcium-dependent chloride channels (570); the calcium-dependent cysteine proteases known as calpains and their varied activities (571) (including involvement in diabetes (572), long-term potentiation (573), and genetically linked muscular dystrophy (574)); a large number of calcium-dependent enzymes, including the venerable protein kinase C and calcineurin (see Chapter 5); activation of potassium channels (575); sodium/calcium-exchanging channels (576); possibly Alzheimer's disease (577); and, of course, pathologies of bone formation and maintenance such as osteoporosis (578). Most of these kinds of investigations should be amenable to the application of CB methods using aequorin, but the details will undoubtedly vary with regard to whether aequorin is expressed by the cells of interest or provided, how coelenterazine is introduced, and so on.

Detection of calcium in biomedical research is currently performed almost exclusively by fluorometric or CB methods. Fluorometric means of quantifying calcium are quite advanced and will be dealt with first.

11.2.1   Fluorometric Quantification of Calcium

Many calcium-sensitive dyes have been developed. The fluo-4 AM and fura-2 AM dyes from Invitrogen will serve as examples of this class of reagents, since they are in widespread use not only for GPCR investigations, but also in other applications requiring calcium quantification. Both are supplied as acetoxymethyl esters (hence the AM). These esters are cell membrane permeant but are hydrolyzed by intracellular esterases, releasing charged groups and leading to retention within the cytosol. Fluo-4 AM is a "single-emission" dye, that is, it responds to association with calcium by increasing its fluorescent output by at least 100-fold and is, therefore, suitable for direct quantification of intracellular calcium. Fura-2 AM, in contrast, has an excitation spectrum that changes upon association with calcium, allowing a ratiometric readout of bound versus unbound fluorophore for microscopy applications. These dyes are useful not only for direct quantification of intracellular calcium, but also for obtaining spatial and time-dependent information about calcium gradients and flux.

Figure 11.2 shows the excellent spatial and temporal resolution that is possible with modern fluorogenic dyes. Calcium concentrations are recorded at four points within a single rat myocyte on a high-millisecond/low-second timescale during a calcium "spark." The responses of the four cellular regions are quite distinct. Although one could quibble about possible artifacts due to the necessity of loading the dye, it is hard to imagine how another technique could yield better results than this data set. Whether or not CB technology can produce results of this kind is not yet known.

Figure 11.3 shows the excitation shift of fura-2 fluorescence upon association with calcium, exhibiting a clean isosbestic point at around 355 nm. Obviously, there is

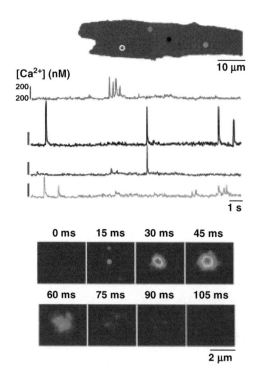

FIGURE 11.2   Bursts of Ca^{2+} release in rat ventricular myocyte labeled with fluo- 4 AM. The outline of a single cell is shown. Courtesy of Invitrogen Corporation. (See the color version of this figure in the Color Plates section.)

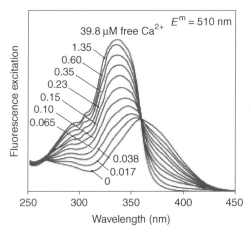

FIGURE 11.3 Fluorescence excitation spectra of dye fura-2 titrated with Ca^{2+}. A clean isosbestic point is seen near 355 nm. Courtesy of Invitrogen Corporation.

considerable fluorescence remaining at 340 nm excitation in the absence of calcium; this enables the production of images such as Fig. 11.4, a false-color photograph showing the variation in the excitation ratios due to the presence of calcium, dramatically revealing details of the heterogeneous distribution in calcium content of this

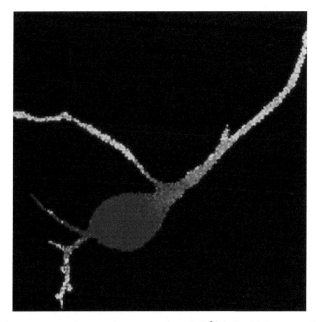

FIGURE 11.4 False-color image of concentration of Ca^{2+} in Purkinje neuron from embryonic mouse cerebellum. Courtesy of Invitrogen Corporation. (See the color version of this figure in the Color Plates section.)

cell type. The value of the demonstration is enhanced by the fact that the unbound dye is shown, as well as bound, making it easier to visualize the calcium gradients.

Despite the quality of these results, the field is moving toward an even higher standard, that is, the no-wash protocol. The FLIPR Calcium-4 Assay Kit from Molecular Devices is a homogeneous formulation. (The FLIPR instrument is described below.) The cells are mixed with the dye and incubated, followed by the readout. Any signal from the extracellular dye is masked by a separate nonpermeant chemical. Fig. 11.5 depicts both the intracellular and extracellular processes involved in this valuable assay in the context of GPCR research.

The FLIPR instrument from Molecular Devices (now part of MDS) is worthy of attention in its own right. It was originally engineered for dedicated calcium quantification and ion-channel work because it had a laser with a fixed wavelength. However, the newer FLIPR TETRA® incorporates an LED array and has much greater

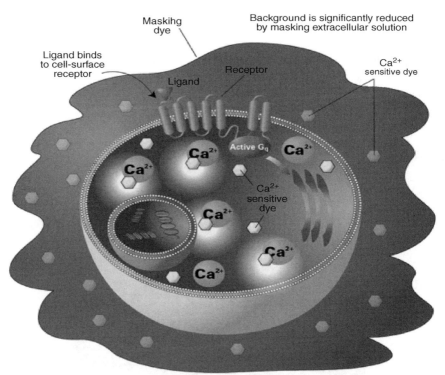

A next-generation no-wash, homogeneous, calcium mobilization assay designed for the most challenging and diffecult GPCR targets

Lower background and increased signal intensity
Significant reduction of ligand or target interference

FIGURE 11.5   Schematic view of the events and strategies underlying the homogeneous FLIPR^TM Calcium-4 Assay. The key to the no-wash format is the masking dye that suppresses the fluorescence of the dye molecules outside the cell membrane. Used by permission of MDS Analytical Technologies. (See the color version of this figure in the Color Plates section.)

wavelength flexibility. Both instruments have HTS-friendly features such as optional internal plate handling, delivery of suspended cells, and easy exchange of 96-, 384-, and 1536-well pipettors. Perhaps the most important feature for our purposes here is the availability of a dual-mode camera that works for both fluorometric and luminometric (aequorin-based) calcium quantifications. However, the instruments are not *general* fluorometers; for example, they do not perform time-resolved fluorescence or fluorescence polarization measurements.

Invitrogen has responded to the challenge of the no-wash fluorometric calcium assay by introducing the Fluo-4 NW Calcium Assay Kit. "NW" means "no wash," which represents an important operational step forward, but this is not necessarily a homogeneous assay, since the vendor still recommends removal of the growth media.

11.2.2   Coupled Bioluminescent Quantification of Calcium Using Aequorin

While the major application of aequorin technology in biomedical research has been and is likely to remain the study of G-protein-coupled receptors (see Chapter 9), aequorin was already being employed as a general calcium sensor in the early 1990s. Cell lines expressing aequorin were developed and shown to respond to cytoplasmic calcium (579, 580), although it was not immediately obvious whether this procedure would exhibit advantages over the use of trappable ester dyes. In contrast to the situation in the cytosol, however, the technical problems associated with specific quantification of calcium in intracellular compartments were severe until a clever means of distinguishing nuclear from cytoplasmic calcium was developed by an Italian group, involving fusion of aequorin cDNA with a translocation signal peptide to ensure the presence of the detector protein primarily within the nucleus (581). Aequorin was also shown in some cases to exhibit superior quantitative response to calcium flux even in the cytosol, for which fluorescent indicators were already available (582), largely because of the dynamic range of the aequorin response and the absence of the undesirable buffering effects associated with the fluorophores. The single-turnover nature of aequorin is an impediment to development of assays requiring more than the data that can be obtained in a single flash reading, but also reconstitution of apoaequorin to the functional protein is not only possible, but also facile, requiring only a 2-h incubation of the expressing cells with 5 μM coelenterazine.

Among the vendors who have taken the greatest advantage of the aequorin possibilities is PerkinElmer, with its AequoScreen products. Among these are various parental cell lines stably expressing apoaequorin (which must, of course, be reconstituted with coelenterazine prior to assay performance) with or without the promiscuous G-protein $G\alpha16$, allowing controlled functional testing of the calcium signaling component of any of a wide range of GPCRs. Other cell lines offered are transfected with specific GPCRs as well. In a separate product line, FroZen^{TM} cells have already been irradiated to halt replication and are ready to be thawed and used directly in an aequorin-based screen or other GPCR assay.

11.2.3   Real-Time Imaging of Calcium Flux with Aequorin

The modern combination of genetic engineering techniques leading to the availability of recombinant aequorin, spatial imaging techniques, and our understanding of the biology of certain metazoan organisms has led to the possibility of making movies of real-time calcium flux. The authors of the report (583) and videoimaging material accessible at the following URL made a record of the successive waves of calcium pulses during the phases of gastrulation of zebrafish embryos: http://www.pnas.org/cgi/content/full/96/1/157/DC1. While quantitative data will always be at the center of scientific inquiry, images such as these greatly enhance our insight into such complex phenomena as development.

11.2.4   Other Uses of Aequorin in Coupled Bioluminescent Assays

Genetic engineering has successfully altered luciferase to fit the strategies of assay developers (see, for example, Chapter 10), and it is not surprising to see the same occurring with aequorin. Among the modifications are extensions of the C-terminus with epitopic peptides for immunoassays (584), extension of a cysteine-free aequorin mutant with an HIV protease substrate to create a luminescent protease assay (557), and the engineering of an aequorin with a unique cysteine residue to enable conjugation of nonpeptide immunoassay targets (585).

It is difficult to generalize about the relative merits of CB and fluorometric methods across this enormous array of potential applications of calcium quantification. The latter techniques are well established. An usual advantage of CB methods, the absence of a labeling step, does not pertain here, since coelenterazine must be added; however, this labeling may be less burdensome and cause less biological perturbation than fluorophore loading. Like other CB methods, aequorin-based assays are often fast and sensitive, but may be subject to interferences due to their coupled nature. The assay developer is encouraged to consider the CB alternative along with other methods.

12

COUPLED BIOLUMINESCENT REPORTER ASSAYS

12.1 INTRODUCTION TO REPORTER ASSAYS

Some five decades after the astonishing Barbara McClintock revealed her profound insights into the mechanisms whereby DNA would not merely specify the structures of proteins but also direct the rate, timing, and circumstances of their synthesis, biologists are fully engaged in the study of *regulation*, the processes that control the activities of genes, and more specifically the production of messenger RNA. The synthesis of this mRNA can be measured in various ways, from tedious and finicky Northern blots to mass spectroscopy, but the most practical approach is to allow the mRNA to be translated into proteins, many of which provide usable signals of their own. Modern techniques of genetic engineering can couple a promoter of interest to expression of virtually any protein that can serve as a readout for promoter activation or quiescence. The luciferases have proven to be the most useful signaling system for this work, although several other reporter enzymes are also in use. (Fluorescent proteins are frequently superior in tissue-imaging applications, which are not covered here.) In the canonical experiment, a promoter of interest is first manipulated into a position "in front of" (or "5′ -to" in molecular-biology jargon) a gene encoding a luciferase. When the promoter is activated, an RNA polymerase associates with the DNA and luciferase messenger RNA molecules are made. Finally, the luciferase protein is expressed by the translation apparatus, and, if appropriate substrates and cofactors are present, light is emitted. The entire process can be carried out *in vitro* if materials for *in vitro* transcription and translation are provided. This simple idea has yielded many thousands of valuable research projects and journal articles and

Coupled Bioluminescent Assays: Methods and Applications, Michael J. Corey
Copyright © 2009 by John Wiley & Sons Inc.

established the careers of vast numbers of scientists. In short, it is by far the most productive area of coupled bioluminescent assays to appear so far.

Superficially, these assays would appear to have little in common with the coupled enzyme assays described elsewhere in this volume, but many of the same principles apply. All the general advantages of CB assays mentioned in the early chapters pertain, as well as all of the caveats. When reporter data are interpreted, care is required in analyzing the kinetics of the reactions, as in any coupled system. The nature of the luciferase (or luciferases, in the case of recently developed methods yielding multiple, independent readouts) is of critical interest, as we shall see. Finally, these are in fact enzyme assays. What is being measured is not only the luciferase activity, but also the catalytic behavior of RNA polymerase as well, and often still other enzymes are involved. As is the case with other CB assays, each of the reactions needed to generate the readout has its own requirements, and the constraints on the system involve both separate requirements and any stresses introduced by combining them. Nevertheless, these methods have enjoyed enormous success, and continuing technical developments guarantee them a promising future.

12.1.1 Brief History and Development of Reporter Assays

Here, we do not consider isotopic and radioactive assays (e.g., (586)), which are obsolete and are not competitive with fluorescent and luminescent methods. The most important competitor for the luciferases in today's reporter assays is green fluorescent protein (GFP), the jellyfish product that was named by molecular biologists in such an uncharacteristically descriptive manner. GFP and its genetically engineered progeny (587, 588) are of inestimable value in such fields as tissue localization of gene expression (589) and studies of vector-transduction efficiency (590). Luciferases are often impractical for such studies, simply because delivery of the requisite substrates and cofactors is inconvenient or difficult. However, from a historical point of view, the advent of GFP as an expression marker is even more recent than the burgeoning of CB assays. The early development of reporter assays instead involved yet another system, known to the biology world as *lacZ*.

LacZ is one of the three structural genes of the *lac* operon, the relatively simple yet highly representative transcriptional unit that helped Jacob and Monod earn their richly merited Nobel Prize for elucidating the fundamental nature of gene regulation. Although *lacZ* is bound to the two other structural genes in natural systems, it was isolated long ago for independent use, because it encodes the enzyme β-galactosidase or "β-gal." Moreover, in the natural system, the substrate lactose, along with a number of other related carbohydrates, acts to "derepress" expression of β-gal (this process is complex, involving a metabolite of lactose that associates with the *lac* repressor protein and inactivates it (591)). Again in nature, β-gal catalyzes the cleavage of the disaccharide lactose into its component monosaccharides, but, like the repressor activity and the gene itself, this process has long ago been manipulated by clever chemistry to serve the needs of biologists. In one of the earliest convenient systems for evaluating gene expression, the synthetic molecule isopropyl β-D-1-thiogalactopyranoside or IPTG was made; this entity serves to inactivate the repressor and stimulate expression of β-gal, but IPTG is not a β-gal substrate and is therefore not consumed. Meanwhile, the enzymatic activity of β-gal is frequently co-opted to cleave either "X-gal" (5-bromo-4-chloro-3-indolyl-β-D-galactoside) or ONPG (orthonitrophenyl-β-galactoside), compounds that yield products with colors in the visible range when cleaved by β-gal.

Combinations of these methods, especially with X-gal, allow exquisitely sensitive selections for expression of β-gal. Moreover, engineered mutants of *E. coli* enable such tricks as detection of fusion proteins and selection of repressor mutations. β-gal has already had a distinguished "career" as an expression marker, and recently, things have come full circle with the development of luminescent substrates for this enzyme (289, 592); thus, a major competitor of bioluminescent assays is now a light-generating enzyme itself. Perhaps a sign of the degree to which the future belongs to luminescent methods is the development of these substrates for an enzyme whose chromogenic activities have played a central role in so many discoveries in the life sciences.

Unfortunately, β-gal is a large protein (1023 residues) and is hard to extract from cells in active form. The current luminescent methods require such extraction. β-gal is also readily denatured and almost impossible to refold into an active form (593). Thus, there are practical limitations to the use of β-gal as a luminescent reporter. Nevertheless, the β-gal enzyme "scaffold" has proven useful in studies of posttranslational modification, one of many clever specialized uses to which the principles of reporter assays can be applied.

12.1.2 Other Non-Luminescent Reporter Proteins

12.1.2.1 Green Fluorescent Protein
Whenever a protein can be harnessed to the service of biology as an information transducer, it probably will be, and it is likely to endure genetic mutilation as well. Green fluorescent protein is no exception. The original source of this extraordinary machine for photon manipulation is the jellyfish *Aequorea victoria*, which naturally fluoresces. This would be enough to render any organism remarkable, but the coelenterate also yields *aequorin*, a calcium-sensitive luminescent enzyme and a major player in assays and screening technologies in its own right. While the light-transducing systems of *Aequorea* and related organisms were elucidated in detail over 30 years ago (594, 595), the subsequent development of biotechnological applications employing these proteins required advanced genetic engineering methods, as well as structural information from protein crystallography (596, 597), and took much longer. Aequorin in calcium-sensing systems is treated in Chapter 11, but its role as an expression reporter enzyme is described below (see Section 12.3).

GFP is a uniquely useful reporter protein for *in vivo* and intracellular applications. It has almost entirely superseded fluorescent small molecules such as fluorescein in this application because it can be produced genetically and because its toxicity is much lower than that of most small fluorophores, which are generally very hydrophobic. If GFP can be cloned behind a cell-type-specific promoter, it becomes possible to follow that cell type visually *in vivo*. Because the light output is strong and reliable, this process can be automated. Thus, GFP has proven to be an extraordinary asset in studies of localization of expression as well as other phenomena.

However, as a rival for luciferases and other luminescent proteins in *in vitro* promoter studies, the future of GFP is uncertain. The light-producing proteins have also been successfully manipulated by the techniques of modern protein engineers, yielding enzymes with tuned wavelengths and improved structural qualities. GFP advocates have responded with the point mutant "enhanced" GFP (EGFP), and subsequently with mGFP (598), or "monster" GFP, consisting of a heavily modified construct originally cloned not from the jellyfish, but from *Montastrea cavernosa*, a coral. Engineered alterations improve light output and stability and reduce apparent cytotoxicity. Nonetheless, all the advantages of luminescence over fluorescence mentioned in Chapter 2 pertain to this situation.

12.1.2.2 New Fluorescent Expression Reporters   The coral protein DsRed showed early promise as an expression reporter complementary to GFP, but technical difficulties related to its biology arose, including robust tetramerization (as with other fluorescent proteins), limiting its breadth of use (599). The anemone protein eqfp611, a similar protein that fluoresces in the far red, was engineered to address related issues (600). It is virtually inevitable that after a "shakeout" period, proteins that can act in concerted fashion in dual-label experiments with GFP and its derivatives will be identified and commercialized. A recent addition to the repertoire of strategies for use of dual-label expression reporters is the possibility of studying alternate splicing phenomena by these methods (601).

12.1.2.3 Chemiluminescent Reporter Assays   Some reporter assays utilize enzymes that are not naturally luminescent but are capable of hydrolyzing chemiluminescent substrates (602–604). This approach has all the advantages and disadvantages of chemiluminescence discussed in Chapter 1. The assays are not in widespread use in the marketplace, although Tropix, now part of PerkinElmer, has actively promoted the concept.

12.2   LUCIFERASES AS REPORTERS OF PROMOTER ACTIVITIES

The use of cloned firefly luciferase as a reporter for promoter activity was reported by Stephen Howell's group as early as 1987 (605). Their expression construct incorporated the same elements as the modern assay, despite the fact that cumbersome *Bal*31 digestion of the promoter of interest and a clever but obsolete method of screening for in-frame products were employed. Cloning difficulties did not prevent the group from creating a number of informative control constructs. The group went on to patent various expression vectors and methods for beetle luciferases and their fusion proteins (606, 607).

The current scene is generally dominated by Promega and its dual-reporter systems, but before addressing those, we turn to the straightforward systems marketed by a number of vendors. Here, it should be mentioned that as of this writing, the

Promega Web site (http://www.promega.com/paguide/chap8.htm) presents a superb and generally useful summary of luminescent reporter technologies, with special points about engineering of luciferases for this application, considerations regarding reporter lifetimes (assays employing long-lived reporter molecules are less sensitive to rapid changes in transcriptional rates), the vectors themselves, and other practical topics not covered herein.

12.2.1 The Modern Luminescent Reporter Assay

Single-enzyme reporter systems are available from many firms. Single-reporter assays requiring cell lysis for measurement are no longer competitive; the enzymes are now mostly secreted. Firefly luciferase can be secreted as well (608), but, perhaps surprisingly, firefly luciferase has been largely superseded in the reporter assay marketplace. The copepod enzymes have generally taken over, partly due to signal strength and convenience considerations. Clontech must have held a naming contest for the Ready-to-Glow™ system, which employs the enzyme from *Metridia* (609), requiring coelenterazine as the prosthetic group. New England Biolabs markets the enzyme from the copepod *Gaussia princeps* (610, 611), which is held to yield much greater luminance than the firefly and *Renilla* enzymes. The assay requires several steps, but the advantage in luminescent intensity appears to be significant; the product literature compares it to the *Renilla* luciferase, which also employs coelenterazine (see Fig. 12.1). The assay appears to be competitively priced as of this writing at ~$0.35/well in bulk.

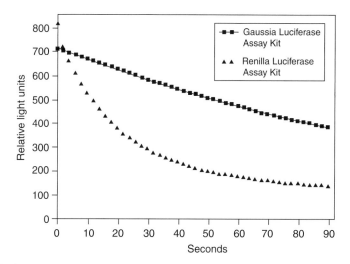

FIGURE 12.1 Kinetics of luminescence of luciferase from *Gaussia* versus *Renilla* enzyme. Commercially available kits were used directly. Used with permission from New England Biolabs, Inc.

To utilize one of these methods, the user first clones the promoter under study into the supplied vector and transiently transfects the cells of interest. A positive control vector contains the same luciferase behind a strong promoter. Expression of the luciferase is sufficient to detect easily after a few hours. With the secreted enzymes, the supernatant can be transferred, mixed with a substrate cocktail, and assayed immediately.

12.2.2 Multicolor Light Emission by Luciferases

Promega's very strong market position in CB reporter assays took considerable time and effort to build. Their patent on mutant luciferases emitting various colors of light, enabling a researcher to distinguish luciferases from different sources (see Section 12.2.3 below), was filed in 1995, but issued only in 2002. This work was based on the earlier observation by Wood et al. that multiple cloned luciferases from the click beetle exhibited varying maximal wavelengths of light emission, from 546 to 593 nanometers, despite their high degree of homology (95–99%) (71). The degree of homology with firefly luciferase was only 48%. This indicated that the beetle luciferases possessed enough structural flexibility to be subject to significant alteration by protein engineering, and, indeed, Sherf and Wood filed a patent in 1994 on a luciferase that had been enhanced in various ways for functionality as a genetic reporter (34). Engineered multicolor luciferases appeared shortly thereafter.

Using luciferases from multiple species, sometimes engineered for stability and emission wavelength as well, yields the ultimate in flexibility for dual-mode assays. Today, the firefly/*Renilla* dual-reporter system is dominant, although click-beetle luciferases emitting different colors are also used (Chroma-Luc™). This intellectual property is also assigned to Promega.

12.2.3 Dual-Mode Reporter Assays in Practice

Dual-reporter systems enable the user to measure separately the expression driven by two independent promoters in a single cell type. This not only yields the potential for studies of gene interactions and potential causal relationships among gene transcriptional patterns, but also allows very simply for a well-controlled reporter assay of a single promoter. Expression of the second transcript can be used for normalization. Both of these options are widely practiced in the research community. To give a few examples: the Promega dual-reporter systems have recently been used for analysis of newly discovered promoter-like sequences (612, 613), elucidation of signaling-pathway transduction mechanisms (614, 615), investigation of the molecular effects of drug candidates (616), and even nutritional studies (617). The Dual-Luciferase® Reporter Assay System is intended for sensitive detection of both luciferase activities, while the Dual-Glo™ Luciferase Assay System, using the same enzymes and readout scheme, provides the rapid data flow needed in high-throughput screening applications.

The Promega method requires lysis of the cells. The disadvantage of this can be significant, but it can also be overstated, especially because Promega has invested considerable research into how to achieve lysis without affecting the luciferase signals. According to their product literature, the process takes 15 min, with a warning that certain cell types or overgrown cultures may require longer periods. The luciferase activities in the lysates are reported to be stable for many hours.

Taking the readings is fast and easy. The product literature cites a total assay time of 30 s: 3 s for mixing, 12 s to measure firefly luciferase activity, 3 s for addition and mixing of the second reagent cocktail, and 12 more seconds for the *Renilla* readout. The second cocktail contains both a highly effective stop reagent for firefly luciferase and (presumably) coelenterazine, the cofactor for the *Renilla* enzyme. The limits of detection are in the attomole (low femtogram) range for firefly and ~30 fg for *Renilla* luciferase. Clearing of the cell debris following lysis is not necessary to obtain a readout, though it may still have to be done if other procedures are planned.

The use of enzymes from related species, of which one is usually engineered to luminesce at a different wavelength, represents a useful alternative to these methods. The Chroma-Luc™ Luciferase Assay System from Promega provides two click-beetle luciferases, one emitting the native green signals, and the other engineered to produce red light. The great advantage is that these photons can be measured without further reagent addition, simply by changing filters. This means of course that one must have the filters, and many luminometer users do not, but the utility of the assay may repay the expense.

The Promega assays are moderately pricey, with a cost as of this writing of over $1 per well in bulk. It should be noted that while the Promega offerings are widely used, the firm is far from possessing a monopoly on the concept of luminescent dual-reporter assays. For example, a firefly/*Gaussia* system was developed independently for *in vivo* assessment of microRNA expression (618).

12.3   AEQUORIN AS A REPORTER ENZYME

Given the extent to which luciferases have dominated the luminescent assay, it is interesting that other enzymes are also under serious consideration as reporters. The fascinating story of aequorin and its actual and potential value in CB assays is discussed in Chapters 9 and 11; here it suffices to mention that aequorin is also used as an expression reporter (619). However, the strategy of these assays is more complicated than that of luciferase reporter assays. Primer extension is first carried out to incorporate biotinylated dUTP, which then associates with a streptavidin–aequorin conjugate. Thus, improvements in protein-modification chemistry, as well as the advent of novel biological lamps such as aequorin, have enabled the uncoupling of the expression and luminogenic steps to yield improved flexibility. In a sense, this trend points to the convergence of fluorescent and luminescent technologies: if the molecular transducers are truly interchangeable, the actual form that the readout takes gradually assumes less and less consequence to assay designers, since later modifications can readily change its nature.

12.4   VECTORS FOR USE IN REPORTER ASSAYS

Unlike most other CB procedures, reporter assays often begin with a cloning step. The objective of this step is to accomplish the physical linkage of the promoter under study to the reporter gene so that activation of the promoter will lead to transcription of the gene, yielding a signal that reflects the stimulus and/or the strength and nature of the promoter. A review of the vast array of historical and modern cloning procedures is beyond the scope of this volume; several laboratory manuals are available for this purpose (620, 621). Our objective here is to convey the range of options currently available to the scientist. One of these options is to construct one's own vector, but many constructs incorporating defined promoters and luminescent (or other) reporter molecules are commercially available.

12.4.1   Systems for Study of Specific Transcriptional Modulators

The androgen receptor (AR) is a transcription factor of fundamental importance in sexual development (622, 623). The DNA target of AR is the androgen response element (ARE) (624). The MMTV-Luc vector is simply a fusion of the ARE to a luciferase gene, enabling measurement of the extent of AR-directed transcriptional activity with a luminescent readout (625). In certain cases, commercial vectors already contain response elements of interest, such as the cAMP-response element CRE, which is available in the pGL4.29 vector from Promega (626)) and the nuclear factor of activated T cells response element (NFAT-RE) in Promega's pGL4.30 vector (627, 628). Promega also supplies HEK293 cells that already contain these vectors, so that responses that require only native pathways may be studied entirely without cloning or transfection.

A tandem series of thyroid response elements has been cloned upstream of a luciferase gene under the overall control of an SV40 promoter, yielding a construct with utility in both medical and environmental evaluation of thyroid receptor activation (629). Many other response elements, structures containing promoters as well as binding sites for signal-transducing proteins, are available commercially or from academic laboratories; the assay developer is encouraged to search for what is needed before expending effort in creating it.

12.4.2   Bacterial Reporter Systems

Despite the fact that *lacZ* has gotten the spotlight, bacterial luciferase was actually one of the earlier luminescent reporter genes. Older applications include screening for antibacterials with defined mechanisms of action. For this purpose, the bacterial luciferase from *Vibrio harveyi* was cloned into *E. coli*, yielding an organism that, with some biochemical tuning and physical manipulation, gave light at a roughly constant rate over several hours (630). (For a modern approach to a similar theme, see Reference 631.) These techniques have also been used to study the frequency of translation errors at high resolution (632) and to study *in vivo*-coupled luminescence and its interferents via the *lux* operon, which yields the bacterial luciferase (633).

The biological goals are distinct from those of researchers working in mammalian systems, but from a conceptual point of view the methods differ little. However, the metabolic state of the prokaryote can influence the luminescent signal, a factor that is especially important with bacterial luciferase, since the obligatory cofactor is itself a reducing agent (634).

12.4.3 Viral Vectors

Gene therapy is rapidly becoming a reality, and the commercial development of appropriate vectors is being accelerated to keep up with the demand. Any group that wishes to test a novel therapeutic idea based on performance of a specific promoter is likely to construct a viral vector incorporating the promoter in tandem with a reporter gene (e.g., see Reference 635). Again, the principles of the assay itself are similar to what we have seen, but the details of vector construction can be dramatically different. For one thing, the production of biological active viruses often requires spontaneous recombination of the elements, rather than simple cut-and-paste PCR and ligation. However, the benefits of the technology have driven the effort, and creation of viruses for CB reporter assays (or other reporter functions) has become almost commonplace (636–638).

12.5 SUMMARY

The important theoretical edge that luminescent methods have over the alternatives, including the every-photon-is-signal advantage, is fully expressed in the arena of liquid-phase reporter assays. Engineered luciferases respond to transcriptional modulation by emitting light of tuned wavelengths, simultaneously yielding information about multiple intracellular events and pathways. Extraordinary sensitivity is achieved without the need for labeling steps, radioactive substances, or even a lamp. The dominance of CB assays in this field is unlikely to diminish for many years.

13

COUPLED BIOLUMINESCENT ASSAYS: REGULATORY CONCERNS

13.1 INTRODUCTION

Entering the commercial biomedical enterprise from academia often entails a considerable degree of culture shock, much of which relates to the set of arcana known
as "GxP." Growing out of GMP (good manufacturing practices), the GxPs now encompass everything from horticulture to hotels. Most of the philosophy of GxPs is in
keeping with the dictates of common sense, encompassing procedures that lower the
risk of unsafe products, ensure proper divisions between functions that are inherently
adversarial in nature, and lessen the probability of fraudulent practices and cover-ups.
"GLP" refers to good laboratory practices, "GCP" to good clinical practices, and so
on. There are even "good policing practices." The reader is encouraged to consult
the many excellent books, articles, and web pages devoted to particular topics, such
as the Wikipedia page on GxP (http://en.wikipedia.org/wiki/GxP), which directs the
visitor to many of the various "x"s.

The term "good practices" in this context refers to methods that are logical, trace-
able, accountable (i.e., both well documented and assignable to an individual), and
tested (usually in multiple ways). The desirability of these characteristics for any
procedure is self-evident, but the profound implications of these concepts are not
obvious to those who have not made a study of the field. A PMA or "premarketing
approval" application to the U.S. Food and Drug Administration (FDA) for a drug
candidate often runs to hundreds of thousands of pages, documenting every conceivable aspect of discovery and development, *in vitro* testing, animal testing, clinical
testing, formulation and dosage, assay development and validation, the results of the

Coupled Bioluminescent Assays: Methods and Applications, Michael J. Corey
Copyright © 2009 by John Wiley & Sons Inc.

required production of three successive batches that pass all required quality tests, and even such organizational aspects as separation (both physical and hierarchical) of manufacturing from quality assurance/quality control functions, along with hundreds of other topics. Each substance used in manufacturing must be "GMP" in its own right; each instrument must be within its calibration period; each worker must be appropriately trained. From the bunny suits in the filling suite to the identification sticker on the pH meter, everything has to "be in compliance," because exceptions may lead to rejection of the PMA and the loss of many millions of dollars, or even abandonment of a project of medical and financial promise.

However, while the regulations and requirements appear burdensome at times, and the regulators often seem to take on an adversarial role (whether imagined or real), it is both important and valuable, at least from the point of view of morale, to remember that the regulatory acts and their enforcement represent measured responses to a real set of problems that have by no means disappeared in the face of stricter rules. The notorious "483 letter" from the FDA often contains allegations capable of shocking the conscience, describing events that can only be explained, but not excused, by the tremendous cost-cutting pressures that exist in the biotechnology and pharmaceutical industries. ("483" is not the designation of the actual letter; the letter is written as a follow-up warning in response to one or usually multiple violations described on "Form FDA 483, Inspectional Observations," which in turn is generally created after a physical inspection of a manufacturing or other site.) Although many or most violations are unintentional, such phrases as the following have appeared in FDA warning letters within the past few years:

"... you continued to use components and intermediates that failed your in-process limit for bioburden and presence of pathogenic microorganisms .... For example, a Too Numerous to Count TNTC bioburden result with the presence of *Pseudomonas fluorescens* was obtained for ... lot ... " (http://www.fda.gov/foi/warning_letters/s6520c.htm)

"... study personnel were aware that cyclosporine testing was infrequently performed for some subjects, yet you did not require compliance with the protocol and simply ignored this protocol requirement ... " (http://www.fda.gov/foi/nooh/law.htm)

"... submitted data from sputum samples that did not belong to the subjects identified with the samples ... the results demonstrate that sputum specimens that were purportedly obtained from 26 of these 35 subjects actually came from 3 individuals ... " (http://www.fda.gov/foi/nooh/deabate.htm)

A perusal of the web page of FDA warning letters, currently found at http://www.fda.gov/foi/warning_letters/, is a fascinating exercise that more than repays the time expenditure, in terms of both appreciation of the human potential for deceit and ineptitude and a deeper understanding of failure modes in general. The 483 letter begins with the amicable phrase "Dear Dr. [X]," but the tone tends to change rather abruptly.

The terms "cGMP" and "cGLP" refer to the current practices. The point is a critical one, since the thrust of all the regulatory agencies is to drive continual improvements in quality and impose more stringent requirements. Regulatory compliance as a field of study resembles academic science in at least these respects: one must not only possess a core body of knowledge but also keep up with the "literature."

Obviously, the assay developer in the GxP environment is primarily concerned with testing of the "product," which may be a candidate drug, a manufacturing method or facility, an instrument, or an assay technique itself. A thorough discussion of assay verification and validation for regulatory purposes is far outside the scope of this volume. Instead, we will concentrate on concepts that are related to CB assays, with some general material on successful development of compliant, validated assays. Moreover, our discussion will be focused on biochemistry and the relevant statistics, with little attention to the physics of instrument calibration or the administrative issues of record keeping and GLP-compliant notebooks. Finally, while other regulatory agencies, such as the International Conference on Harmonization (ICH; www.ich.org); the U.S. Pharmacopeia (USP; www.usp.org); and the International Organization for Standardization (paradoxically known as the ISO, because its name derives not from an acronym, but from the Greek word *isos* or "equal"), which awards the ISO 9000 and ISO 9001 certifications, have critical roles in the approval process, the FDA is the primary domestic "gatekeeper" in the United States and its regulations are our primary focus.

This chapter is neither intended to serve as legal advice nor does it constitute GMP or GLP "training," which can be provided only by a much deeper and more extensive engagement with the material. Before commencing an enterprise that will entail interactions with the regulatory bodies, it is imperative that one consult with individuals who are educated (and preferably highly experienced) in compliance and approval issues.

13.2   REGULATORY ASPECTS OF ASSAY DEVELOPMENT

At this point a brief introduction to the central concepts of developing compliant assays is necessary. Here, we will deal primarily with GLP requirements; when an assay is transferred to GMP personnel for routine analysis of materials in the product stream, the goal is to reproduce as nearly as possible the procedures developed and validated by the assay development group.

13.2.1   Standard Operating Procedures

Faithful replication of the tested and established procedures is the purpose of the standard operating procedures (SOPs), which are formal documents, requiring approval by appropriate department heads, that describe, ideally with great precision, the exact steps involved in the method and the standards to be used in determining whether a particular assay run has "passed" its quality procedures. Since the point is to enable the procedure for operators who may have no contact with the originator, the level of detail is often extreme, down to visual or photometric inspection of each input reagent prior to use, or the length of time on the stir plate for adequate dissolution.

13.2.1.1 Special Considerations for CB Assays For luminometric assays, critical parameters in SOPs might include preliminary testing of instrument performance, using well-defined standards; the necessity, if any, of protecting reagents or mixtures from light and the means of doing so, including any measures required for reservoirs used by automatic injectors; precise mixing instructions, especially if small volumes are being transferred; and procedures for data reduction, especially any special method for handling outliers due to environmentally derived contamination. For bioluminescent assay methods, the best standard is one that replicates most closely the biochemical nature and conditions of the actual test. For luciferase-based assays, a chemiluminescent standard such as a dioxetane/alkaline phosphatase reaction is not a good choice, but still better than a nonenzymatic reaction. If one is going to measure a CB reaction, then the standard should also be a CB reaction, preferably the same one, if possible. For example, the standard for the Promega Kinase-Glo^{TM} reaction could be a kinase of known specificity and activity; the same kinase to be tested would be an even better choice. The light production of the standard can then be compared with that observed in previous runs and deviations can be investigated. Many operators run other standards as well, such as a simple luciferase/ATP reaction. Failure of this standard to meet specifications is readily traced to a problem with the instrument, a problem with the ATP standards, or decay of the luciferase activity, indicating a potential storage or handling issue.

13.2.2 Philosophy of Assay Validation

Validation of an assay is an essential step prior to its use in any phase of drug development. However, what constitutes "validation" depends critically on the intended use of the assay. As a simple example, consider the aCella-TOX^{TM} CB cytotoxicity assay described in Chapter 3. Researcher X might reasonably make use of this method in screening a targeted compound library, say, for enhancement of a cytotoxic effect. Such an application would likely not require validation in itself, since the output of the process is the identity of the lead compound(s), rather than any specific performance characteristics. However, even at this stage, an understanding of the statistical principles underlying assay validation will be of benefit to the enterprise. To continue the example, Dr. X has now proceeded to *in vitro* testing of the lead compound to support an investigational new drug (IND) application, using the same CB cytotoxicity assay. The method now requires validation, but even in-house validation by accepted standards may not be sufficient to satisfy the regulatory agencies. (Regulatory acceptance is discussed further in Section 13.2.4.) Finally, Dr. X has licensed her novel method to Dr. Y, who plans to offer the assay as a service to cancer patients: patient tumor samples will be sent to his commercial site, where a panel of cancer drugs will be tested for cytotoxic activity against the cells to inform treatment decisions. (The wisdom of this approach, given the heterogeneity of cancerous tissues, will not be commented upon.) Because it is to be used directly in medical practice, this is a highly critical application of the assay method and will assuredly receive the most intense level of scrutiny from regulators. While in the ideal sense, the assay validation process should resemble the one performed at Dr. X's site, in practice the tolerance for errors and faulty record keeping will be even lower, and even a basic understanding of the liability exposure of such an enterprise should drive management toward a punctilious insistence on very high levels of quality and accountability. Again, Dr. X's awareness of the essential elements of assay validation from the beginning of the process will greatly ease the way.

Verification may be thought of as strong evidence that an assay works; validation establishes that it is working. One of the products of the assay development phase ought to be a good understanding of the capabilities and limitations of an assay. The validation protocol is written with these known characteristics in mind. To be validated, the assay must meet all of the predetermined requirements set out in the protocol, and any variances from the protocol must be explained and clearly justified. Knowledge of the whole procedure and its "deliverables" aids greatly in writing this protocol. There is usually a "window" of quality between the minimal standards to which the assay must conform to serve its medical and/or business purpose and its actual capabilities; in general, it makes strategic sense to write the protocol with the "must-conform" standards in mind, but if the parameters are set too close to this limit and the required performance changes for some reason, then the standard operating procedure (SOP) may no longer reflect the requirements of the organization. Moreover, setting the parameters too flexibly may lead to "false negatives," in the sense that something is fundamentally wrong with the way the assay is being performed, but because of lenient standards, the error is not identified. Under these circumstances, the error can worsen and lead to costly failures on critical occasions (possible examples are an enzyme degrading with time or a buffer contaminated with live organisms). Thus, setting the validation parameters is an exercise in balancing the desired high probability of success against expensive and potentially dangerous failure modes down the road (including adverse regulatory decisions).

Some parameters are required for all quantitative assays, while others depend on situation. Moreover, the intended use of the assay directly impacts the importance of the various parameters. For example, let us consider again Dr. X's CB cytotoxicity assay. As long as she uses it to determine the cytotoxic activities of various compounds against cancer cells, linearity is a very important criterion, since it is highly desirable to be able to extrapolate (or interpolate) observed data to higher, lower, or intermediate concentrations of the compound to predict its effects. However, one day Dr. X has an idea. She is going to use the same assay to measure undesirable toxicity against normal cells, which she expects to be very low. While this is a specificity issue from the point of view of the compound and its stakeholders, from the point of view of assay development, the limit of detection has now become a critical parameter. The organization's goals can be met only by distinguishing among low levels of toxicity, or even by concluding that none is detectable; but this latter determination will be convincing only if the limit of detection is very low indeed, and if the controls and standards are well considered and also yield expected results. Thus, it may be necessary to reoptimize the assay for this application in a separate development phase.

After this is accomplished, another validation protocol must be written, with the revised parameters in mind, and the assay must be validated again for its new purpose.

13.2.3 Parameters for Assay Validation

The following paragraphs relating to several of the major parameters used in validation protocols are intended to provide insight regarding the use of these parameters with CB assays. We will follow the seminal 2005 ICH Q2(R1) document, which is simply an amalgamation of the 1996 Q2B Supplement with the 1994 Q2A standard, by looking into specificity, linearity, range, accuracy, precision, detection limit, "quantitation" limit (herein "quantification limit"), robustness, and system suitability in the context of CB assays. Many of these metrics are not independent, yet they are often reported separately. (The Q2(R1) document is available on the Internet at http://www.ich.org/LOB/media/MEDIA417.pdf (639).)

While this chapter provides an introduction to the principles involved, the best policy in developing a new assay that will require validation is to acquire (or hire) in-house expertise and to look at similar past cases and their outcomes. Review of the data and conclusions by a biostatistician with a background in assay validation is generally required.

13.2.3.1 Coefficient of Variation   Although it is not specified in the ICH document, the coefficient of variation (CV) is treated first because it is related to several of the other metrics, especially precision. The term "coefficient of variation" refers to the standard deviation of measurements of a given value, expressed as a percentage of the mean of that value. Thus, an assay yielding a value of 200 ± 30, where 30 is the standard deviation (reported errors are assumed to be standard deviations throughout this chapter) is said to have a CV of 15 (the percentage sign is usually omitted). The CV is essentially a measure of precision, but it is not the only useful measure of precision.

Many factors affect the CV of conventional assays, and several additional factors perturb the CV of CB assays. To give the obvious considerations first, calibration of volumetric devices, care in preparation of buffers and other reagents (along with attention to their age and decay properties), training and experience of personnel in measurement techniques, avoidance of cross-contamination, temperature control, warming up the instrument, and a long list of similar technical points can all affect the CV of virtually any biochemical assay. However, CB assays involve special considerations, some of which have already been discussed in Chapter 2 and are repeated here because of their relevance to the CV. The impingement of light on some CB reactions can cause backward reactions and/or degradation of reaction components, either of which can increase the CV (and decrease accuracy as well, as a result of systematic effects over a range of samples). In validating an assay, one should not only simply conclude that "the assay is sensitive to light," but also determine how sensitive, whether dimming room lights helps or hurts, whether the use of opaque shields over each microplate is sufficient (and whether the shields come off or cause other problems, such as shedding of fibers into the wells), and whether the order of addition and component preparation need to be adjusted to avoid exposure of critical component mixtures to light. (For example, enzymatically catalyzed back reactions cannot proceed if the enzymes are separated from their substrates until the plate is enclosed in the light chamber and the reaction is then initiated by injection.) The quality of the plates must also be considered. Not surprisingly, white plates yield higher signals, but occasionally more cross talk among samples, than black plates. Since sensitivity is usually not an issue with CB assays, black plates may be preferred for this reason. In contrast to the situation with fluorometric assays, in which samples usually do not generate photons spontaneously, for CB assays opaque plates are an absolute requirement. Even with absolutely opaque materials, however, cross talk may be observed, increasing the CV, and potentially spoiling accuracy, linearity, and so on. One's first thought upon observing cross talk of samples is operator-borne cross-contamination, and this should be investigated; an obvious way is to request the operator to load a large number of samples, and then instruct a second operator to transfer adjacent samples to widely separated wells in another plate prior to reading. However, one should not be too quick to blame the operator. The dark chamber of a luminometer is subject to the laws of physics, and if the light path from the well to the photomultiplier tube (PMT) is not sealed against photons impinging from other wells, or if reflective material is present somewhere, photon contamination and cross talk may result. For validation purposes, the protocol should be tested with luminance exceeding the highest anticipated readings adjacent to blank wells, which should exhibit luminance within or very near to the range of background variation.

Measurement precision has obvious implications for the CV, but an additional factor in CB and other coupled assays is the potential for superlinear effects. If the rate of a reaction depends on a single limiting component, the flux generally varies in a linear fashion with the concentration of that component (in the absence of special effects such as cooperativity, hysteresis (640, 641), aggregation, etc.) However, if there are two or more limiting components whose interaction is required for the reaction to proceed, then a multiplicative effect of the measurement errors of both components may be observed. This results in a cumulative error, a contribution to the CV that is potentially far greater than the error in any single volumetric transfer. Sometimes, it is necessary to perform a CB assay with multiple limiting reagents because one is the reagent under test, while another must be held limiting because of contamination with another background-inducing substance; thus, the situation can occur even in a well-designed assay. However, attention to the nature of the particular reaction sequence and the nature of the limiting reagents can suggest a solution: for example, one of the critical reagents can be adjusted to a concentration that allows easy and precise measurement, such as 10 μL with a P-20 or 50 μL with a P-200 pipetter. Linearity is discussed in detail in Section 13.2.3.5.

Other sources of error of concern with certain CB assays are clothing and the human body. As mentioned in Chapter 3, skin cells can cause serious "spikes" in such assays as the aCella-TOX method, which is highly sensitive to the presence of even a single cell. This sensitivity is due to its use of the limiting enzyme G3PDH, which is abundantly present in all cells, in the readout process. The use of gloves is mandatory for all assays in which G3PDH is a limiting component, but gloves alone may not be sufficient. If spike problems are not readily resolved, gowning and the provision of a clean environment with aseptic or sterile equipment should be considered.

13.2.3.2 Precision Precision in the ICH standard is defined as "the closeness of agreement (degree of scatter) between a series of measurements obtained from multiple sampling of the same homogeneous material under the prescribed conditions" (639, p. 8). Three levels of precision determinations are described: repeatability (also known as intra-assay precision); intermediate precision, which reflects variation associated with different days, operators, and/or equipment; and reproducibility, which describes lab-to-lab variation. Precision is preferably determined with a homogeneous sample of the type to be used in production.

We will use Dr. X's cytotoxicity assay as an example once again and assume that it is to be employed as a potency assay for a product that enhances the attack capabilities of cytotoxic T-cells (CTLs; see Chapter 3). The standard method of measuring potency is the ^{51}Cr release assay, but Dr. X wishes to validate her approach for use with these T-cell enhancers and eventually replace the ^{51}Cr method. This example is chosen because assessment of the potency of CTLs is especially complex and illustrates many of the relevant points. For this example, we will assume the use of the no-spin supernatant cytotoxicity method described in Chapter 3 in connection with actual CTL assays.

Assuming high levels of quality in equipment and technique, repeatability is likely to be good. Single-point CVs may well be under 5. Factors potentially influencing precision are cell clumping, which may result in the delivery of extra target cells or, conceivably, CTLs to a well (although CTLs are generally not clumpy); uneven settling, perhaps due to mechanical disturbance of the plate or thermal heterogeneities among the well, the latter potentially causing convective flow; or excessive delays between initiation and readout, which could lead to uneven inactivation of G3PDH. However, these problems are rarely observed, and only the last is specific to the assay method.

Intermediate precision is a much more problematic issue, no matter what means of assessing CTL potency one is dealing with. Here, the issue is whether personnel can replicate results from earlier days, perhaps using different equipment and certainly using different materials. However, even batches of CTLs thawed from the same frozen aliquot set do not behave in a uniform manner, probably because of the minute yet nonnegligible differences in handling and timing. One may observe 70% maximum lysis one day and 40% the next, without any known variations in technique or materials. The assay developer must foresee and account for this reality in formulating the analytical methods and in writing the SOP. If, as in Dr. X's case, the goal is to assess a means of enhancing CTL activity, then the activity observed without enhancement represents a possible normalization standard. It is true that less enhancement can be seen if the control is 70% of maximum than if it is 40%, but this is what the development and optimization processes are all about. In most CTL researchers' hands, "70% of maximum lysis" represents (or should represent) the higher asymptote of the four-parameter data reduction process described in Chapter 3, implying that many data from lower parts of the regime are available. It should be possible in some

manner to obtain a CTL-specific rate, together, ideally, with a cooperativity parameter. The rate should more closely indicate CTL potency than the observed maximum
(although the latter is frequently used—unfortunately, in the author's view), and this
rate, which is not subject to asymptotic limitation, as is the maximum, may represent
a superior normalization standard. However, regression calculations of the rate and
cooperativity may require a much higher level of statistical sophistication than direct
observations of the asymptotes. This emphasizes once again the need for statistical
expertise in the assay validation team, especially with a complex assay.

Reproducibility is the highest level of precision testing. Ideally, it should be possible to use materials developed independently with operators trained in different ways
and a different set of instruments and obtain substantially identical results. This is
not the place to go into the vagaries of T-cell culture or target cell preparation, but if
those sources of variability as well as operator training are set aside, we are left with
equipment issues and the materials associated with bioluminescent reactions. Clearly
the data should be normalized to just how much luminance the luciferase is producing
in response to a picomole of ATP on a given day, and the target cells under test should
represent a separate normalization standard. These issues should reduce variability
associated with materials (as has been observed to date with aCella TOX), but there
remains the question of equipment. While the instrumentation market is in constant
flux, it is likely to remain true that dedicated luminometers are simply less noisy and
more sensitive to light than multiple-use instruments, as discussed at greater length
in Chapter 2. The author has observed differences in equipment sensitivity as great as
30-fold, even without conducting a broad survey. The inescapable implication is that
some phenomena will be observable on some instruments, but not others, and even if
they can be assessed on the weaker equipment, the data quality will be considerably
poorer. These differences in sensitivity are generally not observed with fluorometry
and other competing technologies, and as a result the assay developer trying to explain reproducibility problems to a regulator may face a special challenge. Clearly,
it is important to specify performance standards for any instrument to be used in
interlab comparisons, and the standards may have to specify parameters other than
sensitivity. For example, an instrument that is incapable of reading the plate as rapidly
or injecting fluids as accurately as one's "home" instrument may be unsuitable for
some experiments.

13.2.3.3 Accuracy

In plain language, precision (see below) refers to how closely
(i.e., within what error tolerance) one knows the value reported by an assay, while
accuracy specifies how close the observed value is to the "correct" value, which may
refer to an outcome that is known in an absolute sense (such as how many cultures
actually exhibit microbial outgrowth at a later observation) or to a "gold standard"
that is assumed to be correct. (It is hardly surprising that there have been cases in
which a new assay was more accurate than the current gold standard, causing it to
be rejected, at least temporarily, in favor of assays yielding poorer true accuracy
but in better agreement with the standard; unfortunately, the introduction of truly
novel assay methods, even if superior, can require years of "shakeout" and the steady
accumulation of supporting evidence.) It is quite possible to observe a high degree

of accuracy along with poor precision. A simple example is an assay that is well calibrated yet highly sensitive to conditions, reagent aging, or operator errors; the mean value of the measurements is true, but the CV is excessive.

Nevertheless, the meaning of "accuracy" can vary with the type of project. In diagnostic tests, the only form of accuracy that really matters is the ability of a test to distinguish values that indicate the condition of interest (the presence of cancer, liver damage, etc.) from those that do not; however, since data on the biological end point may not be available for comparison, the gold standard may be used as a surrogate. Returning to Dr. X's attempt to enhance CTL activities, the true measure of accuracy of her potency assay would be the degree of agreement between her assay results and the actual *in vivo* activity of the compound in patients, but it is improbable that she will ever have an opportunity to determine this activity unless she can show that her compound is potent in an *in vitro* assay and that the results of her assay correspond to those of an established assay such as the ^{51}Cr release or LDH release assay.

An approach that a regulator might find convincing would be to run the CB and established assays side by side with the same reagents and conditions and provide a graph comparing the results of the two assays, one on each axis, with a linear relationship indicating perfect agreement. Still, one must be careful in applying even such a conceptually ideal approach, because different assay methods have different strengths and weaknesses. It is important to choose concentration regimes that are of biological interest and that also highlight what an assay can do. For example, the ^{51}Cr release assay requires prior labeling of the target cells with radioactive sodium chromate. Since only the target cells are labeled, only they can deliver a signal. Thus, rather odd protocols are sometimes observed in which the CTL "effector" cells are titrated up to enormous numbers in relation to the target cells. These data are usually meaningless in the ^{51}Cr release assay, but they also do no harm. However, in the context of a nonlabeling assay, in which the readout can arise from both lysis of target cells and leakiness of effector cells, very large numbers of effector cells are undesirable, since they do little to enhance the lytic rate while providing a large source of background signal. In the case of a highly sensitive CB assay, the researcher should instead concentrate on the lower range of both CTL and target cell concentrations. This shows what the cells (and any modulating compounds) can actually do, on the one hand, and it reduces the background problem, on the other. This principle is illustrated more fully in Chapter 3. Here, we need only point out that the concentration regimes for comparing two assays with a view to accuracy determination and validation should be selected with regard to readouts of biological interest, and not to bias the determination in favor of one assay or the other.

It is of interest to note that, according to the ICH Q2(R1) document, accuracy may be determined by evaluating performance with a known quantity of pure analyte, determined by comparison with a gold standard method, or inferred from the combination of precision, linearity, and specificity (639, p. 13). The concepts invoked here, like many of those that appear in documents from regulatory agencies regarding analytic procedures, are primarily directed to quantitative determinations of substances, rather than biological potency assays, and the first method is not relevant to Dr. X's challenges. However, the second and third methods both suggest useful approaches. We turn now to specificity and its significance for CB assays.

13.2.3.4 Specificity In analytical terms, specificity refers to the ability of a procedure to confirm the identity of a compound and determine its concentration in the presence of other components. The issue of purity is related but separate, since by its nature it involves substances that are not the compound of interest and may require separate assay methods.

CB assays have both advantages and disadvantages in the area of specificity compared to other methods. The advantages stem from the pseudobiological nature of the reactions. Many enzymes are highly specific for their substrates, rejecting, for example, enantiomeric molecules that are difficult to distinguish from the biological active isomers by most analytic methods. CB phosphate assays, for example, respond specifically to phosphate, and there is little danger that other ions will enter the pathway and successfully produce a light reaction. However, the vulnerable aspect of CB assays is their coupled nature. By definition, multiple enzymatic reactions proceed in series to yield the luminescent readout, and each of these enzymes has its own peculiar features. If one reaction in a coupled series is inhibited, the process may emit little or no light, even if the other enzymes are fully active. A simple example is a nucleoside analog based on adenine. If one can learn enough about the G3PDH enzymes from trypanosomes and humans and identify differences in their association with NAD^+, one can perhaps develop drug candidates for treatment of these pathogenic protozoa (642); however, if one chooses to use a CB assay to assess the activities of these compounds against G3PDH, one must consider the possibility that they will inhibit luciferase as well (and perhaps other enzymes in the coupled pathway). In general, inhibition observed in a CB assay cannot be considered specific to the enzyme under test until the other enzymes in the pathway have been checked independently. This is in keeping with the general principle of assay development that virtually all findings should be confirmed by an orthogonal method. Indeed, the ICH document takes this directly into account with the following statement: "Lack of specificity of an individual analytical procedure may be compensated by other supporting analytical procedures(s)."

Taking CB phosphate assays as a distinct example (see Chapter 5), these methods, as pseudobiological procedures, will generally respond to phosphate and not to physically similar ions such as vanadate and arsenate; however, these ions may interfere with the assay, due to the same biological nature. Thus, while they do not actually raise specificity issues, they may prove to be interfering substances and represent limitations on the achievable specificity, potentially yielding erroneous results in regard to the true phosphate concentration.

Finally, in dealing with a complex assay situation such as the one faced by Dr. X with her cytotoxicity measurements, the concept of substance specificity is largely irrelevant, since one is actually dealing with phenomenological specificity. The real question is: does the readout reflect actual modulation of the CTL activities of interest or is it perturbed (or entirely caused) by other events? A thorough analysis of this question must involve a certain amount of "modular" thinking in regard to the assay process. How can one establish that each step of a complex assay is doing its job?

Suppose that Dr. X's method is being applied to CTL attack on target cells, which are induced to express a target antigen by viral transfection. The transfection is an essential component of the assay procedure, and as such should be independently checked for specificity. Suppose the compound is observed to enhance lysis; does this occur via CTL activity or, if serum is present, via stimulation of complement? The point is not that these failure modes are likely; the point is that these possibilities must be eliminated to ensure the assay procedure is specific.

13.2.3.5 Linearity Linearity is of paramount importance in measuring analyte concentrations. The range of such an assay is usually taken as the concentration range over which linearity is regularly observed, rather than the range over which a statistically interpretable response can be obtained. A linear response to analyte concentration implies that the analyte itself is fully limiting and the response reflects the concentration. Virtually all analytical methods have an asymptotic "decay" regime as well, higher than the linear regime, in which the response to analyte is damped but still greater than zero. The causes of damping are many, including exhaustion or saturation of a reagent, bleaching or other physical saturation of the light detection apparatus, and chemical changes in the solution due to the assay reaction (such as an alteration in pH). If the method involves an enzymatic reaction, further damping modes include product inhibition, a heterogeneous mixture of substrate forms that have different reaction rates, and damage to the enzyme caused by the analyte or the analytical reaction. Most of these potential causes are difficult to replicate quantitatively, making the use of the nonlinear portion of the range for analysis even more risky. However, many commercial assays are nonetheless marketed for use well above their linear ranges. If the causes of nonlinearity are well characterized and controlled by appropriate standards, such assays can be informative, but their results are likely to receive close regulatory scrutiny if used in an approval application. An exception is the very common ELISA method, which frequently yields nonlinear responses (both superlinear and sublinear) over a considerable portion of its range, but whose characteristics are considered to be so well understood that these data can be acceptable.

The most common and important causes of nonlinear responses to analyte in CB assays are reagent exhaustion and enzyme saturation. These phenomena usually result from decisions made by the assay designer, as explained in Chapter 2; in brief, the nature of coupled assays is such that a contaminant in one component frequently leads to a background signal in one of the reactions. This restricts the useful concentration of the component in question, although the situation may be eased if a fairly stable background signal can be deemed acceptable and/or if the component can be further purified. Some of the points discussed in Chapter 2 apply, but, in a regulated environment, the linearity of an assay in the concentration range of interest is generally a more important consideration than expense or convenience.

The ICH document specifies that linearity should be assessed first by visual inspection (639, p.12). This recommendation is entirely in keeping with the possible erroneous analyses discussed in Chapter 2. No single statistical reduction of putative linear data can assure the assay developer or the regulator of the quality of the response throughout the range. However, a combination of the regression statistic (properly known as the Pearson product moment correlation coefficient, which, when squared, is held to represent the degree to which the variation of the dependent variable depends on the variation of the independent variable according to a linear model) with a residual plot is generally sufficient to be convincing regarding the validity of a given response over a range, although this combination does not of course obviate the need for human judgment. The residual plot (discussed at greater length in Chapter 2) is an excellent way of determining whether observed deviations are random or systematic; the latter type of error indicates that the dynamics of the response is not entirely captured by the linear model and that erroneous conclusions from the data are possible if the cause of the systematic deviation is exacerbated on another occasion. In the words of the ICH: "... an analysis of the deviation of the actual data points from the regression line may also be helpful for evaluating linearity" (639, p.12). What is helpful to the ICH is often essential to the assay developer.

These days, of course, such regression calculations are virtually never performed by humans. Statistical packages, including freeware such as the eminent R (available at http://www.r-project.org/), can do the work many orders of magnitude more rapidly than the most prolific human calculators. However, statistics is an area in which Pope's admonition regarding the dangers of "a little learning" applies even more strongly than in most cases. Using a statistical program without a general understanding of its purpose and function can easily lead to erroneous conclusions. In particular, a basic understanding of residuals and their significance is essential to successfully perform regression analysis (see Section 13.2.3.6).

We now turn to assays that are inherently nonlinear. Enzyme catalysis, receptor–ligand association, and the lytic activity of CTLs, among many biochemical and biological processes, are fundamentally saturable phenomena that (ordinarily) exhibit hyperbolic decay as the independent variable is forced to sufficiently high levels. In most studies of these phenomena, whether for regulatory submission or otherwise, it is desirable or even essential to capture the entire response, including the saturation process. This does not imply, however, that the demand for linearity is meaningless in these systems. With very few exceptions, it is possible to observe linearity by adjusting the concentration regime to lower and lower levels until the saturation disappears. (The exceptions occur when the association is so tight or the activity is so rapid that saturation occurs at a vanishingly small concentration. An extreme case is chemical catalysis, which can occur so rapidly that no curvature can ever be observed. With proteins and cells, this is typically found in cases when the catalyst, receptor, or cell is so precious that it cannot be provided in sufficient quantity to observe partial saturation.) A demonstration of a linear response at these low-concentration regimes goes a long way toward assuring that the assay readout truly reflects the activity of interest. Moreover, it may be impossible or mathematically intractable to extract certain kinetic parameters from the whole data set, but they may be accessible from the linear portion of the response (see Reference (92) for an example).

When an assay has undergone considerable development, one is often faced with the problem of exactly what range to claim. If the goals of the organization specify that the assay "must perform" to a given standard, that is, up to a specified concentration, then either the developer must declare that linearity has been achieved to that concentration or development must continue; however, organizations have been known to demand the unachievable, or to insist that certain features are necessities, only to back down when the cost of providing the features becomes evident. Thus, what perhaps ought to be an algorithmic process (e.g., we will continue to work on this assay until 1 μg/mL is in the linear range) becomes yet another exercise in judgment (e.g., do we already have enough advantages over the competition to market this product without extending the linear range? Or, will the agency accept this upper point as part of the linear range?) Linearity, like so much else, depends on one's point of view.

13.2.3.6 Residuals Although linear and nonlinear regressions are fundamentally different, the treatment of residual plots in the two cases is similar and they will be treated together here.

Residuals are simply deviations of the data from the best available fit of the model proposed, usually as calculated by a regression process. A simple example is provided by Fig. 13.1. Panel a shows a data set of which the variation is captured well by a linear model. Panel b displays the residuals, the differences between the actual data and the predictions of the model, plotted against the independent variable x. Note that residuals are defined as discrepancies between actual and expected values of the y variable, not as distances from the regression line. The point is a critical one, especially if the slope of the line is great, since under those conditions a small error in measuring x, the independent variable (such as the quantity of added analyte), may lead to a substantial change in y.

The value of determining residuals does not lie primarily in learning their magnitude, which constitutes information that is provided in a more convenient form by the correlation coefficient itself. Instead their utility lies in their implications regarding

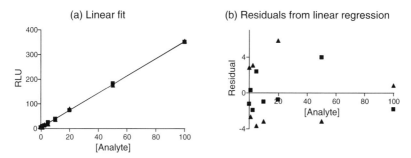

FIGURE 13.1 Constructed data set appropriate for linear regression. (a) The dose–response plot appears to fit the line very well, but this can be deceptive, especially when many of the data are clustered at low values of the independent variable x. However, the r^2 value, the square of the correlation coefficent, is 0.9994, indicating an excellent fit. (b) The residuals plot, which shows the deviation of the actual data from the model for each point, presents a great deal of error-related information that is not available from the primary graph. This plot shows highly random scatter with no trend or pattern that would indicate inadequacy of the model.

the suitability of the model used for the regression. If a particular model is truly the "best" model for the data, then a plot of the residuals should not demonstrate any sort of pattern beyond random variation, apart from the fact that greater errors are sometimes associated with data taken in certain concentration regimes; ideally, this would be accounted for by the model, but this type of nonrandomness in residuals is probably acceptable if no others are seen. An example of nonrandom residuals, contrasting with Fig. 13.1, is shown in Fig. 13.2. If a residuals pattern such as the one seen in panel b is observed, the indications are very strong that the linear model proposed is inadequate to explain the variation encountered. The model may be accounting for a portion of the variation seen, but some of the systematic variation, not due to error, is due to an effect that is not described by the model. If the variation from linearity is slight, the effect may appear subtle in the primary plot, but may be readily observable in the residuals plot; for example, in Fig. 13.3, the inverted U-shaped pattern of the residuals indicates the presence of a sublinear decay effect that is not easily seen in the primary data. (The U-shape is more easily visualized with logarithmic scaling of the x-axis as in panel c.) Note that these are not the only patterns of nonrandom deviation that may be encountered.

If an inappropriate pattern is observed in a residuals plot, the main conclusion is that the model is either insufficient or incomplete with regard to the data. Depending on one's degree of sophistication, one may attempt to draw conclusions from the qualitative nature of the pattern; but one should recall the nature of the plot: the deviations of actual data from the predictions of a model now known to be incorrect. Thus, the primary data themselves are likely to be a superior source of information.

Although nonrandom residuals generally indicate the presence of a problem, it may be premature to throw out the mathematical model itself. As a very simple case,

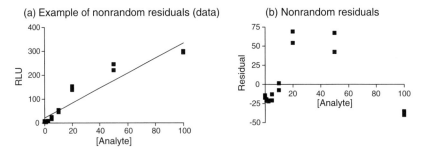

FIGURE 13.2 Constructed data set; typical example of attempt to fit sublinear data to a linear model. While the model captures some information about the variation, it badly fails to explain the sublinear response. The r^2 for this fit is a relatively high 0.8946, showing primarily the flexibility of linear models in fitting nonlinear data. (a) The line may appear to be a reasonable fit. Despite the nonrandom residuals, an estimate made with this model is unlikely to be off by more than 30–50% in the range displayed; however, a point-to-point or hyperbolic decay model would be superior. (b) With the absolute level of the data subtracted and only the deviations from model predictions displayed in the residuals plot, the nonlinearity of the data is rendered obvious.

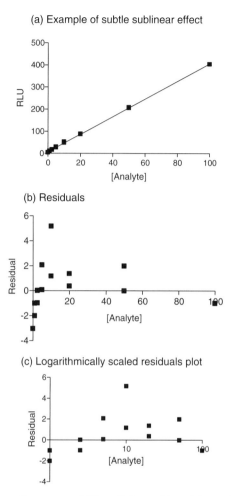

FIGURE 13.3 Intermediate case of "linear" data with subtle nonlinear effect. (a) The primary plot reveals almost no evidence of nonlinear effects. The r^2 value is a very high 0.9998. (b) However, the residuals plot exhibits an obvious bell shape. In such cases, the true linear response is usually represented by the lower portion of the data, while an unidentified damping effects is responsible for the sublinear results at higher values of the independent variable; but this can vary. Note that while the nonlinear effect would be seen in a linear–linear residuals plot, its nature is more obvious in the presented log–linear plot.

consider the asymptotic decay seen in Fig. 13.4. If one attempts to fit these data to a linear model, they will yield a residuals pattern in which positive deviations are observed at intermediate concentrations of analyte, with large negative deviations at low and high concentrations. This is an extreme case of the nonlinear deviation seen in Fig. 13.3. The source of the negative deviation at high concentrations is obviously the decay effect itself, but why are the positive deviations seen? The answer

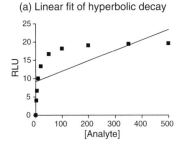
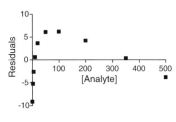

FIGURE 13.4 Attempt to fit hyperbolic decay data with a linear model. The r^2 for this data set is an unacceptable 0.4815. The nature of the mismatch between the data and the model is evident from both the primary plot (a) and the residuals plot (b).

is that regression software can only do what it is designed to do. By commanding the computer to fit this data set to a linear model, one is implicitly telling it to strive to adjust the parameters to make the model fit the data, even though it does not. Under the assumption of normally distributed errors, the software minimizes the sum-of-the-squares deviations from the regression line by adjusting the slope and intercept until no single point is sufficiently far from the line to yield a very large residual. However, this process distorts the apparent residuals of data points that actually exhibit very small deviations from the linear model over a more appropriate range. In other words, the linear-fit process has no way of "knowing" that the mid-range points are more likely to fit the model than the high range. It attempts to incorporate all the data in calculating the model parameters. Here is a clear case of the potential dangers of drawing conclusions from the qualitative nature of the residuals pattern. Although the residuals pattern appears "ugly," the case may be resolved simply by restricting the range of the data to the observed linear regime (in this case, the data from $x = 0$ to about $x = 10$) or, alternatively, by adjusting the conditions to get rid of the decay entirely. After these changes, the model may prove to be appropriate.

If the nonrandom deviations observed cannot be attributed to a simple cause, a different model must be considered. In-depth discussions of nonlinear modeling are outside the scope of this book; a statistics text with relevance to the field of biology under study should be consulted. The F-test is usually an excellent way of determining whether incorporation of an additional parameter in a nonlinear model (or in a linear model, thereby rendering it nonlinear) is justified by the resultant improvements observed in the fit. At some point, small nonrandom variations may be acceptable, but this is more likely to be the case if alternative models have been explored and the statistical analysis is convincing.

Nonlinear regression is a separate topic, worthy of a lifetime of study in itself. It is possible to transform many nonlinear models into linear models and perform linear regression, but this approach is fraught with statistical traps; for example, the Lineweaver–Burke and Eadie–Hofstee plots that were long used as linear transformations of Michaelis–Menten effects were convenient ways of representing the saturation phenomenon in precomputer days, but linear regression using these methods led to errors in weighting the data (errors not immediately evident from a residuals plot), and these transformations should now be considered obsolete. True nonlinear regression (e.g., to the original Michaelis–Menten equation) ordinarily weights the data correctly and is a better approach. However, the transition from linear to nonlinear regression brings additional dangers. Allowing the regression model to curve gives the software the freedom to make errors that are far more egregious and complex than those observed in linear regression. Again, the residuals plot is usually one's best guide to what is going on. A straightforward example of nonlinear regression in the context of CB assays is provided in Chapter 3, but the topic is profound, and statistical consultation should generally be sought.

13.2.3.7 Range   The range of an assay is described in rather circular fashion as the concentration regime over which the assay exhibits adequate accuracy, precision, and linearity. The upper limit of the range of a CB assay is almost always limited by constraints on the concentrations of essential reagents due to contamination, as described above; however, in some cases, range may be limited by the optical response of the reading instrument. With some instruments, it may be possible to overcome this saturation by adjusting the PMT voltage; certain instruments do this automatically.

The specification of range for analysis of drug product content is not one of the more draconian aspects of the regulatory environment. Ranges of 80–120% or 70–130% of expected values are required, depending on the precise application, according to the philosophy that readings outside these ranges are exceptional conditions that require additional analysis by other methods. However, the opposite is true in circumstances under which time-dependent or other complex dosing is part of the claim; the ICH gives an example range of 0–110% of the expected dose (639, p. 8). In testing for impurities, the range must begin at a level commensurate with the reportable concentration and extend to 120% of the specification for the product. Performance, as assessed by accuracy, precision, and linearity, must of course be adequate (as defined in the validation plan) at the limits of the defined range.

13.2.3.8 Limit of Detection   In contrast to some of the other validation parameters, the Limit of Detection (LoD, or "DL" in the ICH document) is refreshingly well defined, and there is general agreement among agencies on the definition. The LoD is a point 3.3 standard deviations (σ) above the mean of the background (639, p.15). This value is chosen to ensure a high degree of confidence that an observed positive is truly different from zero; the assurance that all true positives will be detected is much lower, but this in any case depends on the data distribution of the true positives and cannot be predicted in advance. A well-known example is the PSA marker for prostate cancer (643, 644) and its "gray zone" of 4–10 ng/mL (645), in which prostatic pathology is almost certainly present, but the best course of action depends on many other factors and is not always clear.

Since the LoD depends on the background mean and the standard deviation, these must be known with a high degree of confidence before the LoD is determined. Unfortunately, the standard deviation itself can be subject to large deviations. A superb operator using carefully calibrated pipettes with good seals can measure volumes with a CV of 0.5 or better (i.e., the standard deviation is less than 0.5% of the volume being measured). Another operator might achieve a CV of 3, and yet be certified as adequately skilled. Clearly the standard of deviation of the assay will differ between these operators. This is why assay development and validation require understanding of not only the biochemical system but also the human and physical systems involved. Depending on the goals of the organization, it may be desirable to use the best operator's data, an aggregation of data from all operators, or even data from a less-experienced operator to determine the LoD. Similar considerations apply to determination of the background. Note that the presence of dynamic (time-dependent) background characteristic of many CB assays is much more problematic for these determinations, and will probably require the developer to provide stronger justification for using the CB assay in question.

An additional consideration applies to CB assays used at or near the LoD; this is their extreme sensitivity. Environments and techniques that are adequate for ordinary assays may no longer be usable with assays that are sensitive to the presence of a single cell (PCR is an obvious and well-characterized example). The enzymes present in one mammalian cell are enough to render the results of many types of 96-well CB assays meaningless, and this problem is only exacerbated by the inevitable shift to 384- and 1536-well formats. This is really a different problem from the determination of the LoD, since, at these levels of sensitivity, such contaminants are easily recognized as such; they are in fact outliers, and are addressed separately in the next section. The point is that levels of contamination that might show up as minor fluctuations in less-sensitive methods and concentration regimes can become problematic outliers under CB conditions.

13.2.3.9 Limit of Quantification The limit of quantification or LoQ ("QL" in the ICH document) is defined as 10 standard deviations above the background mean (639, p.16). This definition is based on the observation that many assays exhibit nonlinear effects at their low ends; these can be due to matrix sequestration effects (i.e., absorption of small quantities of reagents by supposedly inert vessel materials), lower levels of precision in measuring small volumes, enzymatic "lag" phases or single-turnover phenomena, biphasic responses due to the presence of multiple isomers or even transient species, and so on. In CB assays, these nonlinear responses can also be caused by back reactions. In any case, the distinction between LoD and LoQ is meant to ensure that exaggerated claims of quantitative accuracy are not made in concentration regimes in which the assay response is not as reliable.

CB assays often perform extremely well at the "low end"; for example, the enzymatic content of a single erythrocyte equivalent was detected by means of the original CB cytotoxicity assay (83), and the data from proximate measurements demonstrate that the result was well within the linear range of detection (see Fig. 13.5). It could be argued that the LoD has not been determined for this assay, but continuing the process far below the single-cell level seemed pointless at the time. If the enterprise had been subject to the pressures of the regulated environment, a different decision might have been made. Nevertheless, extreme sensitivity does not always imply excellent

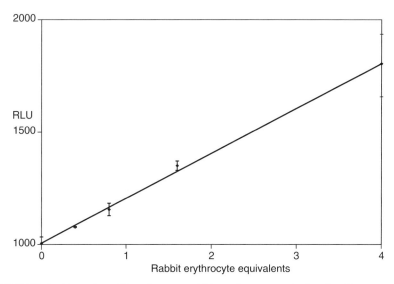

FIGURE 13.5 Detection of less than one rabbit erythrocyte equivalent by CB cytotoxicity assay. The reaction was carried out as in Reference 83, except that the incubation was continued for 3.5 h before luminance was read. The data were taken in duplicate, except blank in quadruplicate. The statistical performance in the low range, with a good linear fit, is strong enough to establish an upper bound on the limit of detection. Reproduced from Reference 83 and used by permission of Elsevier, Inc.

statistical performance, in part due to the reasons mentioned in the previous section, and the separation between LoD and LoQ may be generally appropriate to CB assays as well.

13.2.3.10 Robustness Robustness as a validation parameter is entirely distinct from the others. Here, we are interested in the reliability of an assay in the face of deliberate manipulations of conditions. The reasons that this testing is necessary for procedures involved in drug testing or diagnostics, but not for ordinary laboratory procedures, should be clear; in the former cases, the risks of failure modes involve potential loss of life. Thus, mistakes that one does not necessarily expect, but can sensibly anticipate, should be analyzed for their effects on the result. The goal is not to develop an assay that always "works." Instead the goal is to avoid the production of dangerous misinformation. Thus, a well-designed assay procedure, in the face of technical perturbations that exceed its tolerance, will respond with a notification that "something is wrong" rather than producing incorrect data.

A simple example of robustness is the determination of the effects of normal temperature variation on assay readout. The value of this test is based on the assumption that, despite the fact that most buildings where scientific work is done have controlled temperatures, some variation with season, sun exposure, and the activity of heat-generating equipment is observed. An incubator may be set to slightly higher than normal room temperature, say at 25°C. One does not expect to observe profound effects on a reaction rate as a result of a few degrees' variation in temperature, but at least two phenomena may produce unexpected results: second- or higher order reactions, in which two or more molecules at limiting concentrations must interact via molecular diffusion; and protein denaturation, which ordinarily does not occur readily at or near room temperature, but should be considered nevertheless, especially if a protein constituent is poorly characterized or a recombinant of unknown thermal stability. (The temperature dependence of high-order reactions is greater than that of unimolecular reactions, "pseudo-first-order" reactions in which a linear relationship with concentration of the critical reagent is observed, because the effects of the diffusion rates of multiple species are multiplicative.) If the effects of small variations in temperature are stronger than expected, even if the assay functions according to specifications, the phenomenon should be investigated and characterized and, in the words of the ICH, "the analytical conditions should be suitably controlled or a precautionary statement should be included in the procedure."

The relatively low level of accumulated experience with CB assays suggests that robustness testing, when undertaken, should encompass a wide variety of failure modes. Certain procedures are of obvious importance for all assays, such as spike recovery tests and the effects of diluting samples. Others are especially relevant to CB assays. Does the procedure work with both black and white plates? Do ambient light conditions during set-up make a difference? What are the effects of 5% variations in volumetric measurements? Do independent lots of enzymes perform differently? (The latter is observed fairly frequently, for example, with phosphoglycerokinase in phosphatase tests described in Chapter 5; phosphoglycerokinase is frequently limiting in these reactions.) For reactions in which human enzymes are limiting, the results of intentional contamination by human materials, such as hair and skin cells, might be profitably examined. Other potential targets are minor pH variations, instrument settings (and age), read and incubation times, and order of addition.

13.2.3.11 Outliers   An outlier, by definition, is a measurement that appears to fall outside of a normal error distribution. However, there are really two types of apparent outliers: (1) those due to discrete events, such as reagent mix-ups by the operator, radical but transient equipment failures, gross contamination of a sample well by a dust particle or aerosol, and similar events; and (2) events that are actually statistical extremes along a continuous error spectrum, such as poor volumetric or gravimetric measurements due to poor operator training or fatigue, leaky pipette seals, time-dependent reagent degradation, or combinations of these and like effects. The former type of outlier falls outside any predictive model based on the assumption of a normal error distribution; the latter, in contrast, is an inevitable part of the normal distribution. An ideal method of outlier identification would enable us to discard the "discrete" type as uninformative regarding the quantitative phenomena under study, while retaining the "continuous" type as informative. Unfortunately, numbers do not carry warning flags along with them, and there is no guaranteed algorithm for making this distinction in practice. The most useful procedure in combating outliers is the keeping of comprehensive notes by the operator. If doubts were raised as to sample authenticity, the integrity of transfer procedures, equipment status, and so on before the data are observed, Bayesian inference tells us that one is much less likely to reach false conclusions in eliminating outliers putatively caused by these events. However, the same brand of logic also informs us that an operator who continually speculates in writing as to what may be going wrong to justify later elimination of outliers is not really improving the process. The best approach is honest assessment and recording of actual doubts.

Despite all precautions, unexplained outliers will occur, and CB assays, like other highly sensitive methods, are especially prone to them, particularly during a "break-in" period when personnel have not yet learned how to avoid critical contamination. Occasional outliers will, however, plague data analysis even with an established assay and experienced operators. One must consider this reality during assay development and, especially, in writing the standard operating procedures. The aim is neither exclusively to save time and money nor to demand perfection, but to specify a reasonable and rational course of action to determine the true status of the lot or other system under test. Note that while a priori information about potential failures is best, some failure modes can be convincingly demonstrated after the apparent outlier has appeared, such as intermittent mechanical or lamp failures of an instrument, misreading of a protocol by an operator, and manifest contamination of a reagent. Investigations of potential causes of outliers are likely to prove fruitful at any stage.

As an example of a "discrete" type of outlier, consider the case of a bioburden measurement made by a luminometric assay, such as total ATP content or the live cell assay described in Chapter 14. We assume that a low bioburden is required for a lot to pass inspection. Suppose that in one instance, duplicate measurements are taken, one of which shows a near-zero bioburden and the other yields a high level of luminance, implying the presence of many contaminating microbes in the sample. Now everything depends on how the SOP was written. An assay developer who is used to considering failure modes may have considered this possibility in advance, in which case the SOP should specify the exact follow-up steps: a repetition of the assay with a greater number of data points or, preferably, two or more repetitions. Rarely or never would it be acceptable simply to declare a point an outlier based on a statistical approach and proceed, especially if the sample is critical. However, it should be possible to save the lot by repeating the assay (assuming it is not contaminated), if the SOP writer has foreseen the need. If the SOP simply states that "A reading above 2000 RLU indicates that the lot is contaminated," it is probably too late to rescue the situation. Note that if three or four replicates had been done in the first place, only one of which failed, the assay would still have to be repeated, but the data and reasoning would likely be more convincing to an inspector.

Various statistical methods for identifying and dealing with outliers have been proposed (646). It is not within the scope of this book to discuss Tukey's test, the currently favored Dixon's Q test, and so on. The reader should consult the statistical literature or, preferably, a trained statistician for advice. However, certain general principles are worth stating here. For example, it is unacceptable to fall into a routine in which the presence of unexplained outliers is a common occurrence. However, as assays become increasingly sensitive and complex, the nature of an "outlier" changes.

Certainly, highly sensitive CB assays and other methods capable of detecting a few molecules will yield apparently erroneous readouts as a result of relatively minor failures of operators and equipment, such as a minute light leak or the presence of an aerosol-deposited cell or spore. Thus, the use of a well-supported rule as an inherent part of the validation process and subsequent use may be appropriate; this decision may be affected by assay robustness, as discussed above.

13.2.4   Gaining Regulatory Acceptance of CB Methods

It is quite probable that the use of a method without an established history will cause an agency to request confirmatory data obtained by other means—for example, the ^{51}Cr release assay. If the method can be shown to be concordant over a period of time and under various assay conditions, it may eventually be possible to switch to the more economical method.

In a sense, the FDA Center for Biologics Evaluation and Research (CBER) has already answered the ultimate question of regulatory acceptance by developing and announcing a coupled luminescent assay itself: a reporter-based method of quantifying vaccinia neutralization. Perhaps surprisingly, the β-galactosidase enzyme was chosen as the reporter instead of luciferase; for this reason the method cannot strictly be called a "coupled bioluminescent" assay, since the light-generating reaction is unnatural. However, the distinction makes little practical difference in this case. The odds are good that members of CBER and other such organizations would view CB reporter assays with open minds.

The FDA has also approved a commercial coupled chemiluminescent assay in a clinical diagnostic role (Versant 3.0™ from Panomics). The assay formulation is complex, involving two extenders, two probes, a preamplifier and an amplifier, as well as the chemiluminescent alkaline phosphatase substrate and the conjugated enzyme itself, suggesting that the FDA is more interested it its performance than in its nature. This bodes well for CB assays. However, the failure modes, which are of paramount importance to regulators, are different for chemiluminescent and CB assays. To establish CB assays as sufficiently reliable for clinical use, an exhaustive analysis of all potential failure modes, both in assay performance and in interpretation, will be required. Many of these failure modes are discussed in Chapter 2, but others will appear later and some will depend on the individual nature of the CB methods themselves.

13.3   SUMMARY

This chapter is intended to provide a general introduction to the regulatory field for researchers interested in seeking agency approval of CB assays for clinical use or direct marketing. The material herein is insufficient by itself to enable a researcher either to make a validation plan or to determine whether such a plan is sufficient and successful. However, it is hoped that the many concepts involved in regulatory approval, especially the "failure mode" way of thinking and the critical importance of the intended use of an assay, will be helpful to the researcher who is contemplating these challenges.

PART III

OTHER APPLICATIONS
OF COUPLED BIOLUMINESCENCE

14

COUPLED BIOLUMINESCENT DETERMINATION OF BIOBURDEN AND STERILITY

14.1 INTRODUCTION

Many factors are driving the development of increasingly sensitive means of detecting living organisms and/or their residue. The burgeoning numbers of medical and food products that require sterile preparation and packaging; concerns about newly discovered pathogenic microbes, especially those resistant to multiple antibiotics; and the threat of bioterrorism have all lent urgency to this effort.

The ultimate test of bioburden is a method of amplifying individual molecules of biologically active nucleic acid. The complete absence of such nucleic acids in a truly representative sample of a product is as close to a guarantee of sterility as one is likely to obtain. However, this approach has several drawbacks: it is inappropriate for many products, such as most foods or cell-derived materials; it is usually inconvenient or impractical; it is not always possible, since the nucleic acid content of viruses varies among DNA, RNA, and single-stranded versions of each, which may not respond to standard detection methods; and it is not responsive to the issue of sterility in some cases, since dead organisms that may not present a problem may still leave a residue of intact nucleic acid. (CB methods of working with nucleic acids are discussed in Chapter 8.)

Another preferred bioburden/sterility test is simply the "incubate and watch" approach, with the readout (when plates are used) typically presented in colony-forming units (CFU) and the acceptable limit expressed in the same terms; for liquid cultures, turbidity is monitored, and the readout is usually qualitative. The FDA has a history of approving of this method (see, e.g., the FDA's 1993 "Points to Consider" or

Coupled Bioluminescent Assays: Methods and Applications, Michael J. Corey
Copyright © 2009 by John Wiley & Sons Inc.

"PTC" mycoplasma growth test, and its recent revisions). For routine monitoring of surfaces in laboratories and manufacturing suites, thousands of technicians "swipe" with tens of thousands of wipes every day, followed by overnight incubations in standard media to look for prokaryotic contaminants. This test is also frequently performed on water supplies and other inputs to the product stream. The whole thing sounds relatively simple, but a mere day spent with a bioburden expert convinces one that it is exactly the opposite: culturing of microbes for the purpose of sterility and bioburden determination is a vast field in itself, because there are so many different microbes and so many ways of growing them (or of killing them by accident). If it is the sterility of the actual product, rather than the bioburden level of the environment, that one wishes to determine, matters become complex very quickly. The archetypical case is one in which the presence of a microbe is suspected, but the bug (1) is an obligate anaerobe, (2) has a doubling time measured in weeks, and (3) has poorly defined trace element requirements, possibly with a narrow biochemical survival window as well. One might spend a career learning how to grow such a microbe, and even if one knows how, an absence of growth is rarely enough by itself to dispel suspicion. Fortunately, a subindustry has grown up around bioburden and sterility assessment, and the assay developer may well find it more convenient and less expensive to send samples out for processing by a contract research organization (CRO). However, the manufacturer is ultimately responsible for the sample. Moreover, it may be prohibitively expensive to contract with an outside firm to perform environmental monitoring, which is a skill set that should exist in-house in any case.

For reasons both of expense and, especially, of response time, rapid assessment of the presence of the essential molecules of life has become a useful alternative to these absolutist approaches. Often both are used. In this chapter, we will concentrate on these rapid methods and consider how CB approaches may fit into this picture.

14.2   RAPID METHODS OF BIOBURDEN AND STERILITY ASSESSMENT

Here, we consider "rapid" to mean that the user does not have to wait for growth of the organisms. If very rapidly dividing prokaryotes are the entity of interest and the contamination is substantial, it may be possible to observe turbidity or other effects in 2–3 h, but this is still not "rapid" by CB standards and may not be sufficiently quick to enable many critical decisions related to production, lot acceptance, and other important issues. Thus, we limit the discussion here to biochemical methods that yield results in less than 1 h.

14.2.1   Bioburden Measurement by ATP-Release Assay

The ATP-release assay (ARA) is discussed at length in Chapter 3 as a cytotoxicity/ viability assay. There, it was pointed out that the ARA is much better for measuring viability than it is for cytotoxicity in typical drug screening regimes, because one is, in effect, measuring the number of once-living cells (even though they are now dead).

For this reason, bioburden assessment is one of the best applications of the ARA technology, and the use of this technique in bioburden measurement is expanding at present. An ATP content of 1 fg/mL is roughly equivalent to a CFU readout of 100 prokaryotic organisms per milliliter, but one must keep in mind that to detect the true bioburden by culturing techniques, one must be sure to culture for every organism that might be present—not a simple task. Thus, ARA is much more likely to yield an accurate answer if some of the organisms present are unknown or difficult to culture. An important caveat, however, is that there may be some organisms that are easy to grow, but difficult to lyse, and they may not be known to science.

Firms currently marketing ARAs for bioburden testing of samples of various types include LuminUltra® Technologies, Mo Bio Laboratories, and Scigiene, undoubtedly among many others.

14.2.2 Bioburden Measurement by Protein Assay

Quantitative measurements of protein are rarely used as a means of determining bioburden in drug and biotechnology industries but may be appropriate for food preparation surfaces (647). The method is neither specific for the presence of live organisms nor highly sensitive in this application, and thus is not considered directly competitive with coupled luminescent methods.

14.2.3 Coupled Bioluminescent Methods of Measuring Bioburden

We now take up methods that involve CB detection of enzymes released by dying organisms.

14.2.3.1 Measurements of Bioburden Using Cellular Adenylate Kinase
The only CB assay currently marketed for use in bioburden determination is AKuScreen® from Celsis International (as of this writing, see www.celsis.com). The principle of the assay is the production of ATP from supplied ADP. The ADP is heterolyzed to ATP and AMP by the enzyme adenylate kinase (AK), which is probably ubiquitous in biological cells. Thus, this is a true CB assay, in which one enzymatic reaction is employed to make ATP, which then enters the luminescent luciferase reaction.

One of the standard methods of detecting microbial contamination by ARA involves growth incubation of the sample to increase the number of organisms present prior to the assay. The method recommended by Celsis incorporates this step as well, although it is shortened from 24 to 18 h (but remains 24 h for detection of molds, due to the accurate observation that molds do not grow in convenient, homogeneous fashion in liquid culture and homogenization is required even after the longer incubation). Following the growth incubation, the cells are broken, releasing AK, which is then incubated with vendor-supplied ADP to yield ATP and AMP. This step, which can occur inside the luminometer, is carried out for 40 min at 30°C. The luciferase reagents can then be added and the readouts obtained.

Although it is encouraging to see that a CB assay method has entered this arena, certain aspects of the procedure described are less than ideal. Both the prolonged growth incubation step and the 40-min enzymatic step are slower than one could wish. One suspects that while AK is likely present in virtually all cells, it is not present in large amounts. As a result, the number of cells required for detection is evidently quite large. The Celsis Web site is not helpful in this regard, merely claiming a limit of detection of "one organism per sample." This means, of course, that the method is capable of detecting the organism after it has replicated itself into millions or, likely, billions of its kind. The actual limit of detection cannot be determined from this source, and the associated product literature gives the same information.

The description of the method as presented in the Celsis literature appears to be somewhat misleading. The reaction does not "amplify" ATP present in the sample. Indeed, all of the ATP present in the microbe could be destroyed or removed, and the method would still work, as long as the AK enzyme is not affected. The signal comes from ATP generated *de novo* from the ADP substrate supplied in the kit, not (or only minimally) from ATP originally present in the organisms. It might be possible to construct a system wherein ATP present in limiting concentration is utilized to trigger a cycling reaction with an independent readout, probably one involving NADH (perhaps with a bacterial luciferase). Such a system would "amplify" the signal represented by the ATP present in the original sample; the AK reaction does not.

Generally speaking, this method is enjoying a degree of commercial and regulatory acceptance and, as a CB method, represents an advance in speed and sensitivity over the earlier technology, although the necessity of growing the cells prior to obtaining the readout is a serious drawback in many applications.

14.2.3.2   *The Use of Other Enzymes with CB Technology in Assessing Bioburden*
While at present there appears to be no commercially available CB method apart from AK that is specifically directed to bioburden determination, it should be possible to improve on the AK system by choosing enzymes that are present in larger amounts and/or more catalytically active. Components of major metabolic pathways are of course essential to cell survival and are exploited in schemes such as the aCella-TOX^{TM} method described in Chapter 3, which measures the activity of the highly active and ubiquitous glyceraldehyde-3-phosphate dehydrogenase (G3PDH). Other potential CB assay targets include lactate dehydrogenase, which is already the target of an important coupled fluorometric cytotoxicity assay (see Chapter 3), tricarboxylic acid cycle enzymes, and other glycolytic enzymes. The specific requirements of bioburden measurements include a reasonable degree of enzyme stability, since time may elapse between gathering of the sample and measurement; this issue can be partially addressed by postponing the cell-killing step until immediately before the assay.

The limit of detection of a CB bioburden assay incorporating any of these major metabolic enzymes should be low enough to allow real-time determinations without a growth-incubation step. However, because of the critical nature of these tests, it may be desirable to perform the assay both immediately and after growth incubation. The CB assays of G3PDH is clearly capable of detecting individual eukaryotic cells when properly formulated and performed (83), and the limit of detection for prokaryotes is also very small. This appears to be another case in which the advent of appropriate CB technologies may enable improved decisions based on real-time determinations that were hitherto impossible.

15

ENVIRONMENTAL APPLICATIONS OF COUPLED BIOLUMINESCENT ASSAYS

15.1 INTRODUCTION

The importance of accurate measurements in environmental research and monitoring has never been greater. Given the range of pollutants currently entering the environment; the diverse effects of acid rain, ozone depletion, and global warming; and the absolute requirement of military and antiterrorism professionals for reliable information regarding the presence or absence of biological and chemical threats, it is not surprising that an industry has arisen to supply the assay needs of these workers. Sensitivity and accuracy have been achieved in many cases, but the desired rapidity of response that would allow real-time decision making has generally not yet appeared. As an example, the employees of the U.S. Environmental Protection Agency (EPA) and state agencies looking for eutrophication-promoting nutrients such as phosphate and nitrate in runoff water routinely take samples and send them to central laboratories for analysis, resulting in a response time of several days to several weeks. A rapid, on-site method might enable an immediate response that could stem or stop a source of pollution on the spot. While this might entail something of a paradigm shift in the working approaches of the agencies concerned, the public interest would be well served by such a shift. Obviously, the critical need for a rapid response is even more clear in the case of military or other personnel entering an area of potential biological or chemical threat.

Firms have been addressing the need for rapid detection of pathogens with portable ELISA devices for some time, but the widespread development of advanced technologies for detection and quantification of small molecules is a more recent phenomenon.

Coupled Bioluminescent Assays: Methods and Applications, Michael J. Corey
Copyright © 2009 by John Wiley & Sons Inc.

ELISA methods have advanced a great deal, and a signal may be acquired for many well-defined targets within minutes. Of the many liquid-phase detection technologies available, ELISA is especially amenable to environmental deployment. This suitability is due to several factors, including advancements in technologies for immobilizing antibodies in active states, improved sensors for tiny electrical and optical perturbations, and new ways of transducing the ligand association event. Moreover, the enzymatic component of ELISA may be replaced by mechanical, electrical, or other responses to the binding event; alternatively, enzymes may yield chemiluminescent readouts, or enzyme and antibody may be immobilized separately in a flow system. These ideas gave rise to a set of techniques under development, including quartz crystal microbalance immunosensors, which have been successfully tested in detection of dioxins (648); stabilization of the Fv region on ligand association for the assessment of small molecules (649); the magnetic particle-based immunosupported liquid membrane assay, involving antibodies on magnetic beads suspended in opposing electromagnetic fields (650); and various embodiments of portable flow-based sensors (651).

A central philosophical question, however, is whether the proper detection target is the *analyte* or the *activity*. This issue takes on great importance in the modern era of "designer" chemistry and the associated armamentarium of exotic toxins. Environmental monitoring has long focused on the analyte, largely because the techniques for structural analysis were more widely available, more readily deployed, and better established than those for detection of, for example, an undefined inhibitor of a given enzyme or class of enzymes. However, with the advent of portable and rapid devices employing CB and other rapid methods of detecting enzyme activities, these arguments lose most of their force. The counterargument is a strong one, especially in the age of bioterrorism: an evildoer may well develop and release a molecule with deadly characteristics that does not appear in any structural database, or is even unknown to science. Just because a poison acts as a nerve gas does not mean it automatically appears in the catalog of nerve gases. For the war-fighter facing an invisible and undetermined threat in a theater of action, it is hardly enough to ensure the absence of known harmful agents. What you don't know can kill you.

15.2 CURRENT METHODS FOR ENVIRONMENTAL MONITORING OF WATER QUALITY

Air monitoring bears some similarity to monitoring of drinking water in that a solution must be produced at some point for most agents to be detected. While there are many interesting aspects to collection of test samples from air, their relevance to our topic is limited and we shall concentrate on assays in two types of water sources: environmental waters and drinking water. Here, we use "environment" in a broad sense to refer to the entire human living space, which includes, for example, the interiors of buildings. However, the anticipated contaminants of tap water are obviously quite different from those of stream water or lake water, and are treated in separate sections below.

15.3 METHODS OF MONITORING STREAM WATER AND LAKE WATER

The EPA currently presents an online document entitled "Volunteer Stream Monitoring: a Methods Manual" at http://www.epa.gov/volunteer/stream/index.html. The document is quite sophisticated in terms of the level of the techniques it attempts to impart to the volunteer. Rather than toxic metals and the like, the document emphasizes phosphorus, nitrogen, bacteria, pesticides, and organic nutrients that may lead to consumption of oxygen by microorganisms and partially anaerobic conditions. Some of the methods described are suitable for use "on the spot," just as CB techniques would be expected to be carried out with portable devices. Professional water quality monitors may require higher sensitivity, and they generally collect samples for transfer to a laboratory that performs mass spectroscopy, gas chromatography, or chemical analyses based on UV/spec quantification. Both sets of techniques represent interesting points of comparison for CB methods and other modern approaches.

15.3.1 Monitoring of Phosphate in Freshwater

The inorganic phosphate ion (orthophosphate) is an important contaminant of freshwater because it can serve as a growth-limiting nutrient for a range of microorganisms, especially algae, which in turn can dramatically reduce the oxygen content of the water, adversely affecting fish and other wildlife. It can also disrupt the rate of plant growth, leading to clogging of waterways. The usual source is fertilizer runoff from agriculture, although some industries also produce large amounts of phosphate in their effluent. At this point, measurements of phosphate concentrations in freshwater are limited to a general monitoring, with essentially no reporting of specific problems or enforcement actions; however, the availability of rapid, sensitive methods for use in the field could eventually lead to a change in this approach. Unfortunately, two forms of phosphate that are commonly present are not readily measured by any such rapid approach currently available: polyphosphates, that is, metastable polymers of the orthophosphate ion; and organic phosphates, in which the phosphate group is covalently bound, usually as an ester, to carbon-containing molecules. Detection and quantification of these forms will likely require laboratory analysis, at least until techniques such as mass spectroscopy can be rendered highly rapid, portable, and easy for field personnel. Polyphosphates can be strong chelators and are often found in detergents. They are eventually converted spontaneously to orthophosphate under fresh water conditions. Organic phosphates, however, may persist for long periods.

Two methods for phosphate measurement are presented in the EPA document cited above, both involving a colorimetric method known as "persulfate digestion followed by ascorbic acid" with ascorbic acid in a solution with sulfuric acid, potassium antimonyl tartrate, and ammonium molybdate. A similar method is described on the website of the Hach company, a leading supplier of analytical instruments for environmental monitoring: http://www.hach.com/fmmimghach?/CODE%3APHOSPHORUSTOT_PP_OTH1940%7C1. The online document inserts a 30-min boiling step with potassium persulfate, followed by neutralization with sodium hydroxide, prior to the ascorbic acid procedure. The point is to digest organic forms of phosphate to orthophosphate for quantification. It is likely that the EPA considers this step too dangerous or inconvenient for incompletely trained field volunteers.

Concentrations of phosphate greater than "0.1 mg/L" (actually the concentration of elemental phosphorus as defined) may be assessed by visual comparison with colorimetric standards, while a field spectrophotometer with a 2.5-cm light path is recommended for concentrations down to 0.02 mg/L. This concentration amounts to approximately 600 nM phosphorus (or phosphate), which may be taken as the field limit of detection for this method. Given the graph of standards presented in the document, this concentration of phosphorus would yield an absorbance of 0.02 at a wavelength of 700–880 nm (the exact wavelength to be used is not specified). This is probably the best performance that one can expect in the field from a volunteer without full technical training, although some would argue that a reading of 0.005–0.01 could be meaningful.

The CB method for phosphate quantification described in Chapter 5 has a limit of detection of roughly 10 nM with the currently available reagent set. However, one should allow environmental conditions, as well as the possibility of improved reagents. The reaction time is shorter than that of the assay above (3–5 min), and the assay could be made simpler, with much smaller volumes. (The colorimetric assay requires a large volume to achieve adequate absorbance with the long light path.) On the whole, the CB method, if it were developed further, would likely be highly competitive with the current EPA-recommended field assay.

It may also be useful to compare the potential of CB assays with methods that are still under development. It is perhaps surprising that colorimetric techniques are a player in this context as well. A limit of detection of 1 nM is claimed for such a method, employing a 50-cm path length and analysis at multiple wavelengths (652). The assay is intended for use with ocean water, and the apparatus is unlikely to be adaptable to field measurements in the sense of total portability. Microchip-mounted electrode sensors do not appear to have reached the same range of sensitivity (653), but offer the convenience of small size and often very low cost when mass produced. Amperometric biosensors also fall short on the sensitivity front, but the rapid response (as little as 6 s) is useful (654). With interest in electronic sensors and microfluidics widespread and growing, it is impossible to present an up-to-date review of the field, but these approaches exemplify the possibilities.

15.3.2 Monitoring of Nitrate in Groundwater

Like the phosphate ion, nitrate can be a growth-limiting microbial nutrient and can lead to eutrophication and oxygen depletion of freshwater. Moreover, nitrate salts are generally more soluble than phosphate salts, leading to a greater proportional runoff of this ion into water bodies. Freshwater nitrate is generally considered to reflect any prior pollution by ammonia and nitrite as well, since these are usually oxidized to nitrate under environmental conditions; however, this may not be the case in oxygen-depleted water. Thus, the parameter "total nitrogen" may or may not be reflected accurately by the nitrate concentration.

The two EPA-recognized methods in widespread field use are the cadmium reduction technique and the nitrate electrode; hydrazine has also been used for reduction. Cadmium reduces nitrate to nitrite (thus nitrite in the source is also quantified by this method). This is followed by the Griess reaction described in Chapter 7 and colorimetric quantification of nitrite.

The CB alternative described in Chapter 7 is specific for the nitrate ion, which is coupled into a reaction series involving nitrate reductase and subsequent NAD^+-dependent production of ATP. Its utility therefore depends on the degree to which the nitrate concentration reflects the parameter of interest, which in turn depends on the nature of the source. In many cases, oxidative and microbial processes will process much of the reduced and organic forms of nitrogen to nitrate but this cannot be assumed. The CB assay is very rapid, sensitive, and convenient, and is likely to be highly competitive with other methods if the results adequately address the question of interest.

15.3.3 Monitoring of Pesticides in Freshwater

Detection and quantification of pesticide concentrations are fundamentally different from the analogous procedures with simple ions. Few pesticides are amenable to quantitative assessment by simple chemical procedures followed by colorimetry, largely because such procedures would generally fail to distinguish them from related pesticides or harmless organic molecules. Therefore, monitoring methods for pesticides have mostly relied on structure-based detection by gas chromatography–mass spectroscopy (GC–MS) and high-pressure liquid chromatography (HPLC), sensitive analytical methods that yield highly characteristic molecular fingerprints for pesticides and other molecules, but are slow, expensive, and labor-intensive. However, detection and quantification of specific molecules (as opposed to activity detection) is a suitable approach to pesticide monitoring under many circumstances: when the identity of a contaminant or range of contaminants of interest and their molecular signatures are known, when particular pesticides are in use in a given watershed, or when studies of the fate of a specific molecule are being performed. In fact, it could be argued that structure-based detection is more effective than activity measurement for pesticide detection in freshwater, in contrast to the argument made in Section 15.4.3 regarding protection of personnel from pesticides and nerve agents in more immediate or acute threat situations.

The most complete analysis of pesticides in freshwater in the United States was conducted not by the EPA but by the United States Geological Survey (USGS) under the National Water Quality Assessment (NAWQA) program. The March, 2006 report (revised in February, 2007) on pesticides in freshwater between 1992 and 2001 is currently available online at http://pubs.usgs.gov/circ/2005/1291/ (655). The report summarizes and analyzes data from many thousands of measurements over this extended period, using the methods described. The use of these well-established techniques allowed a high degree of standardization over the long course of this study, providing a clear advantage in terms of our ability to compare and generalize the results.

Despite this nod to the *status quo*, however, it is likely that the near-term future will bring both new needs and novel methods. Ideally, one would not identify a pesticide contamination problem with an assay that is spatially and temporally removed from the source, but near to the source and within minutes of the event. This ideal is currently out of reach both technically and logistically, but if one is willing to wait 90 min or more for a result, one can detect organophosphate pesticides, carbamates, and other acetylcholinesterase inhibitors (see Chapter 6) by means of a microplate assay currently marketed by Envirologix. The QualiPlate^{TM} Kit for cholinesterase employs the conventional acetylthiocholine reaction with Ellman's reagent, yielding a photometric readout. The kit is recommended for use with samples of wine and dried fruit as well as water, though extraction steps (with methanol) are required for detection in the complex matrices. Not surprisingly, the extraction procedure must be optimized for each sample type, and frequently involves aqueous swelling of the sample (e.g., fruit) for several hours before treatment. This would likely also be the case with other detection methods. The claimed limit of detection is as low as 0.5 ppb for some pesticides, such as the notorious chlorpyrifos (656, 657). This figure refers to the concentration in the 80%-methanol extract. This level of sensitivity is quite useful, but it must be assumed that whatever is detected by the Ellman reaction can be matched by CB methods with at least 10–100-fold greater sensitivity. Flow monitors using CB or other methods, described in Section 15.4.3, may be used in the future to obtain a rapid and automated response to a pesticide contamination event in streams and lakes.

15.3.4 Monitoring of Bacterial Content of Freshwater

The problems here are somewhat analogous to those occurring with pesticides, although with bacteria at least a low-cost assay is possible. The obvious method of detecting microbes in an aquatic specimen is to plate an aliquot of sample on growth media and see if anything appears. This approach has the disadvantage that one must wait for many doublings to occur, a matter of a day for most bacteria, though some important species are slower (much longer, of course, for most eukaryotes). The other disadvantage is less evident but of equal consequence. Not all microbes of interest can be cultured so easily. The cultivation of pathogenic, undesirable, and "indicator" microbes is an entire field of study in itself; see Chapter14. It is sufficient here to say that if the presence of a specific organism or range of organisms is suspected, or is of special interest, and is detection by culture is intended, then someone must know how to grow the bug, and that may be far from simple.

Many types of bacteria may be present in freshwater, but the EPA has so far been concerned primarily with those that indicate contamination by sewage: fecal coliforms (including *E. coli*), fecal streptococci, and enterococci. The bacteria are detected by the presence of growth in a tailored nutrient medium, either solid (with agar) or a broth. The medium is inoculated with sample water, which may be concentrated beforehand by using a suitable filter. Colonies, turbidity, or gas production is the readout. Generally, no effort is made to quantify the grown bacteria and relate the number back to the level of contamination of the source; instead, various dilutions of the source water are tested, and a most probable number is calculated based on the analysis of the results using statistical tables. Obviously, this method requires considerable time, effort, and materials (which must be sterile) to obtain information regarding a single source.

The modern strip ELISA technology permits many types of bacteria to be detected very rapidly and conveniently with little apparatus. Envirologix, Kamiya Biomedical, and Teco Diagnostics are among the firms marketing kits and readers for this purpose. However, strip ELISAs are often up to 10-fold less sensitive than plate ELISAs performed in a laboratory, and even the latter methods rarely achieve limits of detection of a few organisms. (Of course, concentrating the sample water by filtration will increase the sensitivity of this or virtually any other method. This process may be assumed when a low limit of detection is required, but even filtration eventually reaches limits in terms of filter clogging (especially with environmental samples) and user exhaustion. Concentrating bacteria on a filter also does not automatically solve the problem of introducing the organisms into the assay mixture, which must usually be accomplished by washing or processing of the whole filter.) ELISA then is an intermediate approach, often enabling sensitive detection of bacteria without an incubation step, but unable in most cases to identify low-level contamination immediately, and requiring separate assay development for each organism to be tested (sometimes for each strain).

Here again, the question of specific versus general detection arises. For an academic study, an identification of the species, or at least the group, may be necessary, and methods such as specific ELISAs may be highly appropriate. However, one can envision a different philosophy of sampling, one more in concert with modern approaches to drinking water monitoring, in which a sudden increase in the count of any prokaryote is of significance. A broad approach based on protein, particulates, conductivity, or enzyme release may be useful under these circumstances.

The general biodetection method of lysing all cells present and quantifying released ATP through luciferase luminescence has several advantages over growth-based methods in this context, the most important of which is the rapidity of detection. Modern alternatives to these methods include CB assays as well as electronic sensors employing piezoelectric (658), amperometric (659), or other advanced principles (660, 661). Moreover, as is the case with small-molecule assessment, the concepts of ELISA can be combined with new technological developments in many ways. These methods are discussed further in Section 15.4.3.

A CB assay for bacterial contamination is likely to rely on a lysis step, followed by enzyme detection. The fluorometric LDH-release assay described in Chapter 3 could also be employed, but this appears to run counter to the goals of maximal sensitivity and speed. The G3PDH-release assay currently recommended marketed as aCella-TOX^{TM} is suitable for this purpose, assuming its limitations are known and acceptable. For example, success of the lysis step cannot be assumed. *E. coli* is a common analytical target, but not all *E. coli* are the same. The author's original lysis protocol for Gram-negatives involved lysis by various combinations of lysozyme, detergents, and polymyxins; this method was found to be highly effective against laboratory strains of *E. coli*, but much less useful against environmentally isolated strains, which have evolved multiple mechanisms for resisting straightforward lysis. As an aside, there are eukaryotic microorganisms, such as the notorious pathogen *Cryptosporidium*, lysis of which without denaturing the internal biomolecules remains a significant scientific challenge. In general, lysis of microorganisms is not always a simple matter and may require its own development phase. The other limitations of CB detection of released bacterial enzymes are likely to be less important. Inhibitors of the coupled system are unlikely to be present in freshwater.

15.4 METHODS OF MONITORING DRINKING WATER

In testing drinking water, although some small molecules are still of interest, the concern generally shifts away from adventitious sources of contamination toward the doings of terrorists and lunatics. Nerve agents and other powerful toxins, bacteria, and eukaryotic pathogens should all be detected rapidly and with very high sensitivity. The field of real-time threat assessment in water is undergoing very rapid development, but few of the measures we would like to see are yet in place, and most municipal water systems still primarily depend on access limitation to prevent catastrophe.

Writing about the subject is complicated by the fact that those who direct and implement safety measures in municipal water are reluctant to reveal their methods or results. At the time of the revelation in 2008 that trace amounts of pharmaceuticals were present in water supplies, reporters encountered difficulty not only in learning about the issue of direct interest but also in getting any answers at all from some water officials. It is quite understandable that these individuals wish to guard sensitive information that might reveal methods of circumvention to evildoers. However, it is also interesting that there was extreme variation among regions of the United States in what information was revealed, with New York city allegedly the least forthcoming.

The simplest way to avoid a reputation as a scientist with information that could be of use to terrorists is never to acquire the information to begin with. The author has carefully followed this policy. Anything written herein, therefore, is both publicly available and judged not to be "sensitive."

15.4.1 Current Drinking Water Reports

Government regulations require water districts serving populations of 100,000 or greater to provide yearly water quality reports. These are usually available on-line as well through the following website: http://www.epa.gov/SAFEWATER/ccr/whereyoulive.html?OpenView#map. The reports typically contain a great deal of reassuring language and conservation advice, but hard data are to be found as well. The author took the unscientific approach of choosing to view reports from three small cities, roughly spanning the nation: the author's hometown of Bellevue, WA; Plano, TX; and Warrington, PA. The Bellevue and Warrington reports were easily found, while the Plano report required several levels of navigation and lucky guesses to locate.

The stated policy in the Bellevue report is to give data only on the few items that were detected at nonzero levels. Therefore, only a few types of data are present in all three reports, including fecal coliform content (low in Plano, zero in the other cities), fluoride (close to 20% of the maximum standard in the two fluoridating communities, zero in the third), nitrate (0–7% of the violation level), barium (5–20% of the violation level in Warrington, near zero in the other two cities), and trihalomethanes (12–80% of the violation in Bellevue, 9–40% in Warrington, reported as unregulated "chloroform" in Plano but present in similar levels; these are expected by-products of chlorination). Some items seem to be location-related. The herbicides atrazine and simazine appear in the Plano report at roughly 0–20% of violation levels, but are not addressed in the other reports. Lead and copper levels resulting from corrosion of pipes in individual households appear in all three reports. Low levels of emitters of β radiation are reported in Plano and Warrington but not in Bellevue. *Cryptosporidium* is reported at near-zero levels in Bellevue and discussed as a hazard, but not reported, in Warrington.

At present there is almost no such thing as a truly rapid response to an environmental contamination event, with the exception of the very visible and emotionally charged oil spill. CB and other advanced methods may eventually change this, but it is also possible that their convenience and rapidity will cause them to be adopted for routine long-term studies as well. The current interest in pharmaceuticals in environmental and drinking water is just one of the trends toward the monitoring of a more extensive set of chemicals and organisms. There are many opportunities ahead for rapid detection technologies, and one day they may have a favorable impact on the annual drinking water report as well.

15.4.2 Biohazard Monitoring in Drinking Water

The distinction between "hazard" and "safety" monitoring is subtle and is meant to convey the difference between detecting an acute threat to health due to a single event, likely a terrorist act, and general assurance of the nation's water quality. By "biohazard" we refer specifically to pathogenic organisms; the other major classes of dangers are viral and chemical. Viral threats cannot ordinarily be detected by CB methods and are not further addressed here, while methods for rapid detection of chemical hazards are discussed. Although the range of substances and organisms that bioterrorists may attempt to introduce into a water supply is broad, it exhibits surprisingly little overlap with the natural and anticipated contaminants from known sources (662).

The volume of thought devoted to the problem of protecting drinking water from intentional contamination by pathogens has increased dramatically since 9/11. The already developed machinery of risk assessment and prioritization is being brought into play, and papers are regularly published on ways of selecting and evaluating threats (663), a significant intellectual exercise in itself. The CDC list of Category-A pathogens, which includes the prokaryotes *Bacillus anthracis* (anthrax), *Yersinia pestis* (plague), *Clostridium botulinum* (botulism), and *Francisella tularensis* (tularemia), along with the smallpox and hemmorhagic fever viruses, is based on the following well-reasoned criteria: potential for illness and death; potential for large-scale production, widespread delivery, and contagion; public perceptions that may give rise to panic and disruption; and the degree of preparedness needed. However, new information on the European *Burkholderia* pathogen led to its inclusion at the highest threat level on a revised list based on more rigorous mathematical principles and published in 2006 (663). The same paper also provided a much longer list of "high threat" diseases, which includes brucellosis, cholera, diphtheria, *Shigella*-caused dysentery, legionellosis, *Neisseria* meningitis, psittacosis (*Chlamydia*), Q fever, Rocky Mountain spotted fever, tuberculosis, typhoid fever, and typhus, along with a host of viral diseases. Detection of each of these pathogens in water samples presents a separate problem in development; the entire range of these methods is outside the scope of this book. Instead, we will simply mention a few modern developments in the rapid detection of *B. anthracis*, the anthrax bacillus, which will serve as examples of how biotechnology is addressing the many threats. There is a rapid immunoassay for anthrax spores (664), as well as the FRET principle applied to spores bound to microbeads and detected by flow cytometry (665).

Other interesting and novel possibilities exist. A time-gated spectrophotometric sensor is based on interaction of the endospore with the Terbium^{3+} ion in the presence of dipicolinic acid (666) (time to detection: about 15 min). Lateral flow microarrays (667) require 2 min for the detection step, but this excludes the critical amplification step, which may require 4 h for high sensitivity (668). A specialized flow system capable of rapid detection uses piezoelectrically actuated cantilevers (658), while a portable fluorometric antibody-based system currently intended for forensic use and oriented toward spore numbers in the thousands requires 15 min to produce a result (669).

Unless used as a secondary detection method (e.g., as the reporter of an ELISA-type procedure), the most obvious way in which CB methods could contribute to this effort is in the area of general biodetection. The background is the same as that presented above for environmental water testing, with the exception that when drinking water is the sample, real-time monitoring becomes even more desirable. Instead of discrete assays, then, the desired system would employ immobilized enzyme molecules, with the sample stream flowing continuously past the enzymes in the presence of appropriate substrates and buffer constituents. In short, the testing would occur continuously in a flow cell. Such an approach could yield an almost instantaneous signal if life were detected. Released enzymes would catalyze the formation of ATP and subsequent light generation within microseconds, triggering an alarm. But can effective CB biodetection methods be developed within the constraints of a flow cell at acceptable cost?

We are not ready to answer this question, but certain aspects of the system are evident. The light-emitting prosthetic group, whether luciferin, coelenterazine, or something else, must be freely diffusible, and must therefore be supplied continuously to the system. This in itself may present problems of expense and practicality. In the case of the specific G3PDH-dependent CB system described in Chapter 3, continuous provision of glyceraldehyde-3-phosphate would also be needed, and this molecule is both expensive and unstable in neutral aqueous solution. The answer to such problems as these may be miniaturization. Expenses tend toward zero as the volumes do so. Flow cells based on nanotubes may be the way to go, and they may prove to be rechargeable and/or disposable. If the unit is small and cheap enough, it may be possible to monitor not just one but many areas of the fluid under test. The microfluidics industry is young and great developments may be expected.

It is also likely that an alternative CB assay formulation will appear that does not require continuous input of expensive and/or labile reagents. One such potential scheme has already been described. At a time when glyceraldehyde-3-phosphate (G3P) was unavailable commercially, the author developed a four-enzyme system incorporating aldolase-catalyzed cleavage of fructose-1,6-bisphosphate (FBP) into dihydroxyacetone phosphate and G3P (143). FBP is much less expensive than G3P, and is also more stable under some conditions, potentially leading to a more practical flow sensor. This is just one additional example; many coupled systems based on enzymes released by lytic treatments are possible. In the author's view, other liquid-phase chemistries are unlikely to compete with CB methods in rapidity and sensitivity. However, the same is not true of the other sophisticated detection methods cited above. The coming decades will yield many surprises in this area, and it is to be expected that several of the advanced methods will play important roles in meeting the threat.

15.4.3 Monitoring of Chemical Hazards in Drinking Water

The dichotomy between the strategies of detecting the chemical structure and detecting activity is most sharply drawn in this area. Again, most of the effort has gone into detection of structures associated with dangerous molecules in internalized databases. The advantage of this concept is that detection is coupled to immediate identification, which may allow more specific countermeasures and provide better information for law-enforcement activities than a general alarm; the serious disadvantage, however, is that not all potentially dangerous molecules have known structures.

Addressing nerve agents and other acetylcholinesterase inhibitors as the most important chemical threats, we find the same structure-based methods we have already seen in use, primarily GC/MS and HPLC. However, the need for activity-based detection methods has not gone unnoticed. The overall capabilities of these traditional methods in the context of bioterrorism and biowarfare have also been criticized on performance grounds (670). A fascinating approach that is receiving a great deal of attention relies on a species of crustacean, *Daphnia magna*, that is highly sensitive to acetylcholinesterase inhibitors (671), with 24-h LC_{50} values in the range of 1.25 μg/L (~4.3 nM) for parathion and comparable responses to other nerve agents (672). Even at lower concentrations, behavioral alterations appear and can be assessed quantitatively. This level of sensitivity is useful, but the system can hardly be considered rapid when it requires at least a day to achieve a readout. Faster alternative approaches include various types of fluorescent sensors, such as those based on enzyme-catalyzed reactions, chemically reactive devices, and the so-called "supramolecular" sensors (all are reviewed in Reference 673); a yeast system for detecting and biodegrading the specific pesticide paraoxon (674); an amperometric biosensor based on whole cells and sensitive only to *p*-nitrophenyl-substituted organophosphate nerve agents (659); a device based on plasmon resonance (660); sensors employing amperometric detection of the products of enzymatic cleavage of the nerve agents themselves, a method that obviously works only for certain agents (675); and a unique type of flow biosensor using that thiocholine substrate with self-assembling acetylcholinesterase in carbon nanotubes and amperometric detection (661).

It is impossible to evaluate these advanced methods against a CB technique that has not yet been rigorously tested against the same threat (a Google search for "aCella AChE" will yield considerable information about the author's patent-pending assay (329)). We do know that CB assays are capable of extraordinary speed and sensitivity, and the method has been shown to be sensitive to subpicomole quantities of various acetylcholinesterase inhibitors with a response time of under 5 min. This and other CB approaches should be considered along with the other modern alternatives in the quest for rapid detection of chemical threats in the environment.

APPENDIX A

ONE-LETTER AMINO ACID ABBREVIATIONS

- A: alanine
- C: cysteine
- D: aspartic acid
- E: glutamic acid
- F: phenylalanine
- G: glycine
- H: histidine
- I: isoleucine
- K: lysine
- L: leucine
- M: methionine
- N: asparagine
- P: proline
- Q: glutamine
- R: arginine
- S: serine
- T: threonine
- V: valine
- W: tryptophan
- Y: tyrosine

GLOSSARY

TERMS

acetylcholine   Critical neurotransmitter

acetylcholinesterase   Enzyme that breaks down critical neurotransmitter acetylcholine

acetylthiocholine   Synthetic substrate for acetylcholinesterase

acridan   Molecule that can yield acridinium ester upon electrical stimulation

acridinium ester   Chemiluminescent molecule with dioxetane structure

ADP   Adenosine diphosphate

AMP   Adenosine monophosphate

ATP   Adenosine triphosphate

aequorin   Photoprotein that generates light in response to Ca^{2+}

apoptosis   Programmed cell death

attomole 10^{-18} mol

***Bal*31**   Enzyme that digests double-stranded DNA

Beer's law $A = ECL$, law relating spectrophotometric absorbance to concentration, where A, absorbance; E, molar extinction coefficient of the solute; C, concentration; and L, path length

bioburden   Quantitative determination of number of microorganisms inappropriately present

---

Coupled Bioluminescent Assays: Methods and Applications, Michael J. Corey
Copyright © 2009 by John Wiley & Sons Inc.

bioluminescence   Generation of light from chemical reactions in which a protein is an essential element

calcineurin   Complex phosphatase that autoactivates

cAMP   Cyclic adenosine monophosphate, important second messenger created from ATP by the enzyme adenylate cyclase, a component of GPCR signaling pathways

cation   Positively charged ion

chemiluminescence   Nonenzymatic generation of light by chemical reactions below the temperature of incandescence

cis   Describes an effect due to an interaction of two parts of the same molecule or system; opposite of *trans*

coelenteramide   Spent form of coelenterazine

coelenterazine   Active form of prosthetic group of many marine luciferases and photoproteins

complement   Complex biochemical system for attacking foreign cells

cross talk   Situation in which one assay sample inappropriately influences the reading of another sample

dication   Ion with two positive charges

1,2-dioxetane   Cyclic structure C_2O_2 that may decompose to yield CO_2 and light

DNA polymerase   Enzyme that catalyzes both synthesis and pyrophosphorolysis of DNA

Ellman's reagent   DTNP, dithionitrobenzene, chemical used in quantification of free –SH groups

exonuclease   Enzyme that degrades the end of a DNA strand to nucleotide monophosphates

Factor I   Regulatory protein of complement

fluorometry   Measurement of fluorescence

gamma counter   Instrument for quantifying gamma radition; may be usable as a luminometer

GPCR   G-protein-coupled receptor, any of hundreds of receptors that transduce extracellular signals to intracellular activities

homogeneous   Refers to an assay method that requires no separations or phase mixing

hydrolysis   Cleavage of a molecule by water

isozyme   Enzyme with very high structural homology to another and virtually the same activity

kinase   Enzyme that transfers the terminal phosphate group of ATP to any of various substrates and can also catalyze the reverse reaction

K_i Inhibitor concentration that yields half-maximal inhibition under appropriate conditions

K_m Substrate concentration at which the enzyme exhibits half-maximal activity

linear regression   Form of regression appropriate for data that vary in linear fashion with the independent variable

luciferase   Enzyme that generates light from chemical energy

luciferin   Prosthetic group of beetle luciferases

luminometer   Instrument for measuring light from liquid-phase reactions

Malachite green   Dye used in quantitative assays of phosphate

mass spectrometry   MS, method of molecular analysis relying on molecular weight of molecules and fragments from the sample

necrosis   Cell death due to damage, disease, or starvation

NOS   Nitric oxide synthase, enzyme that synthesizes nitric oxide (NO)

nitric oxide   NO, reactive small molecule with diverse signaling properties

orthogonal   At right angles; effectively independent

phosphatase   Enzyme that removes phosphate from any of various substrates

pleiotropic   Describes a phenomenon with many diverse effects

pNPP p-nitrophenyl phosphate: common phosphatase substrate

Pol I   DNA polymerase I

protease   Enzyme that cleaves proteins

pyrophosphate P_2O_7, ion generated as a by-product by luciferase-catalyzed hydrolysis of ATP to AMP

quenching   Suppression of fluorescence by a nearby molecule

Renilla   Luminescent sea pansy

regression   Process of altering mathematical model to fit data set

SNP   Single nucleotide polymorphism, a single variable locus of DNA

steady state   A state in which an enzymatic reaction exhibits a constant rate

substrate   A reactant in a chemical transformation catalyzed by an enzyme

trans   Describes an effect due to an interaction of parts of two separate molecules or systems; opposite of *cis*

VX   *O*-ethyl *S*-(2-diisopropylaminoethyl) methylphosphonothiolate

Western blot   Antibody-based method of detecting proteins

Z value   Quality parameter of assay based on separation of signal from background

ACRONYMS

ACh   acetylcholine

AChE   acetylcholinesterase

ATCh   acetylthiocholine

CB   coupled bioluminescent

CCD   charge-coupled device

CRA ^{51}Cr-release assay

CTL   cytotoxic T lymphocyte
CV   coefficient of variation
DAG   diacylglycerol
DiFMUP   6,8-difluoro-4-methylumbelliferyl phosphate
1,3DPG   1,3-diphosphoglycerate
DTNB   dithionitrobenzene
EC$_{50}$ concentration at 50% of observed effect
ECL   electrochemiluminescence
EDTA   ethylene diamine tetraacetic acid
ELISA   enzyme-linked immunosorbent assay
ES   enzyme–substrate
FDP   fluorescein diphosphate
FMN   flavin mononucleotide
FP   fluorescence polarization
FRET   fluorescence resonance energy transfer
G3P   glyceraldehyde-3-phosphate
G3PDH   glyceraldehyde-3-phosphate dehydrogenase
GC/MS   gas chromatography/mass spectrometry
GPL   glyceraldehyde-3-phosphate dehydrogenase/PGK/luciferase
HPLC   high-performance liquid chromatography
HTS   high-throughput screening
IC$_{50}$ concentration of inhibitor at 50% inhibition
IP$_3$ inositol triphosphate
LDH   lactate dehydrogenase
MDCC   *N*-[2-(1-maleimidyl)ethyl]-7-(diethylamino)coumarin-3-carboxamide
MUP   4-methylumbelliferyl phosphate
NAD$^+$ nicotinamide adenine dinucleotide (oxidized form)
NADH   nicotinamide adenine dinucleotide (reduced form)
NADP$^+$ nicotinamide adenine dinucleotide phosphate (oxidized form)
NADPH   nicotinamide adenine dinucleotide phosphate (reduced form)
PCR   polymerase chain reaction
PGK   phosphoglycerokinase
PPE   poly(*p*-phenylene-ethynylene)
PTEN   phosphatase and tensin homolog deleted on chromosome 10
PTP   protein tyrosine phosphatase
QC   quality control
RFLP   restriction fragment length polymorphism
SNP   single nucleotide polymorphism

SPA   scintillation proximity assay
SPR   surface plasmon resonance
TRF   time-resolved fluorescence
TR-FRET   time-resolved fluorescence resonance energy transfer
UV/spec   ultraviolet and/or visible spectrophotometry

BIBLIOGRAPHY

1. Buck, J. and Buck, E. *Sci Am* **234**, 74–85 (1976).

2. Hastings, J. W. *J Mol Evol* **19**, 309–321 (1983).

3. Shimomura, O. *Bioluminescence: Chemical Principles and Methods*. World Scientific Publishing, Singapore (2006).

4. Harvey, E. N. *Bioluminescence*. Academic Press (1952).

5. Levine, E. N. *Physical Chemistry*, 3rd ed. McGraw-Hill (1988).

6. Chantler, P. D., Tao, T., and Stafford III, W. F. *Biophys J* **59**(6), 1242–1250 (1991).

7. Majumdar, D. S., Smirnova, I., Kasho, V., Nir, E., Kong, X., Weiss, S., and Kaback, H. R. *Proc Natl Acad Sci USA* **104**(31), 12640–12645 (2007).

8. Yoon, T. Y., Okumus, B., Zhang, F., Shin, Y. K., and Ha, T. *Proc Natl Acad Sci USA* **103**(52), 19611–19612 (2006).

9. Gavutis, M., Lata, S., and Piehler, J. *Nat Protoc* **1**(4), 2091–2103 (2006).

10. Wang, H., Hammoudeh, D. I., Follis, A. V., Reese, B. E., Lazo, J. S., Metallo, S. J., and Prochownik, E. V. *Mol Cancer Ther* **6**(9), 2399–2408 (2007).

11. Grotenbreg, G. M., Nicholson, M. J., Fowler, K. D., Wilbuer, K., Octavio, L., Yang, M., Chakraborty, A. K., Ploegh, H. L., and Wucherpfennig, K. W. *J Biol Chem* **282**(299), 21425–21436 (2007).

12. White, D., Musse, A. A., Wang, J., London, E., and Merrill, A. R. *J Biol Chem* **281**(43), 32375–32384 (2006).

13. Refsum, H., Smith, A. D., Ueland, P. M., Nexo, E., Clarke, R., McPartlin, J., Johnston, C., Engbaek, F., Schneede, J., McPartlin, C., and Scott, J. M. *Clin Chem* **50**, 3–32 (2004).

14. Szalay, A. A., Hill, P. J., Kricka, L. J., and Stanley, P. E. *Bioluminescence and Chemiluminescence: Chemistry, Biology and Applications World Scientific Publishing Company* (2007).

15. Garcia-Campana, A. M. *Chemiluminescence in Analytical Chemistry* CRC (2001).

16. Weeks, I. *Chemiluminescence Immunoassay (Comprehensive Analytical Chemistry)*. Elsevier Science (1991).

17. Kopecky, K. R. and Mumford, C. *Can J Chem* **47**, 709 (1969).

18. Bronstein, I. Y., Edwards, B., and Juo, R. R. US Patent 5,330,900, (1994).

19. Velasco, J. G. *Bull Electrochem* **10**, 29–38 (1994).

20. Rozhitskii, N. N. *J Anal Chem* **47**, 1288–1301 (1992).

21. Velasco, J. G. *Electroanalysis* **3**, 261–271 (1991).

22. Wilson, R., Akhavan-Tafti, H., Pollet, B. G., de Silva, R., Schaap, A. P., and Schiffring, D. J. *Proc SPIE—Intl Soc Opt Eng* **4414**, 168–177 (2001).

23. WIlson, R., Clavering, C., and Hutchinson, A. *Analyst* **128**, 480–485 (2003).

24. Green, A. A. and McElroy, W. D. *Biochim Biophys Acta* **20**, 170–176 (1956).

25. Harvey, E. N. *Science* **46**(1184), 241–243 (1917).

26. McElroy, W. M. *Proc Natl Acad Sci USA* **33**, 342–345 (1947).

27. Shimomura, O., Goto, T., and Johnson, F. H. *Proc Natl Acad Sci USA* **74**(7), 2799–2802 (1977).

28. Branchini, B. R., Southworth, T. L., Murtiashaw, M. H., Magyar, R. A., Gonzalez, S. A., Ruggiero, M. C., and Stroh, J. G. *Biochemistry* **43**(23), 7255–7262 (2004).

29. White, E. H., Miano, J. D., and Umbriet, M. *J Am Chem Soc* **97**, 198–200 (1975).

30. Shimomura, O. and Johnson, F. H. *Biochem Biophys Res Comm* **44**(2), 340–346 (1971).

31. Airth, R. L., Rhodes, W. C., and McElroy, W. D. *Biochim Biophys Acta* **27**, 519–532 (1958).

32. McElroy, W. D. and Seliger, H. H. *Firefly Bioluminescence*, pp. 427–458. Princeton University Press (1966).

33. Wood, K. V. US Patent 5,283,179 (1990).

34. Sherf, B. A. and Wood, K. V. US Patent 5,670,356 (1997).

35. Squirrell, D. J., White, P. J., Lowe, C. R., and Murray, J. A. H. US Patent 6,265,177 (2001).

36. Wood, K. V. and Gruber, M. G. US Patent 6,387,675 (2002).

37. Kajiyama, N. and Nakano, E. US Patent 5,219,737, (1993).

38. Wood, K. V. and Hall, M. P. US Patent 6,602,677 (1999).

39. O'Brien, M., Wood, K. V., Klaubert, D., and Daily, W. US Patent 7,148,030 (2006).

40. Karkhanis, Y. D. and Cormier, M. J. *Biochemistry* **10**(2), 317–326 (1971).

41. Matthews, J. C., Hori, K., and Cormier, M. J. *Biochemistry* **16**, 85–92 (1977).

42. Cormier, M. J., Hori, K., and Karkhanis, Y. D. *Biochemistry* **9**(5), 1184–1189 (1970).

43. Lorenz, W. W., McCann, R. O., Longiaru, M., and Cormier, M. J. *Proc Natl Acad Sci USA* **88**(10), 4438–4442 (1991).

44. Lorenz, W. W., Cormier, M. J., O'Kane, D. J., Hua, D., Escher, A. A., and Szalay, A. A. *J Biolumin Chemilumin* **11**, 31–37 (1996).

45. Rundlöf, A. K., Carlsten, M., and Arnér, E. S. J. *J Biol Chem* **276**(32), 30542–30551 (2001).

46. Rutter, G. A., White, M. R., and Tavaré, J. M. *Curr Biol* **5**(8), 890–899 (1995).

47. Cormier, M. J. and Strehler, B. L. *J Am Chem Soc* **75**(19), 4864–4865 (1953).

48. Duane, W. and Hastings, J. W. *Mol Cell Biochem* **6**, 53–64 (1975).

49. Hastings, J. W. and Nealson, K. H. *Annu Rev Microbiol* **31**, 549–595 (1977).

50. Zeigler, M. M. and Baldwin, T. O. **12**, 65–113 (1981).

51. Lee, J. *Chemistry and Biochemistry of Flavoenzymes*, pp. 109–151. CRC Press (1991).

52. Baldwin, T. O. and Zeigler, M. M. *The biochemistry and molecular biology of bacterial luminescence*, pp. 467–530. CRC Press (1992).

53. Tu, S. C. and Mager, H. I. X. In *Photobiol Sci its Appl [Proc Int Congr Photobiol]*, Muller, F., editor, pp. 319–328. CRC Press, (1995).

54. Tu, S. C., Lei, B., Liu, M., Tang, C. K., and Jeffers, C. *J Nutr* **130**, 331–332 (2000).

55. Sobolev, A. I., Mazhul, M. M., Malkov, I. A., Danilov, V. S., Gorkin, V. Z., Moskvitina, T. A., and Ovchinnikova, L. N. *Prikl Biokhim Mikrobiol* **26**(5), 700–705 (1990).

56. Hughes, R. J. *Anal Biochem* **131**(2), 318–323 (1983).

57. Huang, W., Feltus, A., Witkowski, A., and Daunert, S. *Anal Chem* **68**(9), 1646–1650 (1996).

58. Raunio, R. P., Leivo, P. V., and Kuusinen, A. M. *J Biolumin Chemilumin* **1**(1), 11–14 (1986).

59. Lavi, J. T. *Anal Biochem* **139**(2), 510–515 (1984).

60. Johnson, F. H. *Nav Res Rev* , 193–266 (1970).

61. Vysotski, E. S., Liu, Z. J., Markova, S. V., Blinks, J. R., Deng, L., Frank, L. A., Herko, M., Malikova, N. P., Rose, J. P., Wang, B. C., and Lee, J. *Biochemistry* **42**(20), 6013–6024 (2003).

62. Deng, L., Markova, S. V., Vysotski, E. S., Liu, Z. J., Lee, J., Rose, J., and Wang, B. C. *J Biol Chem* **279**(32), 33647–33652 (2004).

63. Stepanyuk, G. A., Golz, S., Markova, S. V., Frank, L. A., Lee, J., and Vysotski, E. S. *FEBS Lett* **579**(5), 1008–1014 (2005).

64. Inouye, S., Zenno, S., Sakaki, Y., and Tsuji, F. I. *Protein Expr Purif* **2**, 122–126 (1991).

65. George, C. H., Kendall, J. M., Campbell, A. K., and Evans, W. H. *J Biol Chem* **273**(45), 29822–29829 (1998).

66. Nakahashi, Y., Nelson, E., Fagan, K., Gonzales, E., Guillou, J. L., and Cooper, D. M. F. *J Biol Chem* **272**(29), 18093–18097 (1997).

67. Lin, S., Fagan, K. A., Li, K. X., Shaul, P. Q., Cooper, D. M., and Rodman, D. M. *J Biol Chem* **275**(24), 17979–17985 (2000).

68. Pouli, A. E., Karagenc, N., Wasmeier, C., Hutton, J. C., Bright, N., Arden, S., Schofield, J. G., and Rutter, G. A. *Biochem J* **330**, 1399–1404 (1998).

69. Viviana, V. R. and Bechara, E. J. H. *Photochem Photobiol* **58**, 615–622 (1993).

70. Viviana, V. R. and Bechara, E. J. H. *Ann Entomol Soc Am* **90**, 389–398 (1997).

71. Wood, K. V., Lam, Y. A., Seliger, H. H., and McElroy, W. D. *Science* **244**, 700–702 (1989).

72. Viviani, V. R., Hastings, J. W., and Wilson, T. *Photochem Photobiol* **75**, 22–27 (2002).

73. Harvey, E. N. *J Gen Physiol* **4**, 285–295 (1922).

74. Anderson, R. S. *J Gen Physiol* **19**, 301–305 (1935).

75. Shimomura, O., Goto, T., and Hirata, Y. *Bull Chem Soc Jpn* **30**(8), 929–933 (1957).

76. Miesenböck, G. and Rothman, J. E. *Proc Natl Acad Sci USA* **94**(7), 3402–3407 (1997).

77. Yamagishi, K., Enomoto, T., and Ohmiya, Y. *Anal Biochem* **354**, 15–21 (2006).

78. Wu, C., Kawasaki, K., Ogawa, Y., Yoshida, Y., Ohgiya, S., and Ohmiya, Y. *Anal Chem* **79**(4), 1634–1638 (2007).

79. Shimomura, O. *J Biolumin Chemilumin* **10**(2), 91–101 (1995).

80. Shimomura, O., Masugi, T., Johnson, F. H., and Haneda, Y. *Biochemistry* **17**(6), 994–998 (1978).

81. Tsuji, F. I. and Leisman, G. B. *Proc Natl Acad Sci USA* **78**(11), 6719–6723 (1981).

82. Lundin, A. and Styrélius, I. *Clin Chim Acta* **87**(2), 199–209 (1978).

83. Corey, M. J., Kinders, R. J., Brown, L. G., and Vessella, R. L. *J Immunol Methods* **207**(1), 43–51 (1997).

84. Lundin, A., Richardson, A., and Thore, A. *Anal Biochem* **75**(2), 611–620 (1976).

85. Nyrén, P. and Lundin, A. *Anal Biochem* **151**(2), 504–509 (1985).

86. Hellmér, J., Arner, P., and Lundin, A. *Anal Biochem* **177**, 132–137 (1989).

87. Lundin, A. *Methods Enzymol* **305**, 346–370 (2000).

88. VanCauter, G. C., Osten, D. E., and Tomisek, J. D. US Patent 5,198,670 (1993).

89. Kedar, H., Wallerstein, E. P., and Brown Jr, A. W. US Patent 6,198,577 (2001).

90. Leonard, S. W. and Xu, M. G. US Patent 7,324,202 (2008).

91. Gambini, M. R., Voyta, J. C., Atwood, J., DeSimas II, B. E., Lakatos, E., Levi, J., Metal, I., Sabak, G., and Wang, Y. US Patent 6,518,068 (2003).

92. Corey, E., Wegner, S. K., Stray, J. E., Corey, M. J., Arfman, E. W., Lange, P. H., and Vessella, R. L. *Int J Cancer* **71**(6), 1019–1028 (1997).

93. Fahien, L. A., MacDonald, M. J., Teller, J. K., Fibich, B., and Fahien, C. M. *J Biol Chem* **264**, 12303–12312 (1989).

94. Srivastava, D. K. and Bernhard, S. A. *Curr Top Cell Regul* **28**, 1–68 (1986).

95. Wroblewski, F. and Ladue, J. S. *Proc Soc Exp Biol Med* **91**(4), 569–571 (1956).

96. Bergmeyer, H. U. *Clin Chem* **18**, 1305–1311 (1972).

97. Chang, M. M. and Chung, T. W. *Clin Chem* **21**, 330–333 (1975).

98. Schiffmann, Y. *Mol Cell Biochem* **86**, 19–40 (1989).

99. King, M. M. and Carlson, G. M. *J Biol Chem* **256**, 11058–11064 (1981).

100. Thompson, J., Flichtenthaler, F. J., Peters, S., and Pikis, A. *J Biol Chem* **277**, 34310–34321 (2002).

101. Briggs, G. E. and Haldane, J. B. S. *Biochem J* **19**, 339 (1925).

102. Briggs, G. E. *Biochem J* **20**, 574–579 (1926).

103. Ferscht, A. *Enzyme Structure and Mechanism*, pp. 98–101. WH Freeman (1985).

104. Corey, M. J. and Corey, E. *Proc Natl Acad Sci USA* **93**(21), 11428–11434 (1996).

105. Hahn, K. W., Klis, W. A., and Stewart, J. M. *Science* **248**, 1544–1547 (1990).

106. Marshall, T. H. and Akgün, A. *J Biol Chem* **246**, 6019–6023 (1971).

107. Palomba, S., Berovic, N., and Palmer, R. E. *Langmuir* **22**(12), 5451–5454 (2006).

108. Funck-Brentano, C., Becquemont, L., Kroemer, H. K., Bühl, K., Knebel, N. G., Eichel-baum, M., and Jaillon, P. *Clin Pharmacol Ther* **55**(3), 256–269 (1994).

109. van der Weide, J., van Baalen-Benedek, E. H., and Kootstra-Ros, J. E. *Ther Drug Monit* **27**(4), 478–483 (2005).

110. Stump, A. L., Mayo, T., and Blum, A. *Am Fam Physician* **74**(4), 605–608 (2006).

111. Zhang, J. H., Chung, T. D., and Oldenburg, K. R. *J Biomol Screen* **4**, 67–73 (1999).

112. Corey, M. J. and Kinders, R. J. In *Drug Discovery Handbook,* Gad, S. C., editor, pp. 689–731. Wiley and Sons (2005).

113. Borejdo, J., Gryczynski, Z., Calander, N., Muthu, P., and Gryczynski, I. *Biophys J* **91**, 2626–2635 (2006).

114. Aita, K., Temma, T., Kuge, Y., and Saji, H. *Luminescence* **22**(5), 455–461 (2007).

115. Bronstein, I., Edwards, B., Martin, C., Sparks, A., and Voyta, J. C. US Patent 5,871,938 (1999).

116. Shi, Y. *Mol Cell* **9**(3), 459–470 (2002).

117. Pennell, R. I. and Lamb, C. *Plant Cell* **9**(7), 1157–1168 (1997).

118. Brunner, K. T., Mauel, J., Cerottini, J. C., and Chapuis, B. *Immunology* **14**, 181–196 (1968).

119. Sijts, A. J. A. M. and Pamer, E. G. *J Exp Med* **185**(8), 1403–1412 (1997).

120. Rininsland, F. H., Helms, T., Asaad, R. J., Boehm, B. O., and Tary-Lehmann, M. *J Immunol Methods* **240**, 143–155 (2000).

121. Desem, N. and Jones, S. L. *Clin Vaccine Immunol* **5**(4), 531–536 (1998).

122. Asai, T., Storkus, W. J., and Whiteside, T. L. *Clin Vaccine Immunol* **7**(2), 145–154 (2000).

123. Harris, J. L., Backes, B. J., Leonetti, F., Mahrus, S., Ellman, J. A., and Craik, C. S. *Proc Nat Acad Sci USA* **97**(14), 7754–7759 (2000).

124. Mandelboim, O., Kent, S., Davis, D. M., Wilson, S. B., Okazaki, T., Jackson, R., Hafler, D., and Strominger, J. L. *Proc Natl Acad Sci USA* **95**(7), 3798–3803 (1998).

125. Straathof, K. C. M., Bollard, C. M., Popat, U., Huls, M. H., Lopez, T., Morriss, M. C., Gresik, M. V., Gee, A. P., Russell, H. V., Brenner, M. K., Rooney, C. M., and Heslop, H. E. *Blood* **105**(5), 1898–1904 (2005).

126. Hardwick, A., McMillen, D., Martinez, J., Austin, A., Posey, A., Ave-Teel, C., Maples, P., and Schneider, S. *Bioproc J* **2** 27–31 (2003).

127. Love, W. D. and Burch, G. E. *J Lab Clin Med* **41**(3), 351–362 (1953).

128. Mayer, M. M., Gately, M. K., Okamoto, M., Shin, M. L., and Willoughby, J. B. *Ann N Y Acad Sci* **332**, 395–407 (1979).

129. Mascotti, K., McCullough, J., and Burger, S. R. *Transfusion* **40**(6), 693–696 (2000).

130. Saotome, K., Morita, H., and Umeda, M. *Toxicol In Vitro* **3**(4), 317–321 (1989).

131. Corey, M. J., Kinders, R. J., Poduje, C. M., Bruce, C. L., Rowley, H., Brown, L. G., Hass, M., and Vessella, R. L. *J Biol Chem* **275**, 12917–12925 (2000).

132. Ohmori, H., Takai, T., Tanigawa, T., and Honma, Y. *J Immunol Methods* **147**, 119–124 (1992).

133. Bachy, M., Bonnin-Rivalland, A., Tilliet, V., and Trannoy, E. *J Immunol Methods* **230**, 37–46 (1999).

134. Schäfer, H., Schäfer, A., Kiderlen, A. F., Masihi, K. N., and Burger, R. *J Immunol Methods* **204**, 89–98 (1997).

135. Mosmann, T. *J Immunol Methods* **65**, 55–63 (1983).

136. Korzeniewski, C. and Callewaert, D. M. *J Immunol Methods* **64**(3), 313–320 (1983).

137. Takamatsu, N. *Nippon Ronen Igakkai Zasshi* **35**(7), 535–542 (1998).

138. Vaughan, P. J., Pike, C. J., Cotman, C. W., and Cunningham, D. D. *J Neurosci* **15**, 5389–5401 (1995).

139. Poole, C. A., Brookes, N. H., and Clover, G. M. *J Cell Sci* **106**(2), 685–691 (1993).

140. Schäfer, H., Schäfer, A., Kiderlen, A. F., Masihi, K. N., and Burger, R. *J Immunol Methods* **204**(1), 89–98 (1997).

141. Palgen, K. and Peterson, J. *J Clin Invest* **61**(3), 751–762 (1978).

142. Hou, J. and Zheng, W. F. *J Immunol Methods* **85**(2), 325–333 (1985).

143. Corey, M. J. and Kinders, R. J. US Patent 6,811,990 (2002).

144. Ogbomo, H., Hahn, A., Geiler, J., Michaelis, M., Doerr, H. W., and Cinatl Jr, J. *Biochem Biophys Res Commun* **339**, 375–379 (2006).

145. Jencks, W. P. *Catalysis in Chemistry and Enzymology*. Dover Publications (1987).

146. Branchini, B. R., Magyar, R. A., Marcantonio, K. M., Newberry, K. J., Stroh, J. G., Hinz, L. K., and Murtiashaw, M. H. *J Biol Chem* **272**(31), 19359–19364 (1997).

147. Kalinina, O., Lebedeva, I., Brown, J., and Silver, J. *Nucleic Acids Res* **25**(10), 1999–2004 (1997).

148. Sakati, I. A., Devine, P. C., Devine Jr, C. J., Fiveash Jr, J. G., and Poutasse, E. F. *New Eng J Med* **278**(13), 721–723 (1968).

149. Sharma, P. R., Jain, S., and Tiwari, P. K. *Clin Biochem* **40**(18), 1414–1419 (2007).

150. Ishitani, R., Tanaka, M., Sunaga, K., Katsube, N., and Chuang, D. *Mol Pharm* **53**(4), 701–707 (1998).

151. Muller-Eberhard, H. J. and Schreiber, R. D. *Adv Immunol* **29**, 1–53 (1980).

152. Cooper, N. R. *Adv Immunol* **37**, 151–216 (1985).

153. Fujita, T. *Nat Rev Immunol* **2**(5), 346–353 (2002).

154. Morgan, B. P. *Crit Rev Immunol* **19**(3), 173–198 (1999).

155. Fishelson, Z., Donin, N., Zell, S., Schultz, S., and Kirschfink, M. *Mol Immunol* **40**, 109–123 (2003).

156. Shibuya, A. and N., I. *Hepatol Res* **22**(3), 174–179 (2002).

157. Vyse, T. J., Bates, G. P., Walport, M. J., and Morley, B. J. *Genomics* **24**(1), 90–98 (1994).

158. Rickinson, A. B. *Annu Rev Immunol* **15**, 405–431 (1997).

159. Russell, J. H. and Ley, T. J. *Annu Rev Immunol* **20**, 323–370 (2002).

160. Riddell, S. R. and Greenberg, P. D. US Patent 5,827,642 (1994).

161. Ackers, G. K., Doyle, M. L., Myers, D., and Daugherty, M. A. *Science* **255**(5040), 54–63 (1992).

162. Allison, R. D. and Purich, D. L. *Handbook of Biochemical Kinetics*. Academic Press (2000).

163. Morandat, S. and El Kirat, K. *Langmuir* **22**(13), 5786–5791 (2006).

164. Tan, A., Ziegler, A., Steinbauer, B., and Seelig, J. *Biophys J* **83**(3), 1547–1556 (2002).

165. Ogbomo, H., Hahn, A., Geiler, J., Michaelis, M., Doerr, H. W., and Cinatl Jr, J. *Biochem Biophys Res Commun* **339**, 375–379 (2006).

166. Koechli, O. R., Sevin, B. U., Perras, J. P., Angioli, R., Steren, A., Rodriguez, M., Ganjei, P., and Averette, H. E. *Oncology* **51**, 35–41 (1994).

167. Mossman, T. *J Immunol Methods* **63**, 55–63 (1983).

168. Roehm, N. W., Rodgers, G. H., Hatefield, S. M., and Glasebrook, A. L. *J Immunol Methods* **142**, 257 (1991).

169. Shahan, T. A., Siegel, P. D., Sorenson, W. G., Kuschner, W. G., and Lewis, D. M. *J Immunol Methods* **175**(2), 181–187 (1994).

170. Funk, D., Schrenk, H. H., and Frei, E. *Biotechniques* **43**(2), 178–182 (2007).

171. Wörle-Knirsch, J. M., Pulskamp, K., and Krug, H. F. *Nano Lett* **6**(6), 1261–1268 (2006).

172. Rastogi, N., Potar, M. C., and David, H. L. *Ann Inst Pasteur Microbiol* **137**, 45–53 (1986).

173. Vaara, M. and Vaara, T. *Antimicrob Agents Chemother* **19**(4), 578–583 (1981).

174. Heard, K. S., Diguette, M., Heard, A. C., and Carruthers, A. *Exp Physiol* **83**(2), 195–202 (1998).

175. Druker, B. J. *Trends Mol Med* **8**, 4–14 (2002).

176. Huang, J. M., Wei, Y. F., Kim, Y. H., Osterberg, L., and Matthews, H. R. *J Biol Chem* **266**(14), 9023–9031 (1991).

177. Dhillon, A. S., Hagan, S., Rath, O., and Kolch, W. *Oncogene* **26**, 3279–3290 (2007).

178. Vivanco, I. and Sawyers, C. L. *Nat Rev Cancer* **2**(7), 489–501 (2002).

179. Lesaicherre, M. L., Uttamchandani, M., Chen, G. Y. J., and Yao, S. Q. *Bioorg Med Chem Lett* **12**(16), 2085–2088 (2002).

180. Hu, S., Shively, L., Raubitschek, A., Sherman, M., Williams, L. E., Wong, J. Y., Shively, J. E., and Wu, A. M. *Cancer Res* **56**(13), 3055–3061 (1996).

181. Zuleski, F. R. and McGuinness, E. T. *Anal Biochem* **54**(2), 406–412 (1973).

182. Udenfriend, S., Gerber, L., and Nelson, N. *Anal Biochem* **161**(2), 494–500 (1987).

183. Park, Y. W., Cummings, R. T., Wu, L., Zheng, S., Cameron, P. M., Woods, A., Zaller, D. M., Marcy, A. I., and Hermes, J. D. *Anal Biochem* **269**(1), 94–104 (1999).

184. Beveridge, M., Park, Y. W., Hermes, J., Marenghi, A., Brophy, G., and Santos, A. *J Biomol Screen* **5**(4), 205–211 (2000).

185. Quercia, A. K., Lamarr, W. A., Myung, J., Ozbal, C. C., Landro, J. A., and Lumb, K. J. *J Biomol Screen* **12**, 473–480 (2007).

186. von Ahsen, O., Schmidt, A., Klotz, M., and Parczyk, K. *J Biomol Screen* **11**(6), 606–616 (2006).

187. Huss, K. L., Blonigen, P. E., and Campbell, R. M. *J Biomol Screen* **12**, 578–584 (2007).

188. Lundin, A., Gerhardt, W., Lindberg, K., Lövgren, T., Nordlander, R., Nyquist, O., and Styrelius, I. *Clin Biochem* **12**(6), 214–215 (1979).

189. Wulff, K., Stahler, F., and Gruber, W. US Patent 4,286,057 (1981).

190. Kashem, M. A., Yingling, J. D., Pullen, S. S., Prokopowicz, A. S., Jones, J. W., Wolak, J. P., Rogers, G. R., Snow, R. J., Homon, C. A., and Jakes, S. *J Biomol Screen* **12**(1), 70–83 (2001).

191. McElroy, M. D., McElroy, W. D., Helinski, D. R., Wood, K. V., De Wet, J. R., Ow, D. W., and Howell, S. H. US Patent 5,583,024 (1993).

192. McElroy, M. D., Helinski, D. R., Wood, K. V., Wet, J. D., Ow, D. W., and Howel, S. H. US Patent 5,674,713 (1995).

193. McElroy, M. D., Helinksi, D. R., Wood, K. V., De Wet, J. R., Ow, D. W., and Howel, S. H. US Patent 5,700,673 (1995).

194. Polgar, T., Baki, A., Szendrei, G. I., and Keseru, G. M. *J Med Chem* **48**, 7946–7959 (2005).

195. Koresawa, M. and Okabe, T. *Assay Drug Dev Technol* **2**(2), 153–160 (2004).

196. Tagliati F, B. A., Bosetti, A., Zatelli, M. C., and Uberti, E. E. *J Pharm Biomed Anal* **39**, 811–814 (2005).

197. John, T. A., Ibe, B. O., and Raj, J. U. *Am J Physiol Lung Cell Mol Physiol* **291**, 1079–1093 (2006).

198. Bonnans, C., Fukunaga, K., Keledjian, R., Petasis, N. A., and Levy, B. D. *J Exp Med* **203**, 857–863 (2006).

199. Shay, K. P., Wang, Z., Xing, P. X., McKenzie, I. F., and Magnuson, N. S. *Mol Cancer Res* **3**(3), 170–181 (2005).

200. Langer, T., Vogtherr, M., Elshorst, B., Betz, M., Schieborr, U., Saxena, K., and Schwalbe, H. *Chembiochem* **5**, 1508–1516 (2004).

201. Pargellis, C., Tong, L., Churchill, L., Cirillo, P. F., Gilmore, T., Graham, A. G., Grob, P. M., Hickey, E. R., Moss, N., Pav, S., and Regan, J. *Nat Struct Biol* **9**(4), 268–272 (2002).

202. Ahmed, M., Rocha, J. B., Corrêa, M., Mazzanti, C. M., Zanin, R. F., Morsch, A. L., Morsch, V. M., and Schetinger, M. R. *Chem Biol Interact* **162**(2), 165–171 (2006).

203. Barr, R. K., Boehm, I., Attwood, P. V., Watt, P. M., and Bogoyevitch, M. A. *J Biol Chem* **279**(35), 36327–36338 (2004).

204. Welch, A. R. PCT Patent WO 00/18950 (2000).

205. Wang, Y. and Kent, C. *J Biol Chem* **270**(32), 18948–18952 (1995).

206. Leroy, D., Heriché, J. K., Filhol, O., Chambaz, E. M., and Cochet, C. *J Biol Chem* **272**(33), 20820–20827 (1997).

207. Wang, P., Saraswati, S., Guan, Z., Watkins, C. J., Wurtman, R. J., and Littleton, J. T. *J Neurosci* **24**(19), 4518–4529 (2004).

208. Shizuta, Y., Beavo, J. A., Bechtel, P. J., Hofmann, F., and Krebs, E. G. *J Biol Chem* **250**, 6891–6896 (1975).

209. Smith, E. and Morrison, J. F. *J Biol Chem* **244**, 4224–4234 (1969).

210. Adams, J. A. and Taylor, S. S. *J Biol Chem* **268**, 7747–7752 (1993).

211. Teague Jr, W. E. and Dobson, G. P. *J Biol Chem* **267**(20), 14084–14093 (1992).

212. Ikebe, M. and Hartshorne, D. J. *J Biol Chem* **261**, 8249–8253 (1986).

213. Tomasz, J. *J Chromatogr* **84**(1), 208–213 (1973).

214. Ames, B. N. *Methods Enzymol* **8**, 115–118 (1966).

215. Watanabe, T., Hosyoa, H., and Yonemura, S. *Mol Biol Cell* (2006).

216. Mansuy, I. M. and Shenolikar, S. *Trends Neurosci* **29**(12), 679–686 (2006).

217. Zhu, S., Bjorge, J. D., and Fujita, D. J. *Cancer Res* **67**(21), 10129–10137 (2007).

218. Srivastava, S. K., Bansal, P., Oguri, T., Lazo, J. S., and Singh, S. V. *Cancer Res* **67**(19), 9150–9157 (2007).

219. Knobbe, C. B., Merlo, A., and Reifenberger, G. *Neuro Oncol* **4**(3), 196–211 (2002).

220. Leslie, N. R. and Downes, C. P. *Cell Signal* **14**(4), 285–295 (2002).

221. Reen, D. J. *Methods Mol Biol* **32**, 461–466 (1994).

222. Sugasawara, R. J., Prato, C. M., and Sippel, J. E. *J Clin Microbiol* **19**, 230–234 (1984).

223. Grimshaw, C., Gleason, C., Chojnicki, E., and Young, J. *J Pharm Biomed Anal* **16**(4), 605–612 (1997).

224. Cohen, P. *Annu Rev Biochem* **58**, 453–508 (1989).

225. Fischer, E. H., Charbonneau, H., and Tonks, N. K. *Science* **253**(5018), 401–406 (1991).

226. Tonks, N. K. *Nat Rev Mol Cell Biol* **7**(11), 833–846 (2006).

227. Charbonneau, H. and Tonks, N. K. *Annu Rev Biochem* **8**, 463–493 (1992).

228. Berdy, S. E., Kudla, J., Gruissem, W., and Gillaspy, G. E. *Plant Physiol* **126**(2), 801–810 (2001).

229. Carman, G. M. and Han, G. S. *Trends Biochem Sci* **31**(12), 694–699 (2006).

230. Conklin, P. L., Gatzek, S., Wheeler, G. L., Dowdle, J., Raymond, M. J., Rolinski, S., Isupov, M., Littlechild, J. A., and Smirnoff, N. *J Biol Chem* **281**(23), 15662–15670 (2006).

231. Akimoto, S., Ohki, T., Ichikawa, T., Akakura, K., and Shimazaki, J. *Hinyokika Kiyo* **40**(11), 987–993 (1994).

232. Alvarado-Kristensson, M. and Andersson, T. *J Biol Chem* **280**, 6238–6244 (2005).

233. Stenroos, K., Hurskainen, P., Eriksson, S., Hemmila, I., Blomberg, K., and Lindqvist, C. *Cytokine* **10**(7), 495–499 (1998).

234. Waddleton, D., Ramachandran, C., and Wang, Q. *Anal Biochem* **309**, 150–157 (2002).

235. Seethala, R. and Menzel, R. *Anal Biochem* **253**, 210–218 (1997).

236. Seethala, R. and Menzel, R. *Anal Biochem* **255**(2), 257–262 (1998).

237. Parker, G. J., Law, T. L., Lenoch, F. J., and Bolger, R. E. *J Biomol Screen* **5**, 77–88 (2000).

238. Chen, L., McBranch, D. W., Wang, H. L., Helgeson, R., Wudl, F., and Whitten, D. G. *Proc Natl Acad Sci USA* **96**(22), 12287–12292 (1999).

239. Heeger, P. S. and Heeger, A. J. *Proc Natl Acad Sci USA* **96**(22), 12219–12221 (1999).

240. Sariciftci, N., Smilowitz, L., Heeger, A., and Wudl, F. *Science* **258**, 1474–1476 (1992).

241. Kraabel, B., McBranch, D., Sariciftci, N., Moses, D., and Heeger, A. *Phys Rev B* **50**, 18543–18552 (1994).

242. Rininsland, R., Xia, W., Wittenburg, S., Shi, X., Stankewicz, C., Achyuthan, K., McBranch, D., and David, W. *Proc Natl Acad Sci USA* **101**(43), 15295–15300 (2004).

243. Perrin, D., Fremaux, C., Besson, D., Sauer, W. H., and Scheer, A. *J Biomol Screen* **11**(8), 996–1004 (2006).

244. Rotman, B., Zderic, J. A., and Edelstein, M. *Proc Natl Acad Sci USA* **50**, 1 (1963).

245. Huang, Z., Olson, N. A., You, W., and Haugland, R. P. *J Immunol Methods* **149**, 261 (1992).

246. Park, S. H. and Raines, R. T. *Nat Biotechnol* **18**, 847 (2000).

247. Chiu, D. T., Wilson, C. F., Ryttsén, F., Strömberg, A., Farre, C., Karlsson, A., Nordholm, S., Gaggar, A., Modi, B. P., Moscho, A., Garza-López, R. A., Orawr, O., and Zare, R. N. *Science* **283**, 1892 (1999).

248. Desmarais, S., Friesen, R. W., Zamboni, R., and Ramachandran, C. *Biochem J* **337**(2), 219 (1999).

249. Wang, Q., Scheigetz, J., Gilbert, M., Snider, J., and Ramachandran, C. *Biochim Biophys Acta* **1431**, 14 (1999).

250. Vaz-Velho, M., Duarte, G., and Gibbs, P. *J Microbiol Methods* **40**(2), 147–151 (2000).

251. Gangar, V., Curiale, M. S., D'Onorio, A., Schultz, A., Johnson, R. L., and Atrache, V. *J AOAC Int* **83**(4), 903–918 (2000).

252. Loessner, M. J., Rees, C. E., Stewart, G. S., and Scherer, S. *Appl Environ Microbiol* **62**(4), 1133–1140 (1996).

253. Ramsaran, H., Chen, J., Brunke, B., Hill, A., and Griffiths, M. W. *J Dairy Sci* **81**(7), 1810–1817 (1998).

254. Mountfort, D. O., Kennedy, G., Garthwaite, I., Quilliam, M., Truman, P., and Hannah, D. J. *Toxicon* **37**(6), 909–922 (1999).

255. Taira, A., Merrick, G., Wallner, K., and Dattoli, M. *Oncology (Huntingt)* **21**(8), 1003–1010 (2007).

256. Crowther, J. R., Angarita, L., and Anderson, J. *Biologicals* **18**(4), 331–336 (1990).

257. Fernley, H. N. and Walker, P. G. *Biochem J* **97**(1), 95–103 (1965).

258. Rodems, M. S., Hamman, D. B., Lin, C., Zhao, J., Shah, S., Heiday, D., Makings, M., Stack, J. H., and Pollok, A. B. *Assay Drug Dev Technol* **1**, 9–19 (2002).

259. Brune, M., Hunter, J. L., Corrie, J. E., and Webb, M. R. *Biochemistry* **33**(27), 8262–8271 (1994).

260. Webb, M. R., Brune, M. H., and Corrie, J. E. T. US Patent 5,898,069, (1999).

261. Nishikata, M., Suzuki, K., Yoshimura, Y., Deyama, Y., and Matsumoto, A. *Biochem J* **343**(2), 385–391 (1999).

262. Kokado, A., Arakawa, H., and Maeda, M. *Luminescence* **17**, 5–10 (2002).

263. Hohenwallner, W. and Wimmer, E. *Clin Chim Acta* **45**(2), 169–175 (1973).

264. Gribanov, G. A. and Shchennikova, V. V. *Lab Delo* **4**, 231–233 (1980).

265. Cogan, E. B., Birrell, G. B., and Griffith, O. H. *Anal Biochem* **271**(1), 29–35 (1999).

266. Ng, D. H., Harder, K. W., Clark-Lewis, I., Jirik, F., and Johnson, P. *J Immunol Methods* **179**, 177–185 (1995).

267. Fisher, D. K. and Higgins, T. J. *Pharm Res* **11**(5), 759–763 (1994).

268. Webb, M. R. *Proc Natl Acad Sci USA* **89**(11), 4884–4887 (1992).

269. Upson, R. H., Haugland, R. P., Malekzadeh, M. N., and Haugland, R. P. *Anal Biochem* **243**, 41–45 (1996).

270. Webb, M. R. *Proc Natl Acad Sci USA* **89**, 4884 (1992).

271. Upson, R. H., Haugland, R. P., and Malekzadeh, M. N. *Anal Biochem* **243**(1), 41 (1996).

272. Bronstein, I., Voyta, J. C., Thorpe, G. H. G., Kricka, L. J., and Armstrong, G. *Clin Chem* **35**(7), 1441–1446 (1989).

273. Bronstein, I., Edwards, B., and Voyta, J. C. *J Biolumin Chemilumin* **4**, 99–111 (1989).

274. Kricka, L. J. US Patent 5,306,621 (1994).

275. Kokado, A., Arakawa, H., and Maeda, M. *Luminescence* **17**, 5–10 (2002).

276. Voyta, J. C., Edwards, B., Bronstein, I. Y., and McGrath, P. US Patent 5,145,772 (1991).

277. Dahlquist, F. W., Jao, L., and Raftery, M. *Proc Natl Acad Sci USA* **56**, 26–30 (1996).

278. Hallaway, B. J. and O'Kane, D. J. *Methods Enzymol* **305**, 391–401 (2000).

279. Winston, N. J. and Maro, B. *Biol Cell* **91**(3), 175–183 (1999).

280. Hertzberg, E. L., Sáez, J. C., Corpina, R. A., Roy, C., and Kessler, J. A. *Methods* **20**(2), 129–139 (2000).

281. Winston, N. J. and Maro, B. *Biol Cell* **91**(3), 175–183 (1999).

282. Geiger, R. and Miska, J. W. *J Clin Chem Clin Biochem* **25**, 31–38 (1989).

283. Daum, G., Solca, F., Diltz, C. D., Zhao, Z., Cool, D. E., and Fischer, E. H. *Anal Biochem* **211**, 50–54 (1993).

284. Zhuo, S., Clemens, J. C., Hakes, D. J., Barford, D., and Dixon, J. E. *J Biol Chem* **268**(24), 17754–17761 (1993).

285. Martin, B., Pallen, C. J., Wang, J. H., and Graves, D. H. *J Biol Chem* **260**, 14932–14937 (1985).

286. Perrino, B. A. *Arch Biochem Biophys* **372**, 159–165 (1999).

287. Gribanov, G. A. and Shchennikova, V. V. *Lab Delo* **4**, 231–233 (1980).

288. Koser, B. and Oesper, P. *Anal Biochem* **17**, 119–124 (1966).

289. Martin, C. S., Wight, P. A., Dobretsova, A., and Bronstein, I. *Biotechniques* **21**(3), 520–524 (1996).

290. Leuzinger, W. *Prog Brain Res* **31**, 241–245 (1969).

291. Rosenberry, T. L. *Adv Enzymol Relat Areas Mol Biol* **43**, 103–218 (1975).

292. Castro, A. and Martinez, A. *Curr Pharm Des* **12**(33), 4377–4387 (2006).

293. Seltzer, B. *Expert Opin Drug Metab Toxicol* **1**(3), 527–536 (2005).

294. Porcel, J. and Montalban, X. *J Neurol Sci* **245**, 177–181 (2006).

295. Chouinard, S., Sepehry, A. A., and Stip, E. *Clin Neuropharmacol* **30**(3), 169–182 (2007).

296. Houghton, P. J. and Howes, M. J. *Neurosignals* **14**, 6–22 (2005).

297. Ordentlich, A., Barak, D., Kronman, C., Flashner, Y., Leitner, M., Segall, Y., Ariel, N., Cohen, S., Velan, B., and Shafferman, A. *J Biol Chem* **268**(23), 17083–17095 (1993).

298. Taylor, P., Wong, L., Radic, Z., Tsigelny, I., Bruggemann, R., Hosea, N. A., and Berman, H. A. *Chem Biol Interact* **119**, 3–15 (1999).

299. Berman, H. A. and Leonard, K. *Mol Pharmacol* **41**(2), 412–418 (1992).

300. Gilbert, J. C. *Br Med J* **4**(5987), 33–35 (1975).

301. Juel, V. C. and Massey, J. M. *Orphanet J Rare Dis* **2**, 44 (2007).

302. Perry, E. K., Tomlinson, B. E., Blessed, G., Bergmann, K., Gibson, P. H., and Perry, R. H. *Br Med J* **2**(6150), 1457–1459 (1978).

303. Comfort, A. *Lancet* **311**(8065), 659–660 (1978).

304. Seltzer, B. *Expert Opin Drug Metab Toxicol* **1**(3), 527–536 (2005).

305. Bohnen, N. I., Kaufer, D. I., Hendrickson, R., Ivanco, L. S., Lopresti, B. J., Constantine, G. M., Mathis, C., Davis, J. G., Moore, R. Y., and Dekosky, S. T. *J Neurol* **253**(2), 242–247 (2006).

306. Wang, Z. F., Yan, J., Fu, Y., Tang, X. C., Feng, S., He, X. C., and Bai, D. L. *Cell Mol Neurobiol* **28**, 245–261 (2007).

307. Ordentlich, A., Barak, D., Kronman, C., Ariel, N., Segall, Y., Velan, B., and Shafferman, A. *J Biol Chem* **273**(31), 19507–19517 (1998).

308. Shih, T. M., Koviak, T. A., and Capacio, B. R. *Neurosci Biobehav Rev* **15**(3), 349–362 (1991).

309. Kassa, J. and Bajgar, J. *Acta Medica (Hradec Kralove)* **39**, 27–30 (1996).

310. Lenz, D. E., Yeung, D., Smith, J. R., Sweeney, R. E., Lumley, L. A., and Cerasoli, D. M. *Toxicology.* **233**, 31–39 (2007).

311. Ellman, G. L., Courtney, K. D., Andres Jr, V., and Feather-Stone, R. M. *Biochem Pharmacol* **7**, 88–95 (1961).

312. Hasinoff, B. B. *Biochim Biophys Acta* **704**, 52–58 (1982).

313. Komersová, A., Komers, K., and Zdrazilová, P. *Chem Biol Interact* **157**, 387–388 (2005).

314. Brownson, C. and Watts, D. C. *Biochem J* **131**(2), 369–374 (1973).

315. Reiner, E. and Simeon, V. *Biochim Biophys Acta* **480**, 137–142 (1977).

316. Galzigna, L., Bertazzon, A., Garbin, L., and Deana, R. *Enzyme* **26**, 8–14 (1981).

317. Johnson, C. D. and Russel, R. L. *Anal Biochem* **64**(1), 229–238 (1975).

318. Israël, M. and Lesbats, B. *Neurochem Int* **3**, 81–90 (1981).

319. Israël, M. and Lesbats, B. *J Neurochem* **37**, 1475–1483 (1981).

320. Ternaux, J. P. and Chamoin, M. C. *J Biolumin Chemilumin* **9**(2), 65–72 (1994).

321. Danet, A. F., Badea, M., Marty, J. L., and Aboul-Enein, H. Y. *Biopolymers* **57**, 37–42 (2000).

322. Ozbal, C. C., LaMarr, W. A., Linton, J. R., Green, D. F., Katz, A., Morrison, T. B., and Brenan, C. J. *Assay Drug Dev Technol* **2**(4), 373–381 (2004).

323. Anzai, J. *Yakugaku Zasshi* **126**(12), 1301–1308 (2006).

324. Du, D., Chen, S., Cai, J., and Zhang, A. *Biosens Bioelectron* **23**, 130–134 (2007).

325. Chernyavsky, A. I., Arredondo, J., Karlsson, E., Wessler, I., and Grando, S. A. *J Biol Chem* **280**(47), 39220–39228 (2005).

326. Hadd, A. G., Raymond, D. E., Halliwell, J. W., Jacobson, S. C., and Ramsey, J. M. *Anal Chem* **39**(17), 3407–3412 (1997).

327. Heleg-Shabtai, V., Gratziany, N., and Liron, Z. *Anal Chim Acta* **571**(2), 228–234 (2006).

328. Kiba, N., Ito, S., Tachibana, M., Tani, K., and Koizumi, H. *Anal Sci* **19**(12), 1647–1651 (2003).

329. Corey, M. J. and Dhawan, S. K. US Patent 20080050762 (2008).

330. Palmer, R. M., Ferrige, A. G., and Moncada, S. *Nature* **327**(6122), 524–526 (1987).

331. Palmer, R. M., Ashton, D. S., and Moncada, S. *Nature* **333**(6174), 664–666 (1988).

332. Evans, C. H., Stefanovic-Racic, M., and Lancaster, J. *Clin Orthop Relat Res* **312**, 275–294 (1995).

333. Radomski, M. W., Palmer, R. M., and Moncada, S. *Proc Natl Acad Sci USA* **87**(24), 10043–10047 (1990).

334. Mori, M. *J Nutr* **137**(6), 1616–1620 (2007).

335. Antonova, G., Lichtenbeld, H., Xia, T., Chatterjee, A., Dimitropoulou, C., and Catravas, J. D. *Clin Hemorheol Microcirc* **37**, 19–35 (2007).

336. Kavya, R., Saluja, R., Singh, S., and Dikshit, M. *Nitric Oxide* **15**(4), 280–294 (2006).

337. Dusse, L. M. S. A., Cooper, A. J., and Lwaleed, B. A. *J Thromb Thrombolysis* **23**(2), 129–133 (2007).

338. Cuzzocrea, S. and Salvemini, D. *Kidney Int* **71**(4), 290–297 (2007).

339. Ricciardolo, F. L., Nijkamp, F. P., and Folkerts, G. *Curr Drug Targets* **7**(6), 721–735 (2006).

340. Sydow, K., Mondon, C. E., and Cooke, J. P. *Vasc Med* **10**, 35–43 (2005).

341. Muscará, M. N. and Wallace, J. L. *Am J Physiol* **276**(6), 1316–1316 (1999).

342. Hobbs, A. J., Higgs, A., and Moncada, S. *Annu Rev Pharmacol Toxicol* **39**, 191–220 (1999).

343. Wang, W., Inoue, N., Nakayama, T., Ishii, M., and Kato, T. *Anal Biochem* **227**(2), 274–280 (1995).

344. Robbins, R. A., Hamel, F. G., Floreani, A. A., Gossman, G. L., Nelson, K. J., Belenky, S., and Rubinstein, I. *Life Sci* **52**(8), 709–716 (1993).

345. Devignat, R. *Ann Inst Pasteur* **82**(5), 653–655 (1952).

346. Rasch, M. *Nord Med* **59**(26), 911–912 (1958).

347. Tsikas, D. *Free Radic Res* **39**(8), 797–815 (2005).

348. Lundberg, J. O. and Weitzberg, E. *Arterioscler Thromb Vasc Biol* **25**(5), 915–922 (2005).

349. Feelisch, M. *Eur Heart J* , 123–132 (1993).

350. van Hasselt, M., Weiss, M., and Haase, W. *Curr Med Res Opin* **9**(2), 107–112 (1984).

351. Albert, A. *J Lancet* **81**, 112–114 (1961).

352. Riseman, J. E., Altman, G. E., and Koretsky, S. *Circulation* **17**, 22–39 (1958).

353. Jamal, S. A., Hamilton, C. J., Black, D., and Cummings, S. R. *Trials* **7**, 10 (2006).

354. Arnold, W. P., Mittal, C. K., Katsuki, S., and Murad, F. *Proc Natl Acad Sci USA* **74**(8), 3203–3208 (1977).

355. Rapoport, R. M., Draznin, M. B., and Murad, F. *Nature* **306**(5939), 174–176 (1983).

356. Chae, H. J., Park, R. K., Chung, H. T., Kang, J. S., Kim, M. S., Choi, D. Y., Bang, B. G., and Kim, H. R. *J Pharm Pharmacol* **49**, 897–902 (1997).

357. Evans, D. M. and Ralston, S. H. *J Bone Miner Res* **11**, 300–305 (1996).

358. Jamal, S. A., Browner, W. S., Bauer, D. C., and Cummings, S. R. *J Bone Miner Res* **13**(11), 1755–1759 (1998).

359. Salvemini, D. *Cell Mol Life Sci* **53**(7), 576–582 (1997).

360. Shinmura, K., Xuan, Y. T., Tang, X. L., Kodani, E., Han, H., Zhu, Y., and Bolli, R. *Circ Res* **90**(5), 602–608 (2002).

361. Frumento, G., Piazza, T., Di Carlo, E., and Ferrini, S. *Endocr Metab Immune Disord Drug Targets* **6**(3), 233–237 (2006).

362. Lancaster Jr, J. R. and Xie, K. *Cancer Res* **66**(13), 6459–6462 (2006).

363. Gallagher, K. A., Liu, Z. J., Xiao, M., Chen, H., Goldstein, L. J., Buerk, D. G., Nedeau, A., Thom, S. R., and Velazquez, O. C. *J Clin Invest* **117**(5), 1249–1259 (2007).

364. Cuzzocrea, S. *Curr Pharm Des* **12**(27), 3551–3570 (2006).

365. Raij, L. *J Clin Hypertens* **8**(12), 30–39 (2006).

366. Belge, C., Massion, P. B., Pelat, M., and Balligand, J. L. *Ann N Y Acad Sci* **1047**, 173–182 (2005).

367. Benthin, G., Björkhem, I., Breuer, O., Sakinis, A., and Wennmalm, A. *Biochem J* **323**, 853–858 (1997).

368. Yoshida, K., Kasama, K., Kitabatake, M., and Imai, M. *Int Arch Occup Environ Health* **52**(2), 102–115 (1983).

369. Westfelt, U. N., Benthin, G., Lundin, S., Stenqvist, O., and Wennmalm, A. *Br J Pharmacol* **114**(8), 1621–1624 (1995).

370. McKnight, G. M., Smith, L. M., Drummond, R. S., Duncan, C. W., Golden, M., and Benjamin, N. *Gut* **40**(2), 211–214 (1997).

371. Iijima, K., Henry, E., Moriya, A., Wirz, A., Kelman, A. W., and McColl, K. E. *Gastroenterology* **122**(5), 1248–1257 (2002).

372. Iijima, K., Grant, J., McElroy, K., Fyfe, V., Preston, T., and McColl, K. E. *Carcinogenesis* **24**(12), 1951–1960 (2003).

373. Larsen, F. J., Weitzberg, E., Lundberg, J. O., and Ekblom, B. *Acta Physiol (Oxf)* **191**(1), 59–66 (2007).

374. O'kane, P. D., Jackson, G., and Ferro, A. *Atherosclerosis* **196**, 574–579 (2008).

375. Refsum, H., Nurk, E., Smith, A. D., Ueland, P. M., Gjesdal, C. G., Bjelland, I., Tverdal, A., Tell, G. S., Nygård, O., and Vollset, S. E. *J Nutr* **136**(6), 1731–1740 (2006).

376. Chow, C. K. and Hong, C. B. *Toxicology* **180**(2), 195–207 (2002).

377. Moshage, H., Kok, B., Huizenga, J. R., and Jansen, P. L. *Clin Chem* **41**, 892–896 (1995).

378. Fontijn, A., Sabadell, A. J., and Ronco, R. J. *Anal Chem* **42**(6), 575–579 (1970).

379. Nicholas, D. J. D. and Nason, A. *Methods Enzymol* **3**, 982–984 (1957).

380. Misko, T. P., Schilling, R. J., Salvemini, D., Moore, W. M., and Currie, M. G. *Anal Biochem* **214**(1), 11–16 (1993).

381. Kojima, H., Nakatsubo, N., Kikuchi, K., Kawahara, S., Kirino, Y., Nagoshi, H., Hirata, Y., and Nagano, T. *Anal Chem* **70**(13), 2446–2453 (1998).

382. Kojima, H., Nakatsubo, N., Kikuchi, K., Urano, Y., Higuchi, T., Tanaka, J., Kudo, Y., and Nagano, T. *Neuroreport* **9**, 3345–3348 (1998).

383. Kojima, H., Hirotani, M., Nakatsubo, N., Kikuchi, K., Urano, Y., Higuchi, T., Hirata, Y., and Nagano, T. *Anal Chem* **73**(9), 1967–1973 (2001).

384. Lim, M. H., Xu, D., and Lippard, S. J. *Nat Chem Biol* **2**(7), 375–380 (2006).

385. Cox, R. D. *Anal Chem* **5**, 332–335 (1980).

386. Rowland, I. R., Granli, T., Bøckman, B. C., Key, P. E., and Massey, R. C. *Carcinogenesis* **12**, 1395–1401 (1991).

387. Sen, N. P., Baddoo, P. A., and Seaman, S. W. *J Chromatogr* **673**, 77–84 (1994).

388. Hirata, K. I., Kuroda, R., Sakoda, T., Katayama, M., Inoue, N., Suematsu, M., et al. *Hypertension* **25**, 180–185 (1995).

389. Braman, R. and Hendrix, S. A. *Anal Chem* **61**, 2715–2718 (1989).

390. Myers, P. R., Guerra Jr, R., and Harrison, D. G. *Am J Physiol Heart and Circ Physiol* **256**(4), 1030–1037 (1989).

391. Westphal, A. H., Matorin, A., Hink, M. A., Borst, J. W., van Berkel, W. J., and Visser, A. J. *J Biol Chem* **281**(16), 11074–11081 (2006).

392. Sabbert, D., Engelbrecht, S., and Junge, W. *Nature* **381**(6583), 623–625 (1996).

393. Tally, J. F., Maniscalco, S. J., Saha, S. K., and Fisher, H. F. *Biochemistry* **41**(37), 11284–11293 (2002).

394. Theisen, M. J., Misra, I., Saadat, D., Campobasso, N., Miziorko, H. M., and Harrison, D. H. *Proc Natl Acad Sci USA* **101**(47), 16442–16447 (2004).

395. Taylor, N. E., Maier, K. G., Roman, R. J., and Cowley Jr, A. W. *Hypertension* **48**(6), 1066–1071 (2006).

396. Patel, B. A., Arundell, M., Parker, K. H., Yeoman, M. S., and O'Hare, D. *Anal Chem* **78**(22), 7643–7648 (2006).

397. Levine, D. Z. and Iacovitti, M. *Nitric Oxide* **15**, 87–92 (2006).

398. Moroz, L. L., Dahlgren, R. L., Boudko, D., Sweedler, J. V., and Lovell, P. *J Inorg Biochem* **99**(4), 929–939 (2005).

399. Nakaya, Y., Mawatari, K., Takahashi, A., Harada, N., Hata, A., and Yasui, S. *J Med Invest* **54**, 381–384 (2007).

400. Lang, N., Reppel, M., Hescheler, J., and Fleischmann, B. K. *Cell Physiol Biochem* **20**(5), 293–302 (2007).

401. Whalen, E. J., Foster, M. W., Matsumoto, A., Ozawa, K., Violin, J. D., Que, L. G., Nelson, C. D., Benhar, M., Keys, J. R., Rockman, H. A., Koch, W. J., Daaka, Y., Lefkowitz, R. J., and Stamler, J. S. *Cell* **129**(3), 511–522 (2007).

402. Perfume, G., Morgazo, C., Nabhen, S., Batistone, A., Hope, S. I., Bianciotti, L. G., and Vatta, M. S. *Regul Pept* **142**(3), 69–77 (2007).

403. Xu, L., Okuda-Ashitaka, E., Matsumura, S., Mabuchi, T., Okamoto, S., Sakimura, K., Mishina, M., and Ito, S. *Neuropharmacology* **52**(5), 1318–1325 (2007).

404. Croazza, S., Scarabottolo, L., Lohmer, S., and Liberati, C. *Assay Drug Dev Technol* **4**(2), 165–173 (2006).

405. Palmer, R. M. J., Rees, D. D., Ashton, D. S., and Moncada, S. *Biochem Biophys Res Commun* **153**(3), 1251–1256 (1988).

406. Rees, D. D., Palmer, R. M., Hodson, H. F., and Moncada, S. *Br J Pharmacol* **96**, 418–424 (1989).

407. Palacios, M., Knowles, R. G., Palmer, R. M. J., and Moncada, S. *Biochem Biophys Res Commun* **165**(2), 802–809 (1989).

408. Rees, D. D., Palmer, R. M., Schulz, R., Hodson, H. F., and Moncada, S. *Br J Pharmacol* **101**, 746–752 (1990).

409. Wunder, F., Buehler, G., Hüser, J., Mundt, S., Bechem, M., and Kalthof, B. *Anal Biochem* **363**(2), 219–227 (2007).

410. Iovannisci, D. M., Kupperman, S. O., Lloyd, E. W., and Lammer, E. J. *Genet Test* **6**(4), 245–253 (2002).

411. Liu, R., Paxton, W. A., Choe, S., Ceradini, D., Martin, S. R., Horuk, R., MacDonald, M. E., Stuhlmann, H., Koup, R. A., and Landau, N. R. *Cell* **86**(3), 367–377 (1996).

412. Sim, S. C., Risinger, C., Dahl, M. L., Aklillu, E., Christensen, M., Bertilsson, L., and Ingelman-Sunderberg, M. *Clin Pharmacol Ther* **79**, 103–113 (2006).

413. Lotsch, J., Skarke, C., Liefhold, J., and Geisslinger, G. *Clin Pharmacokinet* **43**(14), 983–1013 (2004).

414. Petronis, A., Gottesman, I. I., Kan, P., Kennedy, J. L., Basile, V. S., Paterson, A. D., and Popendikyte, V. *Schizophr Bull* **29**, 169–178 (2003).

415. Kukanskis, K., Elkind, J., Melendez, J., Murphy, T., Miller, G., and Garner, H. *Anal Biochem* **274**, 7–17 (1999).

416. Afonina, I., Kutyavin, I., Lukhtanov, E., Meyer, R. B., and Gamper, H. *Proc Natl Acad Sci USA* **93**(8), 3199–3204 (1996).

417. Lukhtanov, E. A., Lokhov, S. G., Gorn, V. V., Podyminogin, M. A., and Mahoney, W. *Nucleic Acids Res* **35**, e30 (2007).

418. Ren, B., Zhou, J.-M., and Komiyama, M. *Nucleic Acids Res* **32**(4), 42 (2004).

419. Michikawa, Y., Fujimoto, K., Kinoshita, K., Kawai, S., Sugahara, K., Suga, T., Otsuka, Y., Fujiwara, K., Iwakawa, M., and Imai, T. *Anal Sci* **22**(12), 1537–1545 (2006).

420. Podder, M., Welch, W. J., Zamar, R. H., and Tebbutt, S. J. *BMC Bioinform* **7**, 521 (2006).

421. Rudi, K., Zimonja, M., and Skanseng, B. *Methods Mol Biol* **345**, 111–117 (2006).

422. Blievernicht, J. K., Schaeffeler, E., Klein, K., Eichelbaum, M., Schwab, M., and Zanger, U. M. *Clin Chem* **53**, 24–33 (2006).

423. Kresge, N., Simoni, R. D., and Hill, R. L. *J Biol Chem* **280**, 46 (2005).

424. Klenow, H. and Henningsen, I. *Proc Natl Acad Sci USA* **65**, 168–175 (1970).

425. Bebenek, K., Joyce, C. M., Fitzgerald, M. P., and Kunkel, T. A. *J Biol Chem* **265**(23), 13878–13887 (1990).

426. Longley, M. J., Nguyen, D., Kunkel, T. A., and Copeland, W. C. *J Biol Chem* **276**(42), 38555–38562 (2001).

427. Zhang, X., Xu, L., Wang, P., Wang, Z., and Giese, R. W. *Bioconjug Chem* **13**(5), 1002–1012 (2002).

428. Li, X., Huang, Y., Guan, Y., Zhao, M., and Li, Y. *Anal Chim Acta* **584**, 12–18 (2007).

429. Luedeck, H. and Blasczyk, R. *Tissue Antigens* **50**(6), 627–638 (1997).

430. Duckworth, D. H. and Bessman, M. J. *J Biol Chem* **242**(12), 2877–2885 (1967).

431. Zuiderwijk, M., Tanke, H. J., Sam Neidbala, R., and Corstjens, P. L. *Clin Biochem* **36**(5), 401–403 (2003).

432. Buhl, A., Metzger, J. H., Heegaard, N. H., von Landenberg, P., Fleck, M., and Luppa, P. B. *Clin Chem* **53**(2), 334–341 (2007).

433. Malic, L., Cui, B., Veres, T., and Tabrizian, M. *Opt Lett* **32**(21), 3092–3094 (2007).

434. Newton, C. R. and Graham, A. *PCR. Introduction to Biotechniques*. BIOS Scientific Publishers (1997).

435. McPherson, M. J. and Moller, S. G. *PCR (The Basics)*. Taylor & Francis (2006).

436. Filmore, D. *Mod Drug Discov* **7**, 24–28 (2004).

437. Deshpande, D. A. and Penn, R. B. *Cell Signal* **18**(12), 2105–2120 (2006).

438. Wang, L., Fields, T. A., Pazmino, K., Dai, Q., Burchette, J. L., Howell, D. N., Coffman, T. M., and Spurney, R. F. *J Am Soc Nephrol* **16**(12), 3611–3622 (2005).

439. Li, S., Huang, S., and Peng, S. B. *Int J Oncol* **27**(5), 1329–1339 (2005).

440. Dhanasekaran, N., Tsim, S. T., Dermott, J. M., and Onesime, D. *Oncogene* **17**(11), 1383–1394 (1998).

441. Foerster, K., Groner, F., Matthes, J., Koch, W. J., Birnbaumer, L., and Herzig, S. *Proc Natl Acad Sci USA* **100**(24), 14475–14480 (2003).

442. Neves, S. R., Ram, P. T., and Iyengar, R. *Science* **296**, 1636–1639 (2002).

443. Krebs, E. G. and Fischer, E. H. *J Biol Chem* **218**, 483 (1956).

444. Sutherland, E. W. and Wosilait, W. D. *Nature* **175**, 169–170 (1955).

445. Hescheler, J., Rosenthal, W., Trautwein, W., and Schultz, G. *Nature* **325**(6103), 445–447 (1987).

446. Peralta, E. G., Ashkenazi, A., Winslow, J. W., Smith, D. H., Ramachandran, J., and Capon, D. J. *EMBO J* **6**(13), 3923–3929 (1987).

447. Trumpp-Kallmeyer, S., Hoflack, J., Bruinvels, A., and Hibert, M. *J Med Chem* **35**(19), 3448–3462 (1992).

448. Tyndall, J. D., Pfeiffer, B., Abbenante, G., and Fairlie, D. P. *Chem Rev* **105**(3), 793–826 (2005).

449. Frederiksson, R., Lagerström, M. C., Lundin, L. G., and Schiöth, H. B. *Mol Pharmacol* **63**, 1256–1272 (2003).

450. May, L. T., Avlani, V. A., Sexton, P. M., and Christopoulos, A. *Curr Pharm Des* **10**(17), 2003–2013 (2004).

451. Barnett, A. H., Bain, S. C., Bouter, P., Karlberg, B., Madsbad, S., Jervell, J., and Mustonen, J. *N Eng J Med* **351**, 1952–1961 (2004).

452. McConnell, J. D., Roehrborn, C. G., Bautista, O. M., Andriole, G. L., Dixon, C. M., Kusek, J. W., Lepor, H., McVary, K. T., Nyberg, L. M., Clarke, H. S., Crawford, E. D., Diokno, A., Foley, J. P., Foster, H. E., Jacobs, S. C., Kaplan, S. A., Kreder, K. J., Lieber, M. M., Lucia, M. S., Miller, G. J., Menon, M., Milam, D. F., Ramsdell, J. W., Schnkman, N. S., Slawin, K. M., and Smith, J. A. *N Eng J Med* **349**(25), 2387–2398 (2003).

453. Squire, I. B. and Barnett, D. B. *Br J Clin Pharmacol* **49**, 1–10 (2000).

454. Seeman, P. and Niznik, H. B. *FASEB J* **4**(10), 2737–2744 (1990).

455. Tyndall, J. and Sandilya, R. *Med Chem* **1**, 405–421 (2005).

456. Theis, J. G. W., Dellweg, H., Perzborn, E., and Gross, R. *Biochem Pharmacol* **44**(3), 495–503 (1992).

457. Kondo, J., Nakata, M., Nagano, J., Isono, Y., Nagai, K., and Tamaoki, A. *Chest* **118**, 73–79 (2000).

458. Henahan, J. *JAMA* **253**(5), 617–620 (1985).

459. Badesch, D. B., McLaughlin, V. V., Delcroix, M., Vizza, C. D., Olschewski, H., Sitbon, O., and Barst, R. J. *J Am Coll Cardiol* **43**, 56–61 (2004).

460. Insel, P. A., Tang, C. M., Hahntow, I., and Michel, M. C. *Biochim Biophys Acta* **1768**(4), 994–1005 (2007).

461. Small, K. M., McGraw, D. W., and Liggett, S. B. *Annu Rev Pharmacol Toxicol* **43**, 381–411 (2003).

462. Liggett, S. B., Mialet-Perez, J., Thaneemit-Chen, S., Weber, S. A., Greene, S. M., Hodne, D., Nelson, B., Morrison, J., Domanski, M. J., Abraham, W. T., et al. *Proc Natl Acad Sci USA* **103**, 11288–11293 (2006).

463. Litonjua, A. A. *Curr Opin Pulm Med* **12**, 12–17 (2006).

464. Johnson, J. A. and Turner, S. T. *Curr Opin Mol Ther* **7**(3), 218–225 (2005).

465. Sawa, M. and Harada, H. *Curr Med Chem* **13**, 25–37 (2006).

466. Bonci, A. and Hopf, F. W. *Neuron* **47**(3), 335–338 (2005).

467. Sokoloff, P., Diaz, J., Le Foll, B., Guillin, O., Leriche, L., Bezard, E., and Gross, C. *CNS Neurol Disord Drug Targets* **5**, 25–43 (2006).

468. Galvani, A. P. and Novembre, J. *Microbes Infect* **7**, 302–309 (2005).

469. Becker, Y. *Virus Genes* **31**, 113–119 (2005).

470. Torrecilla, I. and Tobin, A. B. *Curr Pharm Des* **12**(14), 1797–1808 (2006).

471. Presland, J. *Biochem Soc Trans* **32**(5), 888–891 (2004).

472. Oakley, R. H., Hudson, C. C., Cruickshank, R. D., Meyers, D. M., Payne Jr, R. E., Rhem, S. M., and Loomis, C. R. *Assay Drug Dev Technol* **1**(1), 21–30 (2002).

473. Fergusosn, S. S. and Caron, M. G. *Methods Mol Biol* **237**, 121–126 (2004).

474. Ward, W. W. and Cormier, M. J. *Photochem Photobiol* **27**, 389–396 (1978).

475. Yan, Y. X., Boldt-Houle, D. M., Tillotson, B. P., Gee, M. A., D'Eon, B. J., Chang, X. J., Olesen, C. E., and Palmer, M. A. *J Biolmol Screen* **7**(5), 451–459 (2002).

476. Santini, F., Penn, R. B., Gagnon, A. W., Benovic, J. L., and Keen, J. H. *J Cell Sci* **113**, 2463–2470 (2000).

477. Krahn, T., Paffhausen, W., Schade, A., Bechem, M., and Schmidt, D. US Patent 6,420,183 (2002).

478. Marks, K. M., Rosinov, M., and Nolan, G. P. *Chem Biol* **11**(3), 347–356 (2004).

479. Hanson, G. T., Cassutt, K. J., and O'Grady, M. Technical report, (2006).

480. Kost, T. A., Condreay, J. P., and Jarvis, D. L. *Nat Biotechnol* **23**, 567–575 (2005).

481. Anderson, J. M., Charbonneau, H., and Cormier, M. J. *Biochemistry* **13**(16), 1195–1200 (1974).

482. Allen, D. G., Blinks, J. R., and Prendergast, F. G. *Science* **195**, 996–998 (1977).

483. Shimurao, O., Johnson, F. H., and Saiga, Y. *J Cell Comp Physiol* **59**, 223–239 (1962).

484. Shimomura, O., Johnson, F. H., and Saiga, Y. *Science* **140**(3573), 1339–1340 (1963).

485. Button, D. and Brownstein, M. *Cell Calcium* **14**(9), 663–671 (1993).

486. Sheu, Y. A., Kricka, L. J., and Pritchett, D. B. *Anal Biochem* **209**(2), 343–347 (1993).

487. Johnson, P. C., Ware, J. A., Cliveden, P. B., Smith, M., Dvorak, A. M., and Salzman, E. Q. A. *J Biol Chem* **260**, 2069–2076 (1985).

488. Woolkalis, M. J., DeMelfi, T. M., Blanchard, N., Hoxie, J. A., and Brass, L. F. *J Biol Chem* **270**(17), 9896–9903 (1995).

489. Prasher, D., McCann, R. O., and Cormier, M. J. *Biochem Biophys Res Commun* **126**(3), 1259–1268 (1995).

490. Tsuzuki, K., Tricoire, L., Courjean, O., Gibelin, N., Rossier, J., and Lambolez, B. *J Biol Chem* **280**, 34324–34331 (2005).

491. Milligan, G., Marshall, F., and Rees, S. *TiPS* **17**, 235–237 (1996).

492. Bryant, R., Jefferson, F., and Tang, L. Society for Biomolecular Screening (2003).

493. Hemmila, I. *J Biomol Screen* **4**(6), 303–308 (1999).

494. Morin, D., Cotte, N., Balestre, M. N., Mouillac, B., Manning, M., Breton, C., and Barberis, C. *FEBS Lett* **441**(3), 470–475 (1998).

495. Romanelli, A. and van de Werve, G. *Metabolism* **46**(5), 548–555 (1997).

496. Orihuela, P. A., Parada-Bustamante, A., Zuñiga, L. M., and Croxatto, H. B. *J Endocrinol* **188**(3), 579–588 (2006).

497. Hein, L. In *G Protein-Coupled Receptors as Drug Targets,* Seifert, R. and Wieland, T., editors. Wiley & Sons (2005).

498. Seethala, R. and Fernandes, P. B. Technical report (2001).

499. Frang, H., Mukkala, V. M., Syystö, R., Ollikka, P., Hurskainen, P., Scheinin, M., and Hemmilä, I. *Assay Drug Dev Techno* **1**(2), 275–280 (2003).

500. Naor, Z., Benard, O., and Seger, R. *Trends Endocrinol Metab* **11**(3), 91–99 (2000).

501. Wu, E. H., Lo, R. K., and Wong, Y. H. *Biochem Biophys Res Commun* **303**(3), 920–925 (2003).

502. Pi, M. and Quarles, L. D. *J Cell Biochem* **95**(6), 1081–1092 (2005).

503. Rees, S., Martin, D. P., Scott, S. V., Brown, S. H., Fraser, N., O'Shaughnessy, C., and Beresford, I. J. *J Biomol Screen* **6**(1), 19–27 (2001).

504. Wang, P., Yan, H., and Li, J. C. *Biochem Biophys Res Commun* **363**(1), 101–105 (2007).

505. Taylor, C. T., Furuta, G. T., Kristin, S., and Colgan, S. P. *Proc Natl Acad Sci USA* **97**(22), 12091–12096 (2000).

506. Luttrell, D. K. and Luttrell, L. M. *Assay Drug Dev Technol* **1**(2), 327–338 (2003).

507. Ullman, E. F., Kirakossian, H., Singh, S., Wu, Z. P., Irvin, B. R., Pease, J. S., Switchenko, A. C., Irvine, J. D., Dafforn, A., Skold, C. N., and Wagner, D. B. *Proc Natl Acad Sci USA* **91**(12), 5426–5430 (1994).

508. Mao, J., Yuan, H., Xie, W., Simon, M. I., and Wu, D. *J Biol Chem* **273**(42), 27118–27123 (1998).

509. Kitaguchi, N., Takahashi, Y., Tokushima, Y., Shiojiri, S., and Ito, H. *Nature* **331**(6156), 530–532 (1988).

510. Carrell, R. W. *Nature* **331**(6156), 478–479 (1988).

511. Johnston, M. I., Allaudeen, H. S., and Sarver, N. *Trends Pharmacol Sci* **10**(8), 305–307 (1989).

512. Oberg, B. *J Acquir Immune Defic Syndr* **1**(3), 257–266 (1988).

513. Harris, J. O., Olsen, G. N., Castle, J. R., and Maloney, A. S. *Am Rev Respir Dis* **111**(5), 579–586 (1975).

514. Liotta, L. A. *Cancer Res* **46**, 1–7 (1986).

515. Hartley, B. S. *Annu Rev Biochem* **29**, 45–72 (1960).

516. Barrett, A. J. *Ciba Found Symp* **75**, 1–13 (1979).

517. Balls, A. K. and Jansen, E. F. *Advan Enzymol* **13**, 321–343 (1952).

518. Matthews, B. W., Sigler, P. B., Henderson, R., and Blow, D. M. *Nature* **214**, 642–656 (1967).

519. Blow, D. M., Birktoft, J. J., and Hartley, B. S. *Nature* **221**(5178), 337–340 (1969).

520. Strumeyer, D. H., White, W. N., and Koshland Jr, D. E. *Proc Natl Acad Sci USA* **50**, 931–935 (1963).

521. Hunkapiller, M. W., Smallcombe, S. H., Whitaker, D. R., and Richards, J. H. *J Biol Chem* **248**, 8306–8308 (1973).

522. Craik, C. S., Largman, C., Fletcher, T., Roczniak, S., Barr, P. J., Fletterick, R., and Rutter, W. J. *Science* **228**(4697), 291–297 (1985).

523. Gráf, L., Jancsó, A., Szilágyi, L., Hegyi, G., Pintér, K., Náray-Szabó, G., Hepp, J., Medzihradszky, K., and Rutter, W. J. *Proc Natl Acad Sci USA* **85**(14), 4961–4965 (1988).

524. Neurath, H. and Walsh, K. A. *Proc Natl Acad Sci USA* **73**(11), 3825–3832 (1976).

525. Neurath, H. In *Advances in Protein Chemistry*, Vol. XII, Anfinsen, C. B. and Anson, M. L., editors, pp. 320–386. Academic Press (1957).

526. Perkins, S. J. and Smith, K. F. *Biochem J* **295**, 109–114 (1993).

527. Willenbrock, F. and Brocklehurst, K. *Biochem J* **227**(2), 521–528 (1985).

528. Scheer, J. M., Romanowski, M. J., and Wells, J. A. *Proc Natl Acad Sci USA* **103**(20), 7595–7600 (2006).

529. Syrový, I. *Cesk Fysiol* **15**(3), 217–224 (1966).

530. Gomis-Rüth, F. X., Kress, L. F., and Bode, W. *EMBO J* **12**(11), 4151–4157 (1993).

531. Hooper, N. M. *FEBS Lett* **354**, 1–6 (1994).

532. Hughes-Jones, N. C., Pickering, G. W., et al. *J Physiol* **109**(3), 288–307 (1949).

533. Braun-Menendez, E. *Pharmacol Rev* **8**, 25–55 (1956).

534. Swanton, E., Savory, P., Cosulich, S., Clarke, P., and Woodman, P. *Oncogene* **18**(10), 1781–1787 (1999).

535. Kuribayashi, K., Mayes, P. A., and El-Deiry, W. S. *Cancer Biol Ther* **5**(7), 763–765 (2006).

536. Seethala, R. and Menzel, R. *Anal Biochem* **255**, 257–262 (1998).

537. Ma, H., Horiuchi, K. Y., Wang, Y., Kucharewicz, S. A., and Diamond, S. L. *Assay Drug Dev Technol* **3**(2), 177–187 (2005).

538. Maggiora, L. L., Smith, C. W., and Zhang, Z. Y. *J Med Chem* **35**(21), 3727–3730 (1992).

539. Grum-Tokars, V., Ratia, K., Begaye, A., Baker, S. C., and Mesecar, A. D. *Virus Res* **133**, 63–73 (2008).

540. Sun, H., Panicker, R. C., and Yao, S. Q. *Biopolymers* **88**(2), 141–149 (2007).

541. Felber, L. M., Cloutier, S. M., Kündig, C., Kishi, T., Brossard, V., Jichlinski, P., Leisinger, H. J., and Deperthes, D. *Biotechniques* **36**(5), 878–885 (2004).

542. Haugland, R. P. and Zhou, M. US Patent 5,719,031 (1998).

543. Jiang, P. and Mellors, A. *Anal Biochem* **259**, 8–15 (1998).

544. Mancini, F., Naldi, M., Cavrini, V., and Andrisano, V. *Anal Bioanal Chem* **388**(5), 1175–1183 (2007).

545. Kohl, T., Heinze, K. G., Kuhlemann, R., Koltermann, A., and Schwille, P. *Proc Natl Acad Sci USA* **99**(19), 12161–12166 (2002).

546. Préaudat, M., Ouled-Diaf, J., Alpha-Bazin, B., Mathis, G., Mitsugi, T., Aono, Y., Takahashi, K., and Takemoto, H. *J Biomol Screen* **7**(3), 267–274 (2002).

547. Karvinen, J., Hurskainen, P., Gopalakrishnan, S., Burns, D., Warrior, U., and Hemmilä, I. *J Biomol Screen* **7**(3), 223–231 (2002).

548. Kumaraswamy, S., Bergstedt, T., Shi, X., Rininsland, F., Kushon, S., Xia, W., Ley, K., Achyuthan, K., McBranch, D., and Whitten, D. *Proc Natl Acad Sci USA* **101**(20), 7511–7515 (2004).

549. Pinto, M. R. and Schanze, K. S. *Proc Natl Acad Sci USA* **101**(20), 7505–7510 (2004).

550. White, H. E. et al. *J Am Chem Soc* **88**(7), 2015–2019 (1966).

551. Miska, W. and Geiger, R. *J Clin Chem Clin Biochem* **25**, 23–30 (1987).

552. Monsees, T., Miska, W., and Geiger, R. *Anal Biochem* **221**(2), 329–334 (1994).

553. Monsees, T., Geiger, R., and Miska, W. *J Biolumin Chemilumin* **10**(4), 213–218 (1995).

554. O'Brien, M. A., Daily, W. J., Hesselberth, P. E., Moravec, R. A., Scurria, M. A., Klaubert, D. H., Bulleit, R. F., and Wood, K. V. *J Biomol Screen* **10**(2), 137–148 (2005).

555. Kumagai, Y., Konishi, K., Gomi, T., Yagishita, H., Yajima, A., and Yoshikawa, M. *Infect Immun* **68**(2), 716–724 (2000).

556. Leng, J. US Patent 6,890,745 (2005).

557. Deo, S. K., Lewis, J. C., and Daunert, S. *Anal Biochem* **281**, 87–94 (2000).

558. Deo, S. K., Mirasoli, M., and Daunert, S. *Anal Bioanal Chem* **381**(7), 1387–1394 (2005).

559. Deo, S. K. and Daunert, S. *Fresenius J Anal Chem* **369**(3), 258–266 (2001).

560. Wigdal, S. Poster at http://www.promega.com/ (2007).

561. Shimomura, O. *Biol Bull* **189**, 1–5 (1995).

562. Shimomura, O. *J Microscopy* **217**, 3–15 (2005).

563. Prasher, D., McCann, R. O., and Cormier, M. J. *Biochem Biophys Res Commun* **126**(3), 1259–1268 (1985).

564. Inouye, S., Aoyama, S., Miyata, T., Tsuji, F. I., and Sakaki, Y. *J Biochem* **105**(3), 473–477 (1989).

565. Head, J. F., Inouye, S., Teranishi, K., and Shimomura, O. *Nature* **405**(6784), 372–376 (2000).

566. Kurose, K., Inouye, S., Sakaki, Y., and Tsuji, F. I. *Proc Natl Acad Sci USA* **86**(1), 80–84 (1989).

567. Nayak, L. and De, R. K. *J Biosci* **32**(5), 1009–1017 (2007).

568. Liljelund, P., Netzeband, J. G., and Gruol, D. L. *J Neurosci* **20**(19), 7394–7403 (2000).

569. Muller, Y. L., Reitstetter, R., and Yool, A. J. *J Neurosci* **18**, 16–25 (1998).

570. McCann, J. D. and Welsh, M. J. *Annu Rev Physiol* **52**, 115–135 (1990).

571. Murachi, T. *Biochem Soc Symp* **49**, 149–167 (1984).

572. Cox, N. J. *Curr Diab Rep* **2**(2), 186–190 (2002).

573. Tomimatsu, Y., Idemoto, S., Moriguchi, S., Watanabe, S., and Nakanishi, H. *Life Sci* **72**(4), 355–361 (2002).

574. Canki-Klain, N., Milic, A., Kovac, B., Trlaja, A., Grgicevic, D., Zurak, N., Fardeau, M., Leturcq, F., Kaplan, J. C., Urtizberea, J. A., Politano, L., Piluso, G., and Feingold, J. *Am J Med Genet A* **125**(2), 152–156 (2004).

575. Petersen, O. H. and Maruyama, Y. *Nature* **307**, 693–696 (1984).

576. Iwamoto, T., Watanabe, Y., Kita, S., and Blaustein, M. P. *Cardiovasc Hematol Disord Drug Targets* **7**(3), 188–198 (2007).

577. Stutzmann, G. E. *Neuroscientist* **13**(5), 546–559 (2007).

578. Raisz, L. G. *J Clin Invest* **115**(12), 3318–3325 (2005).

579. Button, D. and Brownstein, M. *Cell Calcium* **14**(9), 663–671 (1993).

580. Sheu, Y. A., Kricka, L. J., and Pritchett, D. B. *Anal Biochem* **209**(2), 343–347 (1993).

581. Brini, M., Murgia, M., Pasti, L., Picard, D., Pozzan, T., and Rizzuto, R. *EMBO J* **12**(12), 4813–4819 (1993).

582. Brini, M., Marsault, R., Bastianutto, C., Alvarez, J., Pozzan, T., and Rizzuto, R. *J Biol Chem* **270**(17), 9896–9903 (1995).

583. Gilland, E., Miller, A. L., Karplus, E., Baker, R., and Webb, S. E. *Proc Natl Acad Sci USA* **96**, 157–161 (1999).

584. Deo, S. K., Lewis, J. C., and Daunert, S. *Bioconjug Chem* **12**(3), 378–384 (2001).

585. Lewis, J. C. and Daunert, S. *Anal Chem* **73**(14), 3227–3233 (2001).

586. Yaghoubi, S. S. and Gambhir, S. S. *Nat Protoc* **1**(4), 2137–2142 (2006).

587. Cubitt, A. B., Heim, R., Adams, S. R., Boyd, A. E., Gross, L. A., and Tsien, R. Y. *Trends Biochem Sci* **20**(11), 448–455 (1995).

588. Anderson, M. T., Tjioe, I. M., Lorincz, M. C., Parks, D. R., Herzenberg, L. A., Nolan, G. P., and Herzenberg, L. A. *Proc Natl Acad Sci USA* **93**(16), 8508–8511 (1996).

589. De Giorgi, F., Brini, M., Bastianutto, C., Marsault, R., Montero, M., Pizzo, P., Rossi, R., and Rizzuto, R. *Gene* **173**, 113–117 (1996).

590. Limón, A., Briones, J., Puig, T., Carmona, M., Fornas, O., Cancelas, J. A., Nadal, M., García, J., Rueda, F., and Barquinero, J. *Blood* **90**(9), 3316 (1997).

591. Ptashne, M. *Proc Natl Acad Sci USA* **57**(2), 306–313 (1967).

592. Wehrman, T. S., von Degenfeld, G., Krutzik, P. O., Nolan, G. P., and Blau, H. M. *Nat Methods* **3**(4), 295–301 (2006).

593. Ayling, A. and Baneyx, F. *Protein Sci* **5**(3), 478–487 (1996).

594. Morise, H., Shimomura, O., Johnson, F. H., and Winant, J. *Biochemistry* **13**(12), 2656–2662 (1974).

595. Shimomura, O. and Johnson, F. H. *Proc Natl Acad Sci USA* **72**(4), 1546–1549 (1975).

596. Ormö, M., Cubitt, A., Kallio, K., Gross, L., Tsien, R., and Remington, S. *Science* **273**(5280), 1392–1395 (1996).

597. Yang, F., Moss, L., and Phillips, G. *Nat Biotechnol* **14**(10), 1246–1251 (1996).

598. Corporation, P. Technical report (2007).

599. Baird, G. S., Zacharias, D. A., and Tsien, R. Y. *Proc Natl Acad Sci USA* **97**(22), 11984–11989 (2000).

600. Wiedenmann, J., Vallone, B., Renzi, F., Nienhaus, K., Ivanchenko, S., Röcker, C., and Nienhaus, G. U. *J Biomed Opt* **10**, 14003 (2005).

601. Orengo, J. P., Bundman, D., and Cooper, T. A. *Nucleic Acids Res* **34**(22), 148 (2006).

602. Olesen, C. E., Yan, Y. X., Liu, B., Martin, D., D'Eon, B., Judware, R., Martin, C., Voyta, J. C., and Bronstein, I. *Methods Enzymol* **326**, 175–202 (2000).

603. Culleen, B. R. *Methods Enzymol* **326**, 159–164 (2000).

604. Tannous, B. A., Verhaegen, M., Christopoulos, T. K., and Kourakli, A. *Anal Biochem* **320**(2), 266–272 (2003).

605. Ow, D. W., Jacobs, J. D., and Howell, S. H. *Proc Natl Acad Sci USA* **84**(12), 4870–4874 (1987).

606. McElroy, M. D., McElroy, W. D., Helinski, D. R., Wood, K. V., De Wet, J. R., Ow, D. W., and Howell, S. H. US Patent 5,583,024 (1996).

607. McElroy, M. D., Helinski, D. R., Wood, K. V., De Wet, J. R., Ow, D. W., and Howell, S. H. US Patent 5,700,673 (1997).

608. Chen, X., Ren, S., Jin, Z., and Zhu, S. *Biotechnol Tech* **10**(2), 89–92 (1996).

609. Markova, S. V., Golz, S., Frank, L. A., Kalthof, B., and Vysotski, E. S. *J Biol Chem* **279**(5), 3212–3217 (2003).

610. Verhaegent, M. and Christopoulos, T. *Anal Chem* **74**(17), 4378–4385 (2002).

611. Tannous, B., Kim, D., Fernandez, J., Weissleder, R., and Breakefield, X. *Mol Ther* **11**, 435–443 (2005).

612. Qiu, H., Zhao, S., Yu, M., Fan, B., and Liu, B. *Acta Biochim Biophys Sin* **40**, 85–90 (2008).

613. Chen, W., Wu, W., Zhao, J., Liu, W., Jiang, A., and Zhang, J. *Mol Biol Rep* (2008).

614. Bu, X., Jia, F., Wang, W., Guo, X., Wu, M., and Wei, L. *BMC Cancer* **7**, 208 (2007).

615. Gendron, K., Charbonneau, J., Dulude, D., Heveker, N., Ferbeyre, G., and Brakier-Gingras, L. *Nucliec Acids Res* **36**, 30–40 (2008).

616. Falcetti, E., Flavell, D. M., Staels, B., Tinker, A., Haworth, S. G., and Clapp, L. H. *Biochem Biophys Res Commun* **360**(4), 821–827 (2007).

617. Jin, H., Hwang, S. K., Kwon, J. T., Lee, Y. S., An, G. H., Lee, K. H., Prats, A. C., Morello, D., Beck Jr, G. R., and Cho, M. H. *J Nutr Biochem* **19**, 16–25 (2008).

618. Lee, J. Y., Kim, S., Hwang do, W., Jeong, J. M. Chung, J. K., Lee, M. C., and Lee, D. S. *J Nucl Med* **49**(2), 285–294 (2008).

619. Zerefos, P. G., Ioannou, P. C., Traeger-Synodinos, J., Dimissianos, G., Kanavakis, E., and Christpolous, T. K. *Hum Mutat* **27**(3), 279–285 (2006).

620. Robertson, D., Shore, S. H., and Miller, D. M. *Manipulation and Expression of Recombinant DNA: A Laboratory Manual.* Academic Press (1997).

621. Metzenberg, S. *Working With DNA (Basics S.).* Taylor & Francis (2007).

622. Wilson, J. D., Griffin, J. E., Leshin, M., and George, F. W. *Hum Genet* **58**, 78–84 (1981).

623. Yeap, B. B., Wilce, J. A., and Leedman, P. J. *Bioessays* **26**(6), 672–682 (2004).

624. Prins, G. S. *May Clin Proc* **75**, 32–35 (2000).

625. Ku, C. Y., Loose-Mitchell, D. S., and Sanborn, B. M. *Biol Reprod* **51**(2), 319–326 (1994).

626. Johannessen, M., Delghandi, M. P., and Moens, U. *Ceel Signal* **16**(11), 1211–1227 (2004).

627. Asagiri, M. and Takayanagi, H. *Bone* **40**(2), 251–264 (2007).

628. Wu, H., Peisley, A., Graef, I. A., and Crabtree, G. R. *Trends Cell Biol* **17**(6), 251–260 (2007).

629. Jugan, M. L., Lévy-Bimbot, M., Pomérance, M., Tamisier-Karolak, S., Blondeau, J. P., and Lévi, Y. *Toxicol In Vitro* **21**(6), 1197–1205 (2007).

630. Korpela, M., Mäntsälä, P., Lilius, E. M., and Karp, M. *J Biolumin Chemilumin* **4**, 551–554 (1989).

631. Tilley, L. D., Hine, O. S., Kellog, J. A., Hassinger, J. N., Weller, D. D., Iversen, P. L., and Geller, B. L. *Antimicrob Agents Chemother* **50**(8), 2789–2796 (2006).

632. Kramer, E. B. and Farabaugh, P. J. *RNA* **13**, 87–96 (2007).

633. Galluzzi, L. and Karp, M. *J Biotechnol* **127**(2), 188–198 (2007).

634. Bachmann, H., Santos, F., Kleerebezem, M., and van Hylckama Vlieg, J. E. *Appl Environ Microbiol* **73**(14), 4704–4706 (2007).

635. Zeng, H., Wei, Q., Huang, R., Chen, N., Dong, Q., Yang, Y., and Zhou, Q. *J Androl* **28**(6), 827 (2007).

636. Potts, K. E., Jackson, R. L., and Patick, A. K. US Patent 6,790,612 (2002).

637. Bona, R., Andreotti, M., Buffa, V., Leone, P., Galluzzo, C. M., Amici, R., Palmisano, L., Mancini, M. G., Michelini, Z., Santo, R. D., Costi, R., Roux, A., Pommier, Y., Marchand, C., Vella, S., and Cara, A. *Antimicrob Agents Chemother* **50**(10), 3407–3417 (2006).

638. Richards, K., Rushmore, T. H., and Morsy, M. A. US Patent 6,395,473 (2002).

639. International conference on harmonisation of technical requirements for registration of pharmaceuticals for human use (ICH) http://www.ich.org/LOB/media/MEDIA417.pdf (2005).

640. Nyiri, L. K. and Toth, G. M. *Biotechnol Bioeng* **13**(5), 697–701 (1971).

641. Kane, J. F., Homes, W. M., Smiley Jr, K. L., and Jensen, R. A. *J Bacteriol* **113**, 224–232 (1973).

642. Choe, J., Suresh, S., Wisedchaisri, G., Kennedy, K. J., Gelb, M. H., and Hol, W. G. *Chem Biol* **9**(11), 1189–1197 (2002).

643. Stein, B. S., Vangore, S., Petersen, R. O., and Kendall, A. R. *Am J Surg Pathol* **6**(6), 553–557 (1982).

644. Brawer, M. K., Chetner, M. P., Beatie, J., Buchner, D. M., Vessella, R. L., and Lange, P. H. *J Urol* **137**(3), 841–845 (1992).

645. Luderer, A. A., Chen, Y. T., Soriano, T. F., Kramp, W. J., Carlson, G., Cuny, C., Sharp, T., Smith, W., Etteway, J., Brawer, M. K., et al. *Urology* **46**(2), 187–194 (1995).

646. Barnett, V. and Lewis, T. *Outliers in Statistical Data. Wiley Series in Probability & Statistics.* Wiley (1994).

647. Moore, G., Griffith, C., and Fielding, L. *International Association for Food Protection* (2001).

648. Kurosawa, S., Park, J. W., Aizawa, H., Wakida, S., Tao, H., and Ishihara, K. *Biosens Bioelectron* **22**(4), 473–481 (2006).

649. Ueda, H. *Yakugaku Zasshi* **127**, 71–80 (2007).

650. Tudorache, M., Co, M., Lifgren, H., and Emnéus, J. *Anal Chem* **77**(22), 7156–7162 (2005).

651. Ciumasu, I. M., Krämer, P. M., Weber, C. M., Kolb, G., Tiemann, D., WIndisch, S., Frese, I., and Kettrup, A. A. *Biosens Bioelectron* **21**(2), 354–364 (2005).

652. Adornato, L. R., Kaltenbacher, E. A., Greenbow, D. R., and Byrne, R. H. *Environ Sci Technol* **41**(11), 4045–4052 (2007).

653. Zou, Z., Han, J., Jang, A., Bishop, P. L., and Ahn, C. H. *Biosens Bioelectron* **22**, 1902–1907 (2007).

654. Rahman, M. A., Park, D. S., Chang, S. C., McNeil, C. J., and Shim, Y. B. *Biosen Bioelectron* **21**(7), 1116–1124 (2006).

655. Gilliom, R., Barbash, J., Crawford, C., Hamilton, P., Martin, J., Nakagaki, N., Nowell, L., Scott, J., Stackelberg, P., Thelin, G., et al. Technical report (2006).

656. Whyatt, R. M., Rauh, V., Barr, D. B., Camann, D. E., Andrews, H. F., Garfinkel, R., Hoepner, L. A., Diaz, D., Dietrich, J., Reyes, A., Tang, D., Kinney, P. L., and Perera, F. P. *Environ Health Perspect* **112**(10), 1125–1132 (2004).

657. Rauh, V. A., Garfinkel, R., Perera, F. P., Andrews, H. F., Hoepner, L., Barr, D. B., Whitehead, R., Tang, D., and Whyatt, R. W. *Pediatrics* **118**(6), 1845–1859 (2006).

658. McGovern, J. P., Shih, W. Y., and Shih, W. H. *Analyst* **132**(8), 777–783 (2007).

659. Lei, Y., Mulchandani, P., Wang, J., Chen, W., and Mulchandani, A. *Environ Sci Technol* **39**(22), 8853–8857 (2005).

660. Lin, T. J., Huang, K. T., and Liu, C. Y. *Biosens Bioelectron* **22**(4), 513–518 (2006).

661. Liu, G. and Lin, Y. *Anal Chem* **78**(3), 835–843 (2006).

662. Flowers, L. K., Mothershead, J. L., and Blackwell, T. H. *Emerg Med Clin North Am* **20**, 457–476 (2002).

663. Tegnell, A., Van Loock, F., Baka, A., Wallyn, S., Hendricks, J., Werner, A., and Gouvras, G. *Cell Mol Life Sci* **63**, 2223–2228 (2006).

664. Farrell, S., Halsall, H. B., and Heineman, W. R. *Analyst* **130**(4), 489–497 (2005).

665. Zahavy, E., Fisher, M., Bromber, A., and Olshevsky, U. *Appl Environ Microbiol* **69**(4), 2330–2339 (2003).

666. Yung, P. T., Lester, E. D., Bearman, G., and Ponce, A. *Biotechnol Bioeng* **98**(4), 864–871 (2007).

667. Carter, D. J. and Cary, R. B. *Nucleic Acids Res* **35**, 74 (2007).

668. Baeumner, A. J., Leonard, B., McElwee, J., and Montagna, R. A. *Anal Bioanal Chem* **380**(1), 15–23 (2004).

669. Hoile, R., Yuen, M., James, G., and Gilbert, G. L. *Forensic Sci Int* **171**, 1–4 (2007).

670. Seto, Y. *Yakugaku Zasshi* **126**(12), 1279–1299 (2006).

671. Kikuchi, M., Sasaki, Y., and Wakabayashi, M. *Ecotoxicol Environ Saf* **47**(3), 239–245 (2000).

672. Ren, Z., Zha, J., Ma, M., Wang, Z., and Gerhardt, A. *Environ Monit Assess* **134**, 373–383 (2007).

673. Burnworth, M., Rowan, S. J., and Weder, C. *Chemistry* **13**(28), 7828–7836 (2007).

674. Schofield, D. A., Westwater, C., Barth, J. L., and DiNovo, A. A. *Appl Microbiol Biotechnol* **76**(6), 1383–1394 (2007).

675. Joshi, K. A., Prouza, M., Kum, M., Wang, J., Tang, J., Haddon, R., Chen, W., and Mulchandani, A. *Anal Chem* **78**, 331–336 (2006).

INDEX

Coupled Bioluminescent Assays: Methods and Applications, Michael J. Corey
Copyright © 2009 by John Wiley & Sons Inc.